The Return of Theory in Early Modern English Studies

Also by Paul Cefalu

ENGLISH RENAISSANCE LITERATURE AND CONTEMPORARY THEORY: SUBLIME OBJECTS OF THEOLOGY

MORAL IDENTITY IN EARLY MODERN ENGLISH LITERATURE

REVISIONIST SHAKESPEARE: TRANSITIONAL IDEOLOGIES IN TEXTS AND CONTEXTS

Also by Bryan Reynolds

TRANSVERSAL SUBJECTS: FROM MONTAIGNE TO DELEUZE AFTER DERRIDA

TRANSVERSAL ENTERPRISES IN THE DRAMA OF SHAKESPEARE AND HIS CONTEMPORARIES: FUGITIVE EXPLORATIONS

PERFORMING TRANSVERSALLY: REIMAGINING SHAKESPEARE AND THE CRITICAL FUTURE

BECOMING CRIMINAL: TRANSVERSAL PERFORMANCE AND CULTURAL DISSIDENCE IN EARLY MODERN ENGLAND

CRITICAL RESPONSES TO KIRAN DESAI (*co-editor with Sunita Sinha*)

REMATERIALIZING SHAKESPEARE: AUTHORITY AND REPRESENTATION ON THE EARLY MODERN ENGLISH STAGE (*co-editor with William West*)

SHAKESPEARE WITHOUT CLASS: MISAPPROPRIATIONS OF CULTURAL CAPITAL (*co-editor with Donald Hedrick*)

The Return of Theory in Early Modern English Studies

Tarrying with the Subjunctive

Edited by

Paul Cefalu
Lafayette College, Easton, Pennsylvania

&

Bryan Reynolds
University of California–Irvine

First published 2011 by
PALGRAVE MACMILLAN

Palgrave Macmillan in the UK is an imprint of Macmillan Publishers
Limited, registered in England, company number 785998, of Houndmills,
Basingstoke, Hampshire RG21 6XS.

Palgrave Macmillan in the US is a division of St Martin's Press LLC,
175 Fifth Avenue, New York, NY 10010.

Palgrave Macmillan is the global academic imprint of the above companies
and has companies and representatives throughout the world.

Palgrave® and Macmillan® are registered trademarks in the United States,
the United Kingdom, Europe and other countries.

ISBN 978–0–230–23549–6   hardback

This book is printed on paper suitable for recycling and made from fully
managed and sustained forest sources. Logging, pulping and manufacturing
processes are expected to conform to the environmental regulations of the
country of origin.

A catalogue record for this book is available from the British Library.

Library of Congress Cataloging-in-Publication Data

The return of theory in early modern english studies : tarrying with the
subjunctive / [edited by] Paul Cefalu, Bryan Reynolds.
     p. ; cm.
   Includes index.
     ISBN 978–0–230–23549–6 (hardback)
   1. English literature—Early modern, 1500–1700—History and criticism.
   2. English language—Subjunctive.   3. Historical linguistics.   I. Cefalu, Paul.
   II. Reynolds, Bryan.
       PR421.R49 2011
       820.9'34—dc22
                                                                    2010050721

10   9   8   7   6   5   4   3   2   1
20   19 18 17 16 15 14 13 12 11

Printed and bound in Great Britain by
CPI Antony Rowe, Chippenham and Eastbourne

'The contributors draw on current and emerging fields – cognitive science, the new formalism, political theology – to cast exciting new light on Renaissance texts. This is an essential collection for anyone interested in knowing where the study of early modern literature is headed.'

–Richard Halpern, Johns Hopkins University, USA

'This book makes theory exciting again, both in itself and as a catalyst to re-visioning early modern texts. After a dreary end-of-theory decade, it's good to read essays energized by their engagements with neuroscience, political theology, semantics, and new formalisms. This very smart collection makes a compelling intellectual case for theory's resurgence.'

– Jean E. Howard, George Delacorte Professor in the Humanities, Columbia University, USA

Contents

Part III Rematerialisms

List of Illustrations

Note on Cover

With the jacket design, visual artist Alex Sacui set out to reflect the adventurousness of the theories presented in the book. He altered the landscape photography of ESSJAYNZ to create a dreamlike image of a valley fertile for new and unpredictable thought. Fluid, rhizomatic, and static in different areas, the landscape manifests in-between processes, subjunctive spaces oscillating between recognition and otherworldliness. From its mountainous lush green blend, the childlike illustrations of Sky Reynolds emerge along with the academic text and credits. Given a list of objects that are featured in the book's chapters (feather, crown, sheep, brain, devil creature [Mephistopheles], etc.), six-year-old Reynolds free-associated for thirty minutes in which time she produced the drawings. The idea was to welcome her first impressions so that the drawings represent a synthesis of the imagery that would defy assembling a theory by which to more directly make sense of the book. The result was a surrealist assortment of figures floating as if in outer space, albeit within the rectangular frame of letter-size paper. Sacui relocated the illustrations to serve his multimedia concept, one that implies order, but also, and more powerfully, suggests transversality: dynamic movements outside of established parameters that precipitate surprise and change.

Acknowledgments

Paul Cefalu would like to thank Lafayette College, especially the Academic Research Committee for its support of this project in the form of a research grant to cover indexing and the book jacket. He offers special thanks to Suzanne Westfall, chair of the English Department at Lafayette College for a flexible teaching schedule which allowed time for research; and he extends additional thanks to colleagues Lee Upton, James Woolley, and Lynn Van Dyke for their ongoing encouragement and support. Paul also thanks Jill Greco for creating such a hospitable environment in which to work and create.

Bryan Reynolds is grateful to all the members of the Transversal Theater Company, and especially Michael Hooker, Lonnie Rafael Alcaraz, Luke Cantarella, Gary Busby, Lisa Naugle, Lauren McCue, Adam Bryx, Alex Sacui, Jeanine Nicholas, Oscar Seip, Niels Horeman, Tu Nguyen, Eline van der Velden, Jeanine Nicholas, Chris Marshall, Saskia Polderman, Babette Holtman, Maarten Hutten, Keith Bangs, Matthias Quadekker, Kimberley van Alphen, Jeromy Rutter, Kayla Emerson, Jessica Spaw, and Palmer Jankens for generously supporting his busyness during our rehearsals and tours of *Blue Shade* and *The Green Knight* while he was completing this book. He is also grateful to his colleagues at UC Irvine, especially Robert Cohen, Eli Simon, Stephen Barker, Tony Kubiak, and Ian Munro for their continuous support. His family, Sky, Zeph, and Kris, as always, make it all possible and fun.

Notes on Contributors

Jen Boyle is Assistant Professor and Director of Undergraduate Studies of English at Coastal Carolina University and has held fellowships at the Pembroke Center at Brown University, The Folger Institute, and the Dibner Library. In addition to her work in early modern studies, she has also been an author-collaborator on new media installations. Her book, *Anamorphosis in Early Modern Literature: Mediation and Affect*, a study of mediation and embodiment in early modern literature and techno-science, is in production (forthcoming, November 2010).

Paul Cefalu is Associate Professor of English at Lafayette College, Easton, PA. He is the author of *Moral Identity in Early Modern Literature*; *Revisionist Shakespeare: Transitional Ideologies in Texts and Contexts*; and *Early Modern English Literature and Contemporary Theory: Sublime Objects of Theology*. His essays have appeared in *PMLA*, *ELH*, *Milton Studies*, *The Journal of the Medical Humanities*, and other journals, and he is an Associate Editor of *Literature and Theology*. He is currently working on book on anthropomorphism in early modern religion and literature.

Amy Cook is an Assistant Professor at Indiana University, Bloomington who specializes in the intersection of cognitive science and theories of performance, theater history and dramaturgy, early modern drama, and contemporary productions of Shakespeare. She is the author of *Shakespearean Neuroplay: Reinvigorating the Study of Dramatic Texts and Performance through Cognitive Science* (Palgrave Macmillan, 2010) and her essays have been published in *TDR*, *Theatre Journal*, *SubStance*, and Harold Bloom's Modern Critical Interpretations series (2009).

Gabriel Egan teaches Shakespeare and other English literature as well as the art of hand-printing. His past books include *Shakespeare and Marx* (2004, with a Turkish translation in 2006) and *Green Shakespeare* (2006). His latest book is *The Struggle for Shakespeare's Text* (2010).

Graham Hammill is Associate Professor of English at SUNY-Buffalo. He is the author of *Sexuality and Form: Caravaggio, Marlowe, and Bacon* (2000) and numerous essays in Renaissance studies, critical theory, and queer studies. He has just completed a manuscript entitled *The Mosaic*

Constitution: Political Theology and Literary Imagination from Machiavelli to Milton.

F. Elizabeth Hart is Associate Professor of English (Shakespeare, Renaissance, cognitive literary studies) at the University of Connecticut, Storrs. She is the author of essays on cognitive approaches to early modern literature and culture in the journals *Mosaic, Configurations,* and *Philosophy and Literature*; and the books *The Work of Fiction: Cognition, Culture and Complexity,* ed. Alan Richardson and Ellen Spolsky (2006) and *The Emergence of Mind: Representations of Consciousness in Narrative Discourse in English,* ed. David Herman (forthcoming 2011). She co-edited (with theatre historian Bruce McConachie) *Performance and Cognition: Theatre Studies and the Cognitive Turn* (2006) and is a founding member of the Cognitive Approaches to Theatre and Performance Group of the American Society for Theatre Research (ASTR). Her essays on Shakespeare have appeared in *Shakespeare Quarterly* and *Studies in English Literature.* She is currently working on several projects focused on Shakespeare's late career.

David Hawkes is Professor of English Literature at Arizona State University. He is the author of five books, including *John Milton: A Hero of Our Time* (2009) and *The Culture of Usury in Renaissance England* (2010). His work has appeared in many popular and academic journals, including *The Nation,* the *TLS, ELR, ELH, SEL, HLQ, JHI, Milton Studies, Shakespeare Quarterly* and *In These Times.*

Ken Jackson is Associate Professor of English and Director of Religious Studies at Wayne State University in Detroit. He is the author of *Separate Theaters: Bethlem ("Bedlam") Hospital and the Shakespearean Stage* (2005) and a number of articles on early modern drama, religion, and critical theory. He is currently completing an edited collection of essays with Arthur F. Marotti titled *Shakespeare and Religion: Early Modern and Postmodern Perspectives* and a monograph titled *Shakespeare, Abraham, and the Abrahamic.*

Gary Kuchar is Associate Professor of English at the University of Victoria B.C. Canada. He is the author of *Divine Subjection: The Rhetoric of Sacramental Devotion in Early Modern England* (2005), *The Poetry of Religious Sorrow in Early Modern England* (2008), and articles on, among others, Shakespeare, Donne, and Traherne.

Jerzy Limon is full Professor of English at the English Institute, University of Gdańsk, Poland. His main area of research includes the history of English drama and theater in the sixteenth and seventeenth centuries, and various theoretical aspects of theater. He has published widely on various topics, and his output includes three academic books published in English (*Gentlemen of a Company*; *Dangerous Matter*; and *The Masque of Stuart Culture*), five books in Polish, and over a hundred articles and reviews. His most recent works, published since 2008, include a book on the theory of 'television theater', *Obroty przestrzeni* (*Moving Spaces*), and articles in such journals as *Theatre Research International*, *Shakespeare Jahrbuch*, *Teatr*, *Journal of Drama Theory and Criticism*, *New Theatre Quarterly*, *Poetica*, *Liminalities*, and *Cahiers Élisabéthains*. His new book, *The Chemistry of the Theatre* (on the performativity of time and space), was published by Palgrave Macmillan in 2010. Recently, Limon has also edited a collection of essays, entitled *Theatrical Blends*. He is now preparing a book on King James I's secret service. Limon's literary output includes four published novels and translations of plays by William Shakespeare, Thomas Middleton, Philip Massinger, and Tom Stoppard. He also runs a theater project in Gdańsk, which aims at reconstructing an Elizabethan-in-style theater there, and organizes an annual International Shakespeare Festival.

Julia Reinhard Lupton is Professor of English and Comparative Literature at the University of California, Irvine, where she has taught since 1989. She is the author of *Citizen-Saints*, *Afterlives of the Saints*, and *After Oedipus* (with Kenneth Reinhard), as well as many essays on Shakespeare, religion, and psychoanalysis. She participated with David Pan and Jennifer Rust in the publication of the English edition of Carl Schmitt's *Hamlet or Hecuba*, and she has co-convened several projects on religion and Renaissance literature with Graham Hammill. Her latest book, *Thinking with Shakespeare*, is forthcoming in 2011. She also writes on design with her sister Ellen Lupton.

Ian Munro is Associate Professor of Drama at the University of California, Irvine. He is the author of *The Figure of the Crowd in Early Modern London: The City and Its Double* (Palgrave Macmillan, 2005) and the editor of *A Woman's Answer is Never to Seek: Early Modern Jestbooks, 1526–1625* (2007). Recent publications include 'Knightly Complements: *The Malcontent* and the Matter of Wit' (*English Literary Renaissance*, 2010).

Bryan Reynolds is Chancellor's Fellow and Professor of Drama at the University of California, Irvine. He is also the Artistic Director and resident playwright of the Transversal Theater Company, a collective of American and Dutch artists. Reynolds is the author of four books and many articles in which he develops the social, cognitive, and performance theories, aesthetics, and critical methodology of Transversal Poetics. These are: *Transversal Subjects: From Montaigne to Deleuze after Derrida* (2009); *Transversal Enterprises in the Drama of Shakespeare and his Contemporaries: Fugitive Explorations* (2006); *Performing Transversally: Reimagining Shakespeare and the Critical Future* (2003); and *Becoming Criminal: Transversal Performance and Cultural Dissidence in Early Modern England* (2002). He is also co-editor of *Critical Responses to Kiran Desai* (2009); *Rematerializing Shakespeare: Authority and Representation on the Early Modern English Stage* (2005); and *Shakespeare Without Class: Misappropriations of Cultural Capital* (2000). His plays have been performed in the United States, the Netherlands, Germany, Romania, Poland, the Czech Republic, and Armenia.

William N. West is Associate Professor of English, Classics, and Comparative Literary Studies at Northwestern University. Most recently he published articles on, among other things, the ironies of encyclopedism before and after the Enlightenment, the physics of early modern acting, the authority of the actor's voice, and replaying Renaissance drama. He is currently working on a book about understanding and confusion in the Elizabethan theaters, and another project on Poliziano's anti-theological (in)humanism, from which his essay in this volume is drawn.

Julian Yates is Associate Professor of English and Material Culture Studies at University of Delaware. His first book, *Error, Misuse, Failure: Object Lessons from the English Renaissance* (2003) examined the social and textual lives of relics, portrait miniatures, the printed page, secret hiding places in Renaissance England and was a finalist for the MLA Best First Book Prize in 2003. His recent work focuses on questions of ecology, genre, and reading in Renaissance English literature and beyond.

Tarrying with the Subjunctive, an Introduction

Paul Cefalu & Bryan Reynolds

The idea of a return of theory might give readers pause. When, if at all, did use and development of theory in early modern English literary-cultural studies decline? With the exception of new materialist, posthumanist, and some performance-oriented work, as well as the combined social theory, performance aesthetics, and critical methodology of transversal poetics, the field has suffered a waning interest in theoretically driven approaches that began during the mid-1990s. The radical drop in positions advertised in the MLA Job List for which an emphasis in theory was a primary criterion, the emergence of a large community of scholars whose focus was the history of the book, editing, or philology, and the widespread popularity of Harold Bloom's liberal-humanist account of Shakespeare's genius all exemplified this lag.[1]

The same theorists informed most literary criticism of the 1980s and early 1990s, including the new historicist, cultural materialist, psychoanalytic, and feminist approaches that dominated the study of early modern English literature and culture. The most popular of such theorists were Karl Marx, Sigmund Freud, Jacques Lacan, Michel Foucault, Louis Althusser, Jacques Derrida, Hélène Cixous, Luce Irigaray, and Julia Kristeva. But others, like Antonio Gramsci, Mikhail Bahktin, Raymond Williams, Clifford Geertz, Pierre Bourdieu, Gayle Rubin, and later Judith Butler also considerably influenced early modern English literary-cultural studies. This common genealogy is reflected in much of the work that has become the enduring methodological offspring of new historicism and cultural materialism, namely new materialism and the history of material texts. To a slightly lesser degree, their presentist and posthumanist cousins also reflect this genealogy.[2] This is true for transversal poetics too, but, like posthumanism, it is also significantly indebted to a number of other theorists, like Friedrich Nietzsche, Gilles Deleuze, and

Félix Guattari. Alternatively, presentism shows little positive trace of historical materialism and poststructuralism insofar as it actively avoids the theoretical initiative and thick description that characterizes its predecessors. On the other side of the spectrum is posthumanist work, which tends to be historicized, adventurous, and highly theoretical.[3]

If one accepts that there was, and continues to be, a diminished interest in the poststructuralist interpretations of early modern English culture published in the 1980s and early 1990s, new materialism, transversal poetics, and posthumanism do constitute a return to theory, often manifest in the forming of new theories in explicit response to the existing ones. With this return came enthusiastic attention to other continental theorists (Giorgio Agamben, Alain Badiou, Slavoj Žižek, Jacques Rancière) and to recent developments in certain areas of philosophy, science, and cognitive theory, including cognitive linguistics (Gilles Fauconnier, George Lakoff, Michael Tomasello, Seana Coulson), neurophilosophy (William Bechtel, Patricia Churchland, Paul Churchland, Francis Crick), consciousness studies (David Chalmers, Daniel Dennett, Andy Clark, Roger Penrose), cognitive neuroscience (Vittorio Gallese, V. S. Ramachandran, Giacomo Rizzolatti), neuro-aesthetics (Semir Zeki, Anjan Chatterjee), memetics (Richard Dawkins, Douglas Hofstadter, Richard Brodie, Robert Aunger, Susan Blackmore), and evolutionary psychology (Steven Pinker, Richard Joyce, Richard Byrne, Andrew Whiten). Along with attention to these other theorists and fields emerged new and related topical interests, in some cases interests enabled and affected by the shifts. Performance, religion, and cultural translation surfaced importantly in light of major international events, from the 2001 attacks on the World Trade Center and Pentagon through the wars in Afghanistan, Iraq, and the Congo; and subjectivity and identity formation has become important again. Moreover, scholars are approaching subjectivity and identity formation in radically new ways because of the major changes in political climate globally, the emergence of new modes of experience through the internet, and the groundbreaking research in cognitive neuroscience as furthered by major advances in observational technology, chiefly functional magnetic resonance imaging (fMRI), position emission tomography (PET), and more recently magnetoencephalography (MEG).

Although transversalist work engages and mobilizes with many of the theorists and areas of research that have contributed to the return of theory, we decided not to make transversal poetics a primary focus of this collection because its approach, like that of new materialist work, has already been established within the field, even though it too continues to develop. Instead, we are following transversal poetics' investigative-expansive methodology to feature significant develop-ments that are moving across and outside of longstanding, paradig-matic divisions, even while they sometimes continue to operate from traditional assumptions, or at least have not altogether departed from them. We see the new theoretically driven work featured in this collec-tion as emergent activity.[4] Both in groups and together this work forms distinct and overlapping articulatory spaces, where, in transversal terms, otherwise disparate streams of knowledge-transfer and experience inter-act with both affective presence and eventualization. This means that the work, in effect of the variation of articulatory formations to which it contributes, impacts positively subjectivity and subjective territories, not through specific absence or indirection, but through duration in the mode of a series of events that give the impression of contiguousness. The three sections to this collection each indicates an articulatory space. We have tried to head these sections with as much implicit explanatory power as this naming-function could accomplish while still being both broad enough to include differences and appropriately narrow with regard to research and theoretical focus.

Arguably, of the critical trends mentioned above, the one that has most filled the gap – with both affective presence and eventualization – between the decline of new historicism and the rise of a new genera-tion of theoretically driven work has been the philological turn in early modern English literary-cultural studies. Such work does not eschew high theory as much as it balances it; that is, the 'new' philology merged materialism with salutary attention to early modern textual scholarship, editorial practices, and other media. The 'history of the book' is of course a history of cultural discourse and its means of production and distri-bution, whether of material or abstract substance, and therefore it is a history of the practices motivated by basic cultural-materialist assump-tions of the circulation of social capital. Yet, to a large extent, this new philology sounds like business as usual: early modernists appropriating the most recent research trends, like new media and translation stud-ies, which have produced in such a short time (a decade perhaps) work that is relevant to literary critics in a manner that poststructuralism was relevant some time ago. One might deem this, in uncertain terms, a basic return to theorists (Marx, Foucault, Geertz, Bourdieu, McLuhan, etc.) rather than a return to theory as such, with the implication that the ontological and epistemological principles regarding the nature of subjectivity, materiality, and early modern reading and writing practices have been refined but not paradigmatically changed.

But what most of the essays in this collection reveal is that ontological watersheds have indeed been taking place in the academy, momentous shifts which have invited us to revisit basic presuppositions about consciousness, mind–body relations, subjectivity, linguistic competence, and the very concept of culture. What makes these developments particularly noteworthy is that the most significant developments hail not from continental philosophy or even within the humanities generally, but from the otherwise heterogeneous disciplines of hard science: biology, neurology, and evolutionary psychology, to name the most salient. To a certain extent, we are witnessing the much-heralded consilience of the 'two cultures': literary critics are now participating in the cognitive revolution, partly because some of the seeming bogeys like universalism, essentialism, and eliminative materialism have been shorn of their odious sociobiological implications.

A decade ago it would have been unimaginable for a book like William Flesch's *Comeuppance: Costly Signaling, Altruistic Punishment, and Other Biological Components of Fiction*, which applies the canons of evolutionary psychology to narrative theory, to have received the warm embrace it has among cultural critics.[5] And a wellspring of related critical work has recently been published, such as Dennis Dutton, *The Art Instinct: Beauty, Pleasure, and Human Evolution* (2009), Brian Boyd, *On the Origin of Stories: Evolution, Cognition, and Fiction* (2008), and Maryanne Wolf, *Proust and the Squid: The Story and Science of the Reading Brain* (2008). This body of work offers a compelling case that humanists can gain much from accepting basic notions of embodied cognition.[6] But in order to tell the story of how and to what extent the cognitive revolution has been rendered palatable to scholars of early modern English literature and to cultural critics in general, we should review some of the ontological and epistemological verities which it has worked to displace, and therefore also the official territories of critical practice that the cognitive revolution works to expand and transform.

Culture revisited

Over two decades ago, when a commitment to politicizing early modern literature held sway, Jean Howard and Marion O'Conner edited a timely and influential book, *Shakespeare Reproduced: The Text in History and Ideology*.[7] The editors had set out to counter ahistorical and essentializing interpretations of Shakespeare, to explore instead the 'political functions of the text at specific historical junctures within specific social practices'.[8] Following what had by then become a basic new historicist gambit, 'to place literature in history, rather than as a reflection of it', *Shakespeare Reproduced* conceptualized literature as 'one of a vast ensemble of cultural practices through which constructions of the real are circulated and people are positioned as subjects of ideology'.[9] Paramount to the conceptual territory of this movement, and thus to what became the official territory of new historicism, were two explicitly defended tenets: (1) the idea made popular by Althusser that we can never get outside of ideology, particularly the dominant ideology, because subjectivity is always immersed in and a product of it; and (2) the Foucauldian idea that we are always ever caught in a network of power-knowledge relations from which dissident individual agency is (it seems) impossible to achieve; recall the 'subversion/containment paradigm' of new historicism and the 'entrapment model' of cultural materialism.

Yet underlying the capacious term 'cultural practices' was Geertzian symbolic anthropology which, although often left undertheorized in such appropriations by scholars of early modern English literature and culture, held a quite radical notion of the ways in which individuals relate to their cultural surroundings. In the same ways that semioticians of theater like Jean Alter, Keir Elam, Erika Fischer-Lichte, and Anne Ubersfeld understand theater as a signifying practice in which meaning production occurs in the interrelations of different codes, Geertz understood culture as a semiotic system, an ensemble of 'structures of signification' which shape human action.[10] He vehemently rejected the notion, then fashionable among anthropologists, of a 'critical point' in evolution, a point at which biologically complete hominids invented culture *de novo*.[11] Geertz's response was that culture was an intrinsically prior selective factor that shaped human development. Deeply committed to a rejection of privacy theories of meaning and psychologism, Geertz implied the reverse (or was taken to imply the reverse by literary critics), namely, that culture is entirely social; it contains the mind, constructing individuals *all the way down*.

In response to what has been deemed as reductive social constructivism, cognitive theorists in several fields (anthropology, religious studies, linguistics, psychology) typically have argued that, although cultures are heterogeneous and variable across time and space, they are products of highly adaptive mental schemas that are conducive to human adaptiveness. According to Steven Pinker: 'culture can be seen as ... part of the human phenotype: the distinctive design that allows us to survive, prosper, and perpetuate our lineages'.[12] Cognitive theory, when it eschews extreme neural plasticity, depends on various strains of innatism, and often an attendant modular theory of the mental.

Modularity posits that the brain is an information processor comprised of cooperating parts, each of which plays a role in shaping behavior, and each of which has developed over time to enhance adaptive success. So a 'theory of mind' module helps us to anticipate and interpret the intentions and behavior of others; and a 'cheater detection' module allows us to pinpoint free riders and liars. So much might not be difficult to accept: what might trouble some humanists is the further contention that these modules are often described as hard-wired cognitive structures, many of which are responsible for shaping the culture in which they operate: 'Behavior is not just emitted or elicited, nor does it come directly out of culture or society. It comes from an internal struggle among mental modules with differing agendas and goals.'[13]

How then might we integrate some of the empirical findings of biology with a theory of culture which accepts the implications of embodied cognition but does not embrace biological reductionism? Consider the implications of Robert Boyd and Peter Richerson's notion of 'dual inheritance':

> Culture is neither autonomous and free to vary independently of genetic fitness, nor is it simply a prisoner of genetic constraints. Our rejection of this dichotomy is based on what we call the 'dual inheritance' theory of the interaction of genes and culture ... The essential feature of this theory is that, like genes, culture should be viewed as a system of inheritance. People acquire beliefs, attitudes, and values from others by social learning and then transmit them to others. Human behavior results from the interaction of genetically and culturally inherited information.[14]

Dual inheritance theory should not be confused with memetics. If memetics posits that cultural transmission analogizes genetic transmission (cultural notions, like genes, are selfish replicators) dual inheritance theory more often implies a causal relationship in its assumption that cognitive adaptations are the proximate engines of cultural evolution. Typically, dual inheritance theorists will choose an overriding, species-specific adaptation, and then describe the cultural manifold that evolves in turn: in most of this work, the arrows of causality are such that phylogeny shapes culture, which in turn shapes ontogeny, or historically situated individuals. For example, Michael Tomasello has argued that the genetic and cognitive adaptation that funds much cultural evolution is the phylogenetic capacity of modern humans to identify with conspecifics, to 'understand them as intentional and mental beings like the self'.[15] This entails three types of learning processes (imitative, instructive, and collaborative) that generally do not obtain among non-human primates, and that are responsible for the procession of culture.[16] The adaptive ability to intuit and understand another's intentions (why a particular tool has been developed, for example) often goes by the moniker 'theory of mind' in some quarters; for Tomasello, at least, intentional behavior is the necessary condition for cultural creativity and evolution.[17]

There are several implications of the dual inheritance theory for cultural studies, but the most modest is a methodological one, evidenced in many of the essays in our collection. Perhaps startlingly, the opening critical strategy in several of the essays is neither a culturally specific anecdote, a rich quotation from a treatise, sermon, or play, nor an account of a fateful political and historical event; it is more often an outline of some aspect of the information processing and adaptive ability of the mind, whether that be the role of mirror neurons, as in Jen Boyle's interpretation of Defoe's *A Journal of the Plague Year* and Gabriel Egan's account of early modern drama; or cognitive blending, as in Elizabeth Hart's and Amy Cook's tracking of the iterations of key Shakespearean metaphors.

Where the hard science of cognition is not directly invoked in the essays of this collection (in those essays concerned with the turn to religion, for example) we still get a sense that a robust turn to religion might involve, as Julia Lupton suggests, an understanding of religious concepts which does not reduce such concepts to local contexts.[18] None of this involves a reduction of culture to embodied cognition, but a gestalt shift of sorts, one nicely described by Elizabeth Hart in this volume as a bottom-up approach to interpretation: 'all the knowledge humans can experience is, by definition, *mediated*, first by individual bodies and brains, then by brains and minds, and finally by the aggregates of minds we call "society" or "culture" and without which (in "feedback loop" dependency) individual brains and minds would fail to develop' (23).

New formalisms and materialisms

One manifestation of the use of cognitive theory in the essays below is a renewed interest in formalism, especially cognitive linguistics and theories of metaphor. Several of the essays refer to the concept of conceptual blending developed by the cognitive linguists George Lakoff, Gilles Fauconnier, and Mark Turner. Contending that metaphor is fundamentally semantic and not merely grammatical in nature, cognitive linguists posit the existence of fundamental image schemes on which the brain relies in order to process ordinary and complex figurative language. In a bid to make a science of imagination, Fauconnier and Turner claim that 'characteristic of the human species' is the use of conceptual integration 'in which input spaces are projected into a blended space, and through processes of composition, completion, and elaboration, develop emergent structures in the blend'.[19] To take a simple example: while learning how to ski, one is typically instructed to 'push off' as if one were roller skating. The trainee does not simply duplicate the action of roller skating while skiing, but rather must 'selectively combine the action of pushing off with the action of skiing and develop in the blend a new emergent pattern, known (not coincidentally) as "skating"'.[20] Blending presupposes basic conceptual or 'vital relations' such as change, identity, time, space, cause–effect, part–whole, representation, and analogy.[21] Vital relations are akin to image schemes, described by Hart as 'non-visual mental templates existing prior to the development of concepts'.

Hart's interpretation of *The Merchant of Venice* puts into practice several of the elements of cognitive linguistic theory. The uncovering of an open-ended, 'emergent semantic system' in the play involves tracking semantic iterations such as 'venture', 'forfeit', 'bond', and 'gold', as well as conceptual metaphors such as 'love is money'. In the blended space the elements associated with financial and contractual obligations become intertwined with 'categorical features of intersubjective trust between humans and between humanity and the divine' (34). Similar types of blending occur in *King Lear*, Amy Cook argues, as in Lear's testing of the belief that breath on a looking glass or a stirring feather can mark Cordelia's life. And Jen Boyle's essay relies even more heavily on some of the findings of cognitive science (particularly the role of mirror neurons in empathy) in her transversal approach to mediated perception in Defoe's *A Journal of the Plague Year*. Influenced by the Lucretian presupposition that perception results from the movement of images or simulacra 'through the air and between animate and inanimate objects', Defoe's 'burial pit becomes an exploration of how mediated images transform not only subjective and collective perception, but also the physical and temporal spaces of the city' (82). Defoe's text ultimately offers aesthetic justification for the theory of consciousness implied by the operations of mirror neurons: consciousness is based literally in the mediated exchange of images, 'not material information that our perceptions represent back to us in constructing consciousness, but the traces of the very process of the exchange of these time-images' (71–72). What the cognitivist interpretations in the collection claim is that, as Hart eloquently puts it, 'early modernists must accommodate theories about how human brains and minds function, both in terms of the brains and minds of our historical subjects – the humans who people our histories and whose representations fill our texts – and in terms of ourselves and our own mental habits as critics and historians' (23).

This is not to suggest that cognitive approaches to culture and biology need rely on human exceptionalism. Cognitive ethnology attempts to merge the findings of neuroscience with the traditional, 'ethological' study of animals in their natural environments. With some caveats, this is an approach approved by Donna Haraway, who, along with Katherine Hayles and others has been instrumental in lifting species barriers between humans and the 'companion species'. For example, the cognitive ethologist Stanley Coran has recently demonstrated that canids share a 'theory of mind' with their human companions, that 'dogs do seem to understand that other creatures have their own points of view and mental processes'.[22] Although not directly invoking the scientism of cognitive theory, Julian Yates' essay demonstrates the relevance of posthumanism and distributed cognition to early modern cultural studies. Declaring at the outset that his essay takes 'up the burden that there exists a history of technology, of the machine, the plant, and the animal, that is simultaneously and necessarily also a history of human life', Yates offers a new understanding of the pastoral agenda, asking, for example, whether animal *otium* can exist, whether it bears similarities to human *otium*. For Yates, a bio-politics of pastoral underscores continuities between sheep and shepherds: 'one way of deactivating or stalling the "anthropological machine" and the mutual definition of "sheep" and "not sheep", would be to refuse to treat sheep as "sheep" at all. Instead, like primatologist-turned-sheep farmer Thelma Rowell, you treat them as members of a much broader group' (113).

Political theology and the religious turn

It would be misleading, however, to characterize all of the essays in this volume as directly informed by the cognitive revolution. Several of the essays offer sophisticated reinterpretations of what has become the fashionable topic of political theology, and the theory appropriated in such work is more often philosophical in orientation. We are happy to note that a 'return to theory' does not entail an appropriation of any single theoretical or methodological advance, such as cognitivism, although one can detect the guiding assumption in the essays in this section that religious concepts are not rooted entirely in local contexts, a supposition that emerges in much work in the burgeoning fields of cognitive religious theory and neurotheology.[23]

For some time scholars have been announcing a return to religion in early modern English literary-cultural studies. Perhaps the best overview of the critical implications of this turn can be found in Ken Jackson's and Arthur Marotti's introductory essay to a topical issue of *Criticism* published in 2004.[24] Jackson and Marotti contend that, unlike cultural materialism which strives to locate alterity in Renaissance texts and contexts, usually designating an 'other' subject to ideology critique, the turn to religion, which has often been informed by the late work of Derrida and Levinas, accepts the irreducible quality of alterity in religious contexts: 'Renaissance religion resists our alterity criticism and thereby reveals the aporetic, philosophical problems hardwired into New Historicism and its organizing respect for alterity, its desire to "speak with the dead".'[25]

The implications of a hermeneutics of religion which recognizes the universal pull of many religious concepts is put concisely in Lupton's and Graham Hammill's introduction to a special issue of *Religion and Literature*. After declaring that 'religion is not fully reducible to culture', Lupton and Hammill point out that, while religious habits of thought will shape and be shaped by local formations, religion 'instantiates discourses of value that aim to transcend culture, by creating trans-group alliances and affiliations around shared narratives, commandments, and principles. Unlike forms of national belonging, the singular traits of ascription and prescription around which religious communities form are conceived as coming from outside the groups that adhere to them.'[26]

Employing language that is reminiscent of Richard Dawkins' theory of memetic transmission, Lupton and Hammill claim that, 'like ghosts or viruses, religions leap across groups and epochs, practicing cultural accommodation in order to outlive rather than support the contexts that frame them'.[27] For some anthropologists of religion, one way of explaining the persistence of religious cultures and values is on the analogy of natural selection. Religious notions are cultural units of transmission, memes which, like genes, are high fidelity replicators; those religious memes which have a universal appeal to psychological well-being (the meme for life after death, for example) will naturally survive in the meme pool: 'Religions, like languages, evolve with sufficient randomness, from beginnings that are sufficiently arbitrary, to generate the bewildering – and sometimes dangerous – richness of diversity that we observe. At the same time, it is possible that a form of natural selection, coupled with the fundamental uniformity of human psychology, sees to it that the diverse religions share significant features in common.'[28]

But what of individual agency in advancing particular religious doctrines? Dawkins is keen to point out that some mixture of natural or cultural selection and intelligent design fosters the advancement of religion. So, to the extent that Martin Luther stood behind the Protestant revolution, he was 'not its designer but a shrewd observer of its efficacy'.[29] Whether we agree with Dawkins' undertheorized theory of memes, as well as his lack of explanation as to which constituents of hominid psychology find religion comforting or wish-fulfilling, we find again an emphasis on the ways in which historically recurring religious notions interact with, but at some level are independent of, cultural contexts.

One useful concept that has emerged from the turn to religion, a concept which grows out of early modern contexts, is the 'state of exception'. In his writings on sovereignty and political decisionism, Carl Schmitt grounds his theory of the state of exception (according to which the sovereign may suspend the force of law) on the early modern political theology of Jean Bodin, Samuel Pufendorf, Hobbes, Malebranche, and Locke, among others.[30] Sovereign exceptionalism has its early modern analogues in the sovereign's use (and abuse) of the royal prerogative: as God's vicegerent and the very embodiment of law, early modern sovereigns might declare a state of exception during real or perceived emergency times. Extending Schmitt's decisonism, Giorgio Agamben claims that, under modern regimes, the state of exception is the norm, rather than the rule, and renders law potentially, rather than actively in force. What has emerged from Agamben's work is the useful term 'bare life', a mode of existence that lies somewhere between *zoe*, or purely natural life, and *bios*, life constituted by the *polis*. Bare lives are 'abandoned' to the state not unlike concentration camp victims whose lives are wholly dictated by sovereign exceptionalism.[31]

What this means in part for early modernists is a return to some of the abiding questions of political and intellectual history, and a concern to conceptualize such bare lives which seem uncannily to survive the very negation of cultural accretions. So Lupton's essay in this collection brings attention to the recently translated writings of the Austrian jurist Hans Kelsen, whose work offers 'important constitutional and liberal correctives to the authoritarian arguments of Schmitt' (123).

Distinguishing the group or mass from the state in its ideal form, Kelsen contends that, unlike states, which develop around symbolic ideas, masses converge around charismatic leaders akin to the figure of the father in Freud's primal horde, and so are prone to regressive violence. Still, because the 'idea of the state is inserted in the mesmerizing place of the ego-ideal' (126), it can license the regressive tendencies of the mass, and so Kelsen privileges individual psychology over any substantialization of the body of the state. The ideal state form would be legitimated simply by a system of norms and laws, the basic norm of which, the *Grundnurm*, founds all other norms as the 'normative foundation of God's authority to issue norms' (129). Lupton then ably reconstructs how Kelsen might have interpreted some of the cruxes of *Hamlet*: the injunction to remember his father prompts Hamlet to implement the modes of constitutionalism entailed by elective monarchy. Methodologically, Lupton's essay is typical of many of the essays in the collection in that it partly relies on local texts to raise broader questions, for example, whether there exists a 'political theology native to liberalism and to the broader tradition of constitutionalism to which liberalism belongs' (137).

Hammill, too, is interested in expanding Schmitt's theory of the exception. Pointing out the frequency with which 'constituting' power is opposed by imagination, Hammill contends that the very category of aesthetics is a 'competing figure for constituting power' (145). The earliest edition of Marlowe's *Hero and Leander* – in which Hero decides to consummate with Leander, rather than, as later canonical editions hold, Hero is raped by Leander – points to an unreliable politics of imagination: if imagination enables prudential reasoning, it just as easily misdirects such reasoning toward false conclusions (153). The entanglement of imagination and politics has important consequences for understanding representations of subjectivity in that, according to what Hammill describes as the 'Marlovian sublime', characters defend themselves against the encroachments of imprudent sovereignty by an 'undoing or betrayal of the self' (158). In the case of *Hero and Leander*, Hero's 'ravishment becomes a compromise position in which imagination reclaims the force of making against sovereignty by staging the very effects of sovereign power in and as imagination' (160).

In further exemplifications of the turn to religion, Ken Jackson's and Gary Kuchar's essays more specifically return to Pauline theology. Jackson argues that Leontes' jealousy in *The Winter's Tale* derives from his inability to grasp the full implications of messianic time: since messianic time nullifies existing time, and renders ephemeral chronologically established distinctions of love, Leontes' marriage to Hermione becomes virtually indistinguishable from Hermione's gesture of friendship with Polixenes. The Pauline messianic 'time of now' shares much with Deleuze's notion of the virtual, 'the non-present, i.e., the whole continuum of time itself' (195), and thus locates messianic time immanently, rather than transcendentally.

In his study of Shakespeare's sonnets, Kuchar argues that Shakespeare offers two perspectives on the Pauline notion of the old man becoming new by putting on Christ: Neoplatonic *eros*, on the one hand, and the asymmetries of reformist *agape*, on the other hand. According to the former, the speaker becomes one with his beloved; according to the latter, the speaker is merely imputed to be one with Christ and so still experiences alienation from the beloved. Not only is Kuchar's essay a welcome correction to those approaches to the sonnets that separate theology and psychology; it realizes, as do all of the essays in the book's section on theology, Lupton's contention that a return to religion marks a return to 'concepts, concerns, and modes of reading that found worlds and cross contexts, born out of specific historical situations, traumas, and debates, but not reducible to them' (211).

Rematerialism

What brings so many of the essays in this book together is a return to materialism, and it is here in particular that one again detects a productive convergence of the 'two cultures'. Part of the goal of the new materialist work by early modernists has been to restore the primacy of objects without also assuming that such objects are fetishized by their bearers. In *Renaissance Clothing and the Materials of Memory*, Peter Stallybrass and Ann Rosalind Jones remark that fetishization emerges during the transitional period as a term of economic and religious abuse (in the sense that natives might fetishize trifles) and that a truly materialist inquiry into early modern culture would recognize that material goods such as clothing were markers of a transition from understanding objects as 'social symbols, bearers of social status', to understanding them as 'circulating commodities'.[32]

Materialist inquiries salutarily direct a focus away from timeworn notions of subjectivity and toward a fresh consideration of objects in themselves. However, as Douglas Bruster has argued, materialist inquiries would do well to avoid pulling objects out of their historical contexts and symbolic significance.[33] Furthermore, materialist inquiries run the danger of reductionism when they slide into eliminativism.

One exemplary study that is alive to the dangers of eliminative materialism is Jonathan Gil Harris' *Untimely Matter*, which notes that objects are not only changing, but are constituted by a regress of smaller objects down to subatomic particles, the substrate of which might indeed be complexly physical but is not crudely material.[34]

To the extent that the essays in this collection, particularly those in the last section on theater, extend and refine this new materialism, we note a focus not simply on material objects but rather on material bodies, or even more specifically, material cognition. As Jerzy Limon points out, the elements of postdramatic theater (depsychologized, non-verbal performances which treat the 'human body in a manner analogical to body art') were all key elements of medieval and Renaissance dramaturgy, especially the masque and its attendant machinery of the court theater. Especially interesting is the Renaissance 'postdramatic' theater's project of representing the human body in its own right. One finds a preoccupation with the naked body and with an 'aesthetics of ugliness, deformation, vagueness of anatomy, and sex': in keeping with the masque's project to project harmony out of chaos, 'nakedness is not so much erotic challenge or licentiousness as divinity' (271).

If the postdramatic theater privileges bodies in their own right, spectacular bare lives, as it were, early modern theater also, as Gabriel Egan suggests, proleptically tests the most recent theories of embodied cognition, especially the 'theory of mind'. As mentioned earlier, one adaptive trait that hominids have developed is the ability to track their sense of another's intentions and behavior, evidenced, Egan argues, by the back and forth that one finds between Iago and Othello. And the use of 'cognitive prostheses' or tools such as plot scripts, versification, simple rules governing soliloquies, even the structure of the playhouse itself, all facilitated the efficiency with which actors could master several roles and plays in a short time period.

The early modern theatricalization of bodies as such, as well as embodied cognition, informs Ian Munro's claim that early modern English drama realizes performance theory's concern to privilege the phenomenal or lived body over the abstract, signifying body, one reduced 'to transparent semiotic meaning' (298). Undermining the script as master discourse, theater criticism would do well to consider the theater as a 'place of embodied social practices and the body (especially the actor's body) as a site of durable social dispositions' (301). What should be added to this movement, according to Munro, is a sense of the uncanny and metadramatic moments when audience expectation shifts – as at the end of *Doctor Faustus* – and characters themselves 'register the limitations of the representational mode in which they find themselves' (302).

The essays collected here tarry with the subjunctive, that is, they engage *as-if* and *what-if* hypotheses that have the potential to expand our conceptual territory with regard to, at the very least, the study of early modern English literature and culture. In the positive flow of the essays, we would like to wrap up this introductory chapter by emphasizing the emergent quality of the scholarly activity we have chosen to include. As Reynolds and William West put it in their introduction to the anthology *Rematerializing Shakespeare*: 'No longer can the critical terrain be mapped exclusively in terms of binaries, dialectics, pyramids, squares; critical intervention must offer itself as interconnected and vibrating in sympathy with other interventions, but fully reducible to no other unit or category. The intervention is the emergent product of the forces and discourses of the community, but the properties uniquely associated with the community depend for their existence on the individuals who comprise the community but themselves lack the emergent properties.'[35] To engage the contributions to this collection as not satellite or integrally linked to the communities from which they emerge – their history, networks, and affects – suggests a misunderstanding of, and does a disservice to, the critical intersections at which we find ourselves. To explore reasons to validate the communities – their overlaps, shared interests, and complementary contributions (from peer-reviewed published research, critical exchange, and conference dialogues to the immediacy of classroom pedagogy) – is to acknowledge that we are already participating in both raising critical awareness and the making of theory that hermeneutics require. In so doing, we further the kinds of productive intellectual activism that enables and affects politically and socially purposeful educational enterprises. The future of the field is uncertain, but the return of theory is not: already happening, its various emergences are gaining momentum.

Notes

1. The University of Pennsylvania and Florida State University, which may have the strongest collectives of scholars of early modern English literature, both have programs that emphasize the history of the book, entitled 'Workshop in the History of Material Texts' and 'The History of Text Technologies Program', respectively. This is not to say that scholars in these departments do not employ theory, but rather that theory neither drives nor is necessarily engaged in their work.
2. On presentism, see Terence Hawkes, *Shakespeare in the Present* (London: Routledge, 2002) and Hugh Grady and Terence Hawkes, eds, *Presentist*

Shakespeare (London: Routledge, 2006); for an example of ambience studies, see Bruce R. Smith, *The Acoustic World of Early Modern England: Attending to the O-Factor* (Chicago: University of Chicago Press, 2009), as well as Smith's recent *The Key of Green: Passion and Perception in Renaissance Culture* (Chicago: University of Chicago Press, 2009); for examples of the new materialism, see Jonathan Gil Harris, *Untimely Matter in the Time of Shakespeare* (Philadelphia: University of Pennsylvania Press, 2008); for a recent posthumanist assessment of the species difference in Shakespearean drama, see Laurie Shannon, 'Poor, Bare, Forked: Animal Sovereignty, Human Negative Exceptionalism, and the Natural History of *King Lear*', *Shakespeare Quarterly* 60:2 (2009), 168–196.

3. For many examples of posthumanist work in early modern English studies, see the inaugural issue of *postmedieval: a journal of medieval cultural studies*, entitled 'When Did We Become Post/human?' 1:1/2 (April 2010).

4. On such scholarly 'emergent activity', see Bryan Reynolds and William West, 'Shakespearean Emergences: Back from Materialisms to Transversalisms and Beyond', in Reynolds and West, eds, *Rematerializing Shakespeare: Authority and Representation on the Early Modern English Stage* (Basingstoke: Palgrave Macmillan, 2005), 1–18.

5. William Flesch, *Comeuppance: Costly Signaling, Altruistic Punishment, and other Biological Components of Fiction* (Cambridge, MA: Harvard University Press, 2009).

6. Denis Dutton, *The Art Instinct: Beauty, Pleasure, and Human Evolution* (New York: Bloomsbury, 2008); Brian Boyd, *On the Origin of Stories: Evolution, Cognition, and Fiction* (Cambridge, MA: Harvard University Press, 2009); Maryanne Wolf, *Proust and the Squid: The Story and Science of the Reading Brain* (New York: Harper, 2008).

7. Jean Howard and Marion F. O'Connor, *Shakespeare Reproduced: The Text in History and Ideology* (London: Methuen, 1987).

8. Ibid., 5.

9. Ibid., 8.

10. Cited in Bradd Shore, *Culture in Mind: Cognition, Culture, and the Problem of Meaning* (New York: Oxford University Press, 1996), 32.

11. Ibid., 33.

12. Steven Pinker, *The Blank Slate: The Modern Denial of Human Nature* (New York: Penguin Books, 2002), 60.

13. Ibid., 40.

14. Robert Boyd and Peter J. Richerson, *The Origin and Evolution of Cultures* (New York: Oxford University Press, 2005), 104.

15. Michael Tomasello, *The Cultural Origins of Human Cognition* (Cambridge, MA: Harvard University Press, 1999), 10.

16. Ibid., 5.

17. Ibid., 6. See also Richard Byrne and Andrew Whiten, *Machiavellian Intelligence: Social Expertise and the Evolution of Intellect in Monkeys, Apes, and Humans* (Oxford: Clarendon Press, 1988), for an account of primates having a theory of mind and the implications this has for humans. For engagement with Byrne and Whiten's theory with regard to Shakespeare, theater, and performance studies, see Anthony Kubiak and Bryan Reynolds, 'The Delusion of Critique: Subjunctive Space, Transversality, and the Conceit of

Deceit in Hamlet', in Bryan Reynolds, *Transversal Enterprises in the Drama of Shakespeare and his Contemporaries: Fugitive Explorations* (London: Palgrave Macmillan, 2006), 64–84.

18. For an example of work along these lines, see Andrew Newberg, *Why God Won't Go Away* (New York: Random House, 2002).
19. Gilles Fauconnier and Mark Turner, *The Way We Think: Conceptual Blending and the Mind's Hidden Complexities* (New York: Basic Books, 2002), 89.
20. Ibid., 21.
21. Ibid., 92–102.
22. Cited in Donna J. Haraway, *When Species Meet* (Minnesota: University of Minnesota Press, 2008), 236.
23. See Andrew Newberg, *Principles of Neurotheology* (Burlington, VT: Ashgate, 2010).
24. Ken Jackson and Arthur Marotti, 'The Turn to Religion in Early Modern Studies', *Criticism* 46:1 (2004), 167–190.
25. Ibid., 179.
26. Graham Hammill and Julia Lupton, 'Sovereign, Citizens, and Saints: Political Theology and Renaissance Literature', *Religion and Literature* 38:3 (2006), 1–11.
27. Ibid., 2.
28. Richard Dawkins, *The God Delusion* (New York: Houghton Mifflin, 2006), 220.
29. Ibid., 221.
30. See Carl Schmitt, *Political Theology: Four Chapters in the Concept of Sovereignty*, ed. George Schwab (Chicago: University of Chicago Press, 2006).
31. See Giorgio Agamben, *States of Exception*, trans. Kevin Attell (Chicago: University of Chicago Press, 2005). For a transversalist perspective on this notion of bare life, see Anna Kłosowska and Bryan Reynolds, 'Civilizing Subjects, or Not: Montaigne's Guide to Modernity, Agamben's Exception, and Human Rights after Derrida', in Bryan Reynolds, *Transversal Subjects: From Montaigne to Deleuze after Derrida* (London: Palgrave Macmillan, 2009), 203–61.
32. Ann Rosalind Jones and Peter Stallybrass, *Renaissance Clothing and the Materials of Memory* (Cambridge: Cambridge University Press, 2001), 7–11.
33. Douglas Bruster, *Shakespeare and the Question of Culture: Early Modern Literature and the Cultural Turn* (New York: Palgrave Macmillan, 2003).
34. Jonathan Gil Harris, *Untimely Matter in the Time of Shakespeare* (Philadelphia: University of Pennsylvania Press, 2008).
35. Reynolds and West, 'Shakespearean Emergences', 13.

Part I
New Formalisms and Cognitivism

1
A Paltry 'Hoop of Gold': Semantics and Systematicity in Early Modern Studies

F. Elizabeth Hart

Introduction

In recent years, early modernists across a spectrum of fields have been picking up on the intellectual trends in a variety of other disciplines in their quest for a descriptive vocabulary for the patterns we find in early modern cultures. The objects we are trying to describe – material histories crossing a range of phenomena and levels of analysis – defy description by more traditional and received vocabularies, including the progeny of the various poststructuralisms: deconstruction, psycho-analysis, and Althusserian and Foucauldian materialisms.[1] The behav-iors of these systems, involving trade and economics, exploration and map-making, social and political formations and change, production and cultivation of the arts, and the spread of literate culture throughout Europe, to name only some, invite critics and historians to experiment with the vocabulary of 'complex' systems that has been more readily associated with the social sciences and – undergirding them – the physi-cal sciences.

To describe, for instance, a now-common type of early modern study: An examination of the small-scale trading practices associated with a particular commodity or geopolitical site points to specific material practices or technologies that give way to larger-scale matrices of agents, things, practices, and technologies, all in complex alliance. These matri-ces extend from the subjective agency of individuals to the intertwining agencies of social groups – encompassing the fluctuating conceptual paradigms of such groups – and then eventually to the groups' develop-ment of new forms of materiality, i.e., practices or technologies that in turn shape new forms of subjectivity/agency, often in colonial contexts.[2] Such studies routinely cite the 'emergence' of patterns within such cultural scenarios, their authors sliding between the varying imagery of 'networks', 'webs', and 'circuits' in their attempts to capture the multi-dimensional connections that they sense are motivating these uncanny abstractions. Ironically, or so it has seemed to this author, the harder we focus on such scenarios, deploying the newly refined techniques of materialist history, the more bewildering their complexities become.[3]

It seems apparent that the rhetoric of 'complex' or 'dynamic' systems theory has become increasingly useful within the humanities, but do we know *why*? In this essay, I will suggest that there is a profound correlation between our discovery of the salience of systems rhetoric and the parallel, if modest, rise of cognitive literary and cultural studies, notably in early modern studies of recent years. Cognitive literary theory provides insights into the roles that the brain and mind play in both enabling and constraining social agency and cultural production. As an analytical tool, such theory helps us to model the relays of influence that connect texts and readers – and also performances and spectators – within aggregates of cognition for a redefinition of the all-important 'cultural contexts' of materialist history as dynamic systems of human intersubjectivity.[4] Building on such theory, I would like to draw attention to an idea that is now generating some empirical support from the science of cognition and dynamic systems: Human cognition increasingly impresses cognitive scientists for its resemblance to 'connectionist' computer models that give imagery to a wide variety of complex biological as well as physical systems – from the symmetries of flocking birds to the patterns of insect migrations; from the spread of viral epidemics to the flowerings of household mold spores; or from the geometry of eroding coastlines to the a-rhythms of the healthy human heart (and so on in seemingly numberless examples). Scientists now realize that such systems constitute the *rule* rather than the exception in nature, and some are beginning to test for the possibility that the human conceptual system is included among these dynamic systems.[5]

I argue that, in order for such science to be meaningful to us, that is, to humanists who are uniquely trained to analyze the world from the largest-scale vantage points of social and cultural structures built upon material strata, we humanists must 'flip' our top-down postmodern cultural models to view them *primarily* as bottom-up ones. We must imagine the relay of cultural determinants of texts and performances and their receptions as a system in which aggregates of minds give rise to the cultural contexts in which they are embedded, in processes of accretive creativity that include (but are not exclusive to) the expressive artifacts of texts and performances.[6] And it is only with such an etiology in our toolbox that we may track and articulate a theory of the 'emergence' – the self-organization – of patterns within the systems that it is our business to observe: e.g., trade histories, shifting economies, rises in literacy, reorganizations of knowledge, social and political agency and change, arts production and cultivation, and the many other examples of discursive systems that our materialist investigations now link to specific things or practices in the early modern world.

Mostly in concurrence with the cognitive reading program of Mary Thomas Crane, I contend that early modernists must accommodate theories about how human brains and minds function, both in terms of the brains and minds of our historical subjects – the humans who people our histories and whose representations fill our texts – and in terms of ourselves and our own mental habits as critics and historians.[7] The epistemology that I adopt with Crane marks a subtle but crucial shift from both the traditional, idealist epistemologies of the pre-1980 twentieth century and the array of Marxist and neo-Marxist materialisms that now widely inform critical practice. Based on the condition of 'cognitive (or mind) embodiment', in which minds create cultures along the conduit of a materialist continuum extending from the concrete exigencies of the human body to the very abstract complexities of technological and social systems, a cognitively inflected epistemology inhabits neither the realism of idealist traditions nor the relativism of postmodern ones; rather, it hovers somewhere between the two, accepting some aspects of both while avoiding any absolute commitment to either.[8]

Put a little more colloquially, whereas both realist and relativist epistemologies represent polarized positions ('Reality exists, and humans can know it on its own terms' vs. 'Knowledge is contingent and determined only by the conditions of knowing'), the alternative, cognitive embodiment, succeeds in combining the two ('Reality exists, but humans can only access it through contingent conditions of knowing') and thereby eludes the absolutist traps inherent to both. Cognitive embodiment offers the recognition that all the knowledge humans can experience is, by definition, *mediated*, first by individual bodies and brains, then by brains and minds, and finally by the aggregates of minds we call 'society' or 'culture' and without which (in feedback-loop dependency) individual brains and minds would fail to develop. Some of the implications of cognitive embodiment include the position that the postmodern critique of transcendent idealism remains intact (i.e., there is no such thing as 'Truth') *and* the even more controversial one that human beings, for all their many social contingencies, are constrained enough epistemologically to allow for some same – or similar-enough – knowledge to cross both the transhistorical and the transcultural divides (i.e., some aspects of humanity are, or effectively serve as, universal).[9]

A bottom-up epistemology conjoining mind with culture – in that order – may be useful to critical theory as we search for metaphors to describe what our data increasingly reveal about early modern history: patterns of systematic iteration and replication, what the science calls 'self-similarity', that bespeak puzzling yet demonstrably motivated abstractions of alliances. This is an epistemology, on the one hand, of both non-realism and non-relativism and, on the other hand, of the measure of 'good enough' knowledge as distinct from the zero-sum games of both formalism and poststructuralism (that is, their mutual dedication to the absolute and always-troubling position that 'If something is not perfectly aligned in one-on-one correspondence with something else, then it must be unmeasurable'). Theories derived from philosophical frameworks that do *not* take into account the contemporary science and philosophy of mind engage in what George Lakoff and Mark Johnson have labeled fallacies of 'objectivist' or 'subjectivist' epistemologies, both of which tend to skew analysis into one or another untenable position: toward either the 'God's-eye view' of realism or the 'anything goes' relativism of postmodernism.[10]

I begin this essay – almost word-for-word here – at exactly the point where I left off another, published some ten years ago, about the value for early modern historicism of what at that time I was calling a 'materialist linguistics'.[11] Although the terminology I had at my disposal has since shifted – and my need to sift the discussion through the discipline of linguistics has abated – I feel that some of the conclusions I reached then are still relevant and worth reasserting: Materialist critics have long understood that social ideology and literary representation are twin artifacts of metaphor and of the processes by which humans categorize the world. But we are only beginning to appreciate the important claims made by metaphor theorists and proponents of embodied cognition over the past several decades. Understanding both metaphor and processes of semantic categorization may help us close our theoretical gaps between subjectivity/agency and the cultural systems in which subjectivity/agency is manifested. Metaphor has a material origin in the body and brain; and like all matter in a state of flux, metaphor behaves systematically – first within individual minds, and then in projections outside those minds, in distributions from concrete experience toward the vast abstractions we tag as 'cultural' experience. This model gives imagery to early modern materialist critics' profound suspicion – one often articulated though rarely theorized – that subjectivity, language, text, performance, and culture are bound up within a matrix of very complex causes and effects, each affecting the others and being affected *by* the others in a dynamic of exchange that is both accretive and infinitely circular.

In this essay, I will elaborate on these statements, beginning with a brief overview of the Marxist issues at stake in today's materialist studies, including a review of 'manifesto'-style claims from my previous discussion, and following with a description of some of the theoretical foundations of Crane's cognitive method. I will then turn to examples of the kinds of interpretive difficulties that critics face in trying to reconcile their materialist findings with the daunting complexities of early modern culture, noting, in particular, the problems of contradiction generated within the dialectical tensions of a text like Shakespeare's *The Merchant of Venice*. I will suggest how cognitive theory, inflected by an awareness of dynamic systems theory, helps us reformulate our understanding of such problems as being symptomatic of the brain/mind's own structuring mechanisms and those mechanisms' complex parameters, citing the phenomenon of the semantic 'conflation' that critics describe with insight but lack the vocabulary to explain within a broader view of *how knowledge itself is organized*. Recent work on the economic contexts of *The Merchant of Venice* provides a case history for modeling the culturally constructive forces of cognition.

A 'materialist linguistics'

Now some four decades old, materialist criticism is still grappling with difficulties related to transforming early forms of Marxism into more fluid and encompassing historical theory. Among the problems for materialist studies of early modern culture, and for Shakespeare studies in particular, has been the struggle to model a viable historical materialist dialectic – to establish, as Ivo Kamps described it, a 'precise balance ... between the impact of base and superstructure in the analysis of literary texts'.[12] How, for instance, should critics reconcile the economic determinism at the core of Marxist historical theory with their own instinct to focus on a variety of early modern material practices, some but certainly not all of them economic in character? Is there sufficient theory to account for these instincts and for the tendency they provoke toward a privileging of superstructural effects – be they the religious, mythic, or philosophical objects of 'old' historicism, or the ideological structures of the 'new'? At stake are key issues, the approaches to which still roughly distinguish neo-Marxist, cultural materialist, and New Historicist considerations of early modern texts, performances, and history. Arguably, the most important of these issues still waiting to be resolved relates to the ideological means by which the human subject is constructed and the limits to agency imposed on the subject through its interpellation by and into the cultural system.

For early modern studies to realize the potential of a historical materialist dialectic, it will require a more satisfyingly materialist linguistics, one that realizes key attributes of poststructuralism but does so without maintaining the formalist habits of poststructuralist language theory or the limited definition of systematicity that we find in Saussure, Derrida, and their theoretical descendants (including, and perhaps most influentially, Foucault).[13] The following claims intertwine the concerns of linguists with those of literary theorists because linguistics – however we have chosen to appropriate it – has always been and should continue to be fundamental to the development of any text-based (i.e., historical and literary) methodology. The portrait of the 'materialist linguistics' that we develop would contain – but need not be limited to – the following features:

- Its claims to materiality would include the material processes underlying a unified human brain/mind activity as laid out, for example, by the schools of cognitive and functionalist linguistics.
- Its forms and structures would be shown to *self-organize* in an analogous sense to, but perhaps also literally from, the emergent structures of complex dynamic systems: in a bottom-to-top direction from out of the human cognitive apparatus, giving the production of cognitive and linguistic structures a cause-and-effect schematics operating from a material – indeed, a biological – 'base' toward an increasingly abstract 'superstructure'.
- It would be able to account for the general properties of linguistic systematicity long known and accepted in theoretical linguistics: creativity, or language users' capacity to produce and comprehend an infinite number of new utterances through the use of finite resources; and productivity, the means by which language users transform linguistic resources from level to level and from context to context.
- Contrary to what Noam Chomsky asserted when he launched the first comprehensive theoretical linguistics, a materialist linguistics would assume that the linguistic system cannot be idealized, frozen for purposes of observation. Instead, its analysis begins on the assumption that linguistic forms *emerge* from their environmental and historical contexts, and that no analysis with claims to explanatory

power may be attempted without factoring in the contingencies of the language speaker's social and physical conditions – determinants, in other words, of human subjectivity. Forms emerge from the subject's material situatedness through the mediating presence of a dynamic semantic system. This system's origins, coming from within human conceptual apparatuses, generate the differential relations between human subjectivity, human agency, cultural contexts, and cultural agency that are the cultural theorist's objects of analysis.

- Corollary to the above, the methods of synchronic and diachronic analyses would collapse into a modified version of the diachronic, a byproduct of the materialist model's capacity to figure the one phenomenon that formalisms of all kinds must ignore or deny: change. The collapse of synchrony into diachrony encourages a systems analysis that recognizes the inseparability of evolution – both biological and cultural – from a system's identity.

- Finally, a materialist linguistics would maintain key insights into the characteristics of ideological constructivism achieved through the forty-year advance of poststructuralism. Poststructuralism derationalizes language and consequently elevates metaphor and processes of semantic categorization as mechanisms through which ideology circulates in cultural discourse. A materialist linguistics furthers this poststructuralist ethical project by demonstrating a continuity between, on the one hand, small-scale associative cognition shaped by a subject's historical contingencies and, on the other hand, large-scale ideological processes that function at the level of culture. For a more concise terminology here, we might distinguish between 'cognitive metaphor' and 'cultural metaphor' respectively, but we should preserve in such terms their shared origin in the material determinants of the human body/brain/mind.

Like Crane, I believe that we find the most likely candidate for a linguistics of this description in metaphor studies, a multidisciplinary endeavor whose main goals, since the late 1950s or early 1960s, have been to draw attention to the semantic powers of metaphor and to challenge rationalism's privileging (and indeed its definition) of literal discourse. While metaphor studies have spawned a variety of approaches, I focus here on the particular school of cognitive linguistics, specifically the 'conceptual metaphor theory' (CMT) of Lakoff and Johnson[14] and the conceptual integration theory ('blending') of Gilles Fauconnier and Mark Turner, which have been receiving recognition in recent years across the science and humanities divide.[15] Cognitive linguistics is important because of its investment in embodied epistemology – through which it provides a viable alternative to the realism that poststructuralism continually refutes but cannot replace – and because it has produced some detailed models of the causal links between semantics and other levels of language experience that may apply to cultural analysis more broadly.[16] Useful to scientists but also highly relevant to the social sciences, arts, and humanities, CMT/blending theories now inform a number of scholarly conversations about human thought and language and have begun to enter into mainstream discourses as well.

Cognitive linguistics has the potential to effect a materialist challenge to the formalist assumptions embedded in Chomskyan linguistics that is analogous to the impact of Marxism on liberal theories of history and economics. Its model rewrites the essential difference between linguistic meaning and structure to accommodate another kind of difference altogether: a difference in the degree of structure within a linguistic framework. It describes a difference that holds among elements organized not binarily but along a structural continuum, whose interactions are vituperatively dialectical. In early modern materialist studies, this model has the special potential to contribute to older but still generally ongoing debates over 'the precise nature of the subject/ structural relationship, especially with regard to the subject's (in)ability to impact or subvert the social structure', to quote Kamps again.[17] The 'limited form of human agency' (ibid.) that these critics seek requires a theory that enables them to resist images of totalization and to replace those images with some more in keeping with the historical materialist dialectic. This theory should elaborate the dynamics of a continuum that stretches between the materially situated human subject and the subject's abstract cultural milieu. As Louis Montrose recognized, such a theory would enable critics 'to resist the inevitably reductive tendency to constitute ... terms as binary oppositions, instead construing them as mutually constitutive *processes*'.[18] (Readers already familiar with CMT/ blending theories might wish to skip to the next section.)

As with other inquiries within metaphor studies, cognitive linguistics documents the constructive role of metaphor in human thought and language systems. Philosophers, linguists, and literary critics have traditionally considered metaphor to be a deviant and even opposite form of expression from the literal, but many since I. A. Richards have challenged this view, arguing that the literal and the metaphorical are not dichotomously opposing discourses but varying uses of the same materials operating within the same expressive system. Rather than deviating from or merely supplementing literal meaning (as the rhetorical and literary traditions would have metaphor do), this Nietzschean, postmodern metaphor creates meanings as well as restructures existing ones. Its function, therefore, is semantic as well as grammatical; or, to put it another way, metaphor operates in thought as well as in language. And according to the cognitive linguists, at least, it is semantic (thought-based) before it is grammatical (in linguistic form).

Beginning in 1980 with George Lakoff and Mark Johnson's *Metaphors We Live By*, researchers have collected data that they say point to a causal link between semantics and higher-order levels of language, in which grammatical form develops out of a densely structured semantic system. The cognitive linguists assert or imply a number of important claims deriving from such a view, some now being verified by empirical means. These claims include the idea that human thought and language are interactive but essentially independent products of the same biologically derived cognitive system; that thought is partly structured by metaphor and partly by some form of direct, material exchange between the embodied mind and its environment; that language, unlike thought, is completely structured by metaphor; and that the presence and processes of metaphor in language (and thus in texts) are the result of a complex network of semantic exchanges at the cognitive level – exchanges in which cognitive metaphor plays an integral part, even though only limited aspects of it may be manifested in lexical (e.g., textual) choices.

Consider, for example, the two statements 'The life spilled from her body' and 'Doctors struggled valiantly to keep her flame burning'. Most English speakers would agree that both statements make reference to the concept of life. Most would also agree, however, that the two define life very differently. Each statement subsumes a complex of cognitive-level metaphorical associations that control the speaker's lexical choices: The first uses what Lakoff and Johnson call a 'conceptual metaphor', LIFE IS A FLUID, signaled by the verb 'to spill'. The second, the conceptual metaphor LIFE IS A FIRE, is signaled by the noun 'flame'. The two conceptual metaphors are equally conventional in English in the sense that we commonly use both in constructing our understanding of life, and few speakers of the language would insist that one or the other is actually *wrong* as a descriptor for life. But while many would agree that 'The life spilled from her body' and 'Doctors struggled valiantly to keep her flame alive' are statements that use metaphor, it is not immediately apparent that the concepts founding these statements are themselves metaphoric – that is, that thought itself operates metaphorically. Central insights gained from this recognition include, first, the idea that *no* statement about the abstraction 'life' can escape from essential metaphoricity; and second, that it is this condition and not a preexisting algorithmic rule system (as more formalist linguistic theory would have it) that determines the utterance's meaning.

Such a description of semantic exchange necessitates a theory of semantic category formation. Cognitive linguistics constructs a model of category formation using a bottom-to-top analytical schema in which the embodied mind's most basic interactions with its environment are figured as inhabiting the ground for all successive construction. Conceptual metaphor results from a process of interaction between similar, contrasting, or simply juxtaposed 'source' and 'target' semantic domains. The source domains of cognitive metaphors are constructed from basic-level 'image schemas', uncomplicated abstractions of concrete experience gained through sensory interaction between the human body and its social and material environments. Modeled on the 'conceptual schemes' of Hilary Putnam and other 'experientialist' philosophers – and ultimately on the aesthetics of Kant – image schemas are non-visual mental templates existing prior to the development of concepts. They are representations of the very simple geometries of such physical episodes as collision, release, propulsion, attraction, and compulsion, experience first gained in pre-linguistic infancy and stored afterward in the developing child's (and eventually adult's) memory. Learning is a process by which an image schema exports its structure via 'metaphorical projection' onto the target domain of unstructured experience – a domain that is 'unstructured' precisely because it is lacking in concrete detail. The resulting mixture of concrete and abstract elements within the target domain provides the architecture for wholly new semantic categories and experiential domains as well as the ground for semantic and experiential shifts. Once new semantic material has been created, it gains, in turn, the capacity to project its structure iteratively into even more abstract realms of experience. Each new metaphorical iteration may escalate this learning process toward greater and greater abstraction, thinning – but never entirely shedding – as it goes the bounds and traces of originary perceptual constraints. Basic-level image schemas and metaphorical projections, themselves abstractions from perceptual experience, are thus responsible for generating meaning where none may have existed before. But such 'meaning', it should be stressed, is wholly contingent, relative not to some realm of absolute experience but to representations of situated mind-embodiment manifested within environmentally and historically specific cognitive systems.[19]

Building on this theoretical foundation, the blending theory of Fauconnier and Turner pushes CMT a step further, replacing a relatively static model with one of generative dynamism: A 'blend', as Fauconnier and Turner call it, is an 'online' mental construction composed initially of at least three mental spaces, each one containing semantic features that, when integrated with the others, creates a fourth mental space and new conceptual material. Blends are symptoms of conceptual processing, chiefly utilizing 'working' or short-term memory – hence they operate 'online' – although intermediate and long-term memory are also required for at least part of every blending operation. The process begins when two concepts or domains of experience, called 'input spaces', are juxtaposed, or framed together by linguistic or imagistic means, causing the mind to scan automatically for underlying similarities. If the two have schematic traits in common – some spatial patterning such as verticality, a trajectory of movement, a dominating shape – then the result of the scanning will be the recall from longer-term memory of a third or 'generic' space containing the outlines of such schematic relations. Correspondences to this memory-based generic space prime the mind to project or 'map' connections, resulting in yet a fourth space, the blend itself. Here, not only are the schematic associations between the two inputs 'run' (tested for fitness), but ancillary associations within the two input spaces are imported into the fourth space and similarly 'run', creating within the blend connections that might never have existed in either the real world or in any human's imagination. In other words, blends make possible new meanings and enable changes to existing ones; and they are the basis for humans' imaginative constructions and habitations of 'fictional worlds'.

A blend is a temporary construction, a product of fleeting working memory; but it always has the potential to become a more permanent construction – a holding of a person's longer-term memory – if the 'reality' it constructs seems especially resonant within an individual's social experience. Thus, a blend might form a new semantic category or contribute to the expansion or revision of existing categories, and its salience as such would be a function of its acceptance by a collection of minds. This process of meaning-making – or rather 'reality-making' – is relative only to what brains and minds *understand*, constituting the embodied alternative to both realist and relativist epistemologies noted earlier. That is, the standard for determining salience is not the 'God's-eye view' epistemology of 'objectivist' realism but rather the realities that minds create – the products of dynamic negotiations between single minds' idiosyncratic constructions and multiple minds' negotiations of the salience or 'relevance' of those constructions.[20]

Shakespearean knowledges and complexities

The etiological model that this cognitive approach connotes is not intended to replace the top-down approaches that today's early modern scholars fruitfully deploy in their analyses of cultural dynamics and discourse. Rather, I hope it would serve as a supplement to – and within – our historical models in recognition of the view that even the most abstract systems have *histories* and thus belie the idealism inherent in traditional formalisms. And it would recognize that a cultural system's history manifests through proliferations of individuals' cognitive – and by this I mean the semantic – matter within it, matter-driven conceptualization that we may reconstruct using the contemporary theories of metaphor and blending just described.

Before offering a further description and demonstration, however, I should mention briefly the conversations in early modern criticism and theory to which this model already bears some (sometimes troubled) relation. Recent scholarship in the period has been flirting with – and increasingly engaging directly – cognitive theory and the embodied-mind epistemology inherent to a cognitive approach that makes our adoption of a humanities-inflected systems theory seem especially timely and warranted (see, for examples, Paster, Watson, and Smith).[21] These newer studies, constituting something of a second wave, contribute to an earlier (and ongoing) corpus of works that first began appearing in the early 1990s by a group of critics in 'early modern cognitive studies of literature and culture' – Donald Freeman, F. Elizabeth Hart, Mary Thomas Crane, Ellen Spolsky, Evelyn Tribble, Lalita Pandit, and Amy Cook (the label is Crane's).[22] Less cognitive-oriented but deeply engaged with the vocabulary of complex dynamic systems are studies by Henry Turner and Alison Games, the latter providing a particularly strong example of how the terms of dynamic systems theory may be deployed to recast our understanding of early modern historical change.[23] Among this latter group I also place Valerie Traub's work on 'Mapping the Global Body', which details a highly nuanced archaeology of early modern map-making and -conceptualizing and the co-conceptualizing of maps interactively with categories of the colonized human Other.[24] Although Traub does not identify her methodology as cognitive *per se*, she obviously shares a fascination with the detailed mechanisms of semantic change and creativity that are labeled 'cognitive' by the cognitive critics (Crane et al.) mentioned above – her fascination apparent in her efforts to track instantiations of geopolitical gestalts as they create, situate, shift, dissolve, recreate, resituate, and reshift repeatedly over the long view of the period.

In some of these newer studies (by literary critics like Smith, Turner, and Traub in particular), the poststructuralism still being leaned upon for theoretical backing seems frankly insufficient to explain the bottom-up energies of conceptual cause and effect that they strongly intuit in their embrace of materialism. Specifically, the rhetoric of 'sites' of 'conflations' and the 'discourses' they help create remains unhelpfully rooted – despite these critics' references to instruments like the *OED* to track semantic change – in a linguistics that overly differentiates the materiality of *things* from the materiality of *thought* and by extension language. I believe that this disconnect stems from the residual formalism inherent in poststructuralism that many in our field now recognize and that we hear being worried in insistent debates over issues that poststructuralism was supposed to have rendered moot: authorship, intention, subjectivity, agency, the relationships between history and literary form, all of which are being earnestly revisited as issues in need of better theory. What I find particularly restrictive is the static, realist epistemology that poststructuralism passively inherited from its various precursor structuralisms (literary and linguistic) and that has contributed to its once-enlightening (e.g., in Foucault or Butler) but now limiting top-down methodology. Such embedded realism has the ironic effects of objectifying and reifying the very systems that these critics otherwise *portray* quite persuasively as dynamic and cumulative proliferations of semantic complexity.

Returning now to the etiological cognitive model I am proposing in this essay, let me attempt to demonstrate how an understanding of the history of a literary text's semantic accumulations might render something of genuine critical value. To do this, I must show the relevance of a discussion of literary adaptability – what is lately being offered as 'historical formalism' – for conversations ultimately more concerned with the ethical dimensions of social change. Here I offer a sample reading of the text of Shakespeare's *The Merchant of Venice* to show how a CMT/ blending analysis might contribute to our sense of this text's 'interconnectivities' with its cultural contexts.[25] Such analysis may never fully describe the complexities that historians and critics discern in tracking this play's participation in its original cultural moment, but it may sketch out some of the patterns of human agency – what the CMT/ blending theorists call semantic 'motivations' – in the accumulations from local levels of individual experience to the collective levels that comprise cultural contexts. Specifically, a CMT/blending analysis adds a diachrony of human experientialism to Marxist-inflected readings such as the recent addresses to the play's participation in its economic contexts by Linda Woodbridge, Natasha Korda, Amanda Bailey, and others.[26] Such readings, relying on historical narratives and *OED* snapshots of the play's semantics of usury, contractual bonds, credit-based law, and even accounting practices, help position the play within the changing world of early modern finance; but they lack the dimension of an experientially situated human agency that would tie the play's poetics to its early modern – and indeed its modern – receptions.

My own reading is intended to show how the figurative details of *The Merchant* generate an emergent semantic system whose constrained open-endedness may be viewed from a bottom-up perspective as establishing the basic mental frames within which all its conceptual accumulations – from the most local-level metaphors to the highest-level abstractions of epistemological oscillations – appear to us retrospectively to bear 'self-similar' patterning throughout (to use a term from dynamic systems theory). The criticism on this play is of course vast and comprehensive, and I have space here to provide only a limited set of suggestions building on its modern reception. I can state generally, however, that most of the play's critics tend to notice its distinct habits of semantic iteration – what Bailey usefully terms 'refractions' – of specific words, dozens of whose repetitions across the play can be counted and tracked: to cite only a very few examples, 'venture' (Ralph Berry),[27] 'bond' (Korda and Bailey), 'forfeit' (Bailey), and 'gold' (my reading here).

Critics now routinely remark on this play's conflations of economics with human relationships. The conceptual metaphor LOVE IS MONEY, a species of what Robert Watson cites as elisions of 'flesh and gold', begins right away in the Act 1 dialogue between Antonio and Bassanio. Here the text consistently juxtaposes the domains of MONEY and LOVE in such a way that these two very different aspects of human experience – love an exceedingly abstract semantic domain, money less so and undergoing shifts in the early modern period that highlight both its abstract and its concrete dimensions – combine to form a single gestalt, a blended recategorization, in which the features associated with contracting financial obligations become inextricably intertwined with the categorical features of intersubjective trust between humans and between humanity and the divine. Bassanio states: 'To you, Antonio, / I owe the most, in money and in love, / And from your love I have a warranty / To unburthen all my plots and purposes / How to get clear of all the debts I owe' (1.1.130–134).[28] Personal attachment and financial obligation come together as separate inputs in a conceptual blend that is then reinforced in Antonio's double-use of the word 'uttermost' in his response: '[O]ut of doubt you do me now more wrong / In making question of my uttermost / Than if you had made waste of all I have./ … Try what my credit can in Venice do; / That shall be racked even to the uttermost / To furnish thee to Belmont, to fair Portia' (1.1.155–157, 180–182). The blending of love and money is further reinforced in the lines that end the scene, in which Antonio collapses 'money' and personal 'trust': 'Go presently inquire … / Where money is, and I no question make / To have it of my trust or for my sake' (1.1.183–185).

This blend then recategorizes with new inputs from Act 1, Scene 3, where it makes more coherent the framework within which Shylock gains Antonio's 'friendship' through his sarcastic appeal to the opposing traditions of Christian charity and Jewish usury: 'This is kind I offer' (1.3.140), says Shylock, to which Antonio sneers: 'The Hebrew will turn Christian; he grows kind' (1.3.177). At this point, the effect of the blend is to focus semantic features into 'coupling points' (to use a general cognitive science term) or even 'strange attractors' (another dynamic systems term) through the detailed interactions of its specific entailments, taking forms of concrete materiality – in the text as metonyms, in theatrical performance as stage props. With the entry of Shylock into the play, and additionally with reinforcements through the developing elopement plot involving Lorenzo and Shylock's daughter, Jessica, MONEY as a domain of experience begins to narrow into the more concrete 'prototype' category feature of the ducat.[29] To Shylock, money *is* the ducat; and Jessica, extending this elision, metonymizes her father's ducats with her father. She now expresses her pledge to Lorenzo *and* her anxiety about the authenticity of Lorenzo's love 'promise' (2.3.20) specifically in terms of the ducats that she frantically collects before emerging from the house (and that may very well materialize as stage props). Shylock's parallel conflation of Jessica with his money once he has discovered her missing ('My daughter! O my ducats! O my daughter!' [2.8.15]) becomes imaginatively *viable* – a reality we are able to entertain regardless of its 'objective' possibility – according to the terms of LOVE IS MONEY: father and daughter do indeed 'love' each other by associating each other so consistently with their wealth.

Shakespeare's increasing references to 'gold' add further concrete features to the domain of MONEY and project those features onto

LOVE in a number of ways. Bassanio's Act 1 comparison of Portia to the Golden Fleece of Jason's quest (1.2.170–172), noting too her 'sunny locks', frames his own quest for her as a trophy-hunting venture and also subtly sets up the audience to expect – along with the Prince of Morocco – that her essence must somehow be enclosed within the golden casket in the first of the three casket scenes (2.7). That ducats were forged out of gold (as well as silver) underlines historically – and also cognitively refocuses – the emergent blends of love and money encountered in earlier scenes. Thus, it may seem conceptually jarring, though conceptually not impossible, when Morocco receives the golden casket's stern reprimand: 'All that glisters is *not* gold; ... Many a man his life hath sold / But my outside to behold. / Gilded worms do tombs infold' (2.7.65, 67–69, emphasis added). Interestingly, this inscription instantiates 'gold' with an inside/outside or CONTAINER image-schematics and prepares us to marshal the alternative – and not untraditional, even if surprising – association between gold and moral decrepitude, even death. Hence, the skull is literally tucked *inside* the golden casket ('A carrion Death, within whose empty eye / There is a written scroll!' [2.7.63–64]). Gold as a concept now serves to anchor both MONEY (via ducats) and menacing avarice and so, by blended association, the domain of LOVE. We understand then, if only in a murky series of figurative leaps, that Bassanio's Act 3, Scene 2 chastisement of the golden casket is also an indirect expression of his and his culture's paranoia about Portia's gold-blond ('sunny') beauty:

> ... Look on beauty,
> And you shall see 'tis purchased by the weight,
> Which therein works a miracle in nature,
> Making them lightest that wear most of it.
> So are those crisped, snaky, golden locks,
> Which maketh such wanton gambols with the wind
> Upon supposed fairness, often known
> To be the dowry of a second head,
> The skull that bred them in the sepulchre.

> (3.2.88–96)

The contradictions Bassanio's rebuke encapsulates here carry over but become much more complicated in the Act 4 trial scene and in the ring plot of Act 5. The trial scene, preoccupied with the problems of justice and mercy, places spectators into the conceptual position of being the fulcrum to a great, imaginary scale, as its members weigh the balance between fine moral and ethical distinctions in a process that is eerily similar to the semantic balancing act the play has been requiring in its earlier LOVE IS MONEY associations. Put another way, the mental effort that the audience has put into gauging the positive and negative connotations of 'gold' now expand to encompass the more abstract cultural domains of economic, legal, religious, and philosophical contexts. Such leaps are not at all uncommon in Shakespeare's plays, but in *The Merchant* they are held conspicuously captive to the dialectical discomforts of overly crisp dichotomies: mercy pitted very rigidly against justice, charity against vengeance, authority against vigilantism, and New against Old Testaments of biblical theology.

It is therefore no coincidence that an actual scale – a 'balance' – soon appears on stage, requested by Portia in her guise as doctor of law and promptly produced presumably right out of Shylock's pocket (4.1.253–254). What is termed 'weighing' throughout the scene – taking both concrete and abstract forms of weighing flesh and weighing moral and ethical dilemmas – represents an iteration of the same species of tension already required for processing the LOVE IS MONEY blend, particularly with respect to its emergent – and noticeably binary – meanings of 'gold'. BALANCE is itself a complex conceptual schema, taking its shape from the bilateral symmetries of the human body: our having two arms, two hands, two eyes, two ears, and general front-to-back corporeality, all in carefully orchestrated, embodied proportion in our brains' constant if unconscious efforts to maintain our upright stance against gravity. BALANCE provides the structure of many aspects of basic human embodiment; and because it is so pervasive as an experience, it stamps a framework of proportionality onto a very wide range of our most abstract conceptual systems: argumentation, physical and psychological health, mathematics, relationships, equity (to name only some) – and of course justice.[30]

The ineffability of mercy, and hence its spiritual value, is at least partly owing to the inability of most able-bodied humans – at least after the age of two – to *avoid* balancing, and thus our tendency to cast abstract problems almost invariably in terms of binary oppositions. The trial scene stokes the frustration that an audience might feel in attempting to step outside of BALANCE conceptualization to grasp Portia's 'quality of mercy' (4.1.182), a 'quality' of experience that the characters themselves fail to achieve by the scene's end. Yet it is just this heightened, abstract dimension of 'gold' that Act 5 strives to establish through the ring plot and specifically through the ritualistic re-exchange of wedding rings before the witness of Antonio (whose imbalanced affections for Bassanio, we recall, staged the play's earliest commitments to LOVE IS MONEY). Shakespeare emphasizes the fact that *these rings are made of gold* with Gratiano's seemingly flippant description of his from Nerissa: '[A] hoop of gold, a paltry ring / That she did give me, whose posy was / For all the world like cutler's poetry / Upon a knife, "Love me, and leave me not"' (5.1.147–150).

The ring trick's purpose, at least ostensibly, has been to test (that is, to weigh) the men's sense of marital fidelity: As a loaded form of metonymy (in the text) and stage-prop materiality (in performance) signifying the love 'promise', these rings should have been the focus of all the reverence due to the marriages themselves. Yet the fact that they are made of gold complicates this reverence, not just for Bassanio and Gratiano but for the spectators who have themselves been watching and weighing all along. After all, was it not on the basis of the shift of gold's meaning from its heightened, abstract, and spiritual dimensions down to its more mundane, concrete, and even menacing connotations that Bassanio won his wife in the first place? ('Look on beauty, / And you shall see 'tis purchased by the weight ...' [3.2.88–89]). The rings signify the failure of human bodies in general – here embodied in both the play's characters and its intersubjectively active spectators – to flatten out the contradictions that blends make imaginatively possible and sometimes (actually *always* in Shakespeare) conceptually satisfying and enriching. But because such semantics are not verifiable by any formal system of logic – which, in complex iteration, is *itself* subject to the BALANCE schema – these semantic options remain suspended, still available within the extra-logical processing capacities of copiously blending minds. Such semantic 'bifurcations', I argue, are not just projected onto the cultural field, but are 'scaled up' (to deploy more terms from dynamical systems theory) in the uneasy but insistently festive resolutions to the play's multiple plots. Comic form, with its dependence on the spiritual register of marriage, asserts itself as an overriding authority (whether or not modern audiences are able to accept it as such) in resolving the paradoxes that arise from Act 5's lingering difficulties: the extremes of discomfort to which Bassanio is subjected, Gratiano's gross and inappropriate sexual puns, Antonio's bittersweet dejections, Jessica's shunning, and the absence of any signs of conscience over Shylock's ruination.

Among the many ways in which these text- and performance-scaled energies open up toward the wider-scaled domains of the play's cultural contexts, these tensions richly encapsulate the very *spatial* poetics of the Neoplatonism to which Lorenzo alludes near the beginning of Act 5 (and which today's critics largely ignore in their efforts to historicize the play beyond the history of ideas). Pointedly casting this brief philosophical foray in the binary terms of earth and heaven, human and divine, grossness and perfection, Lorenzo's nod to cosmic harmonics depends for its coherence on the paradoxical sense of radical concreteness coexisting with radical abstraction, paralleling, in a complex pattern of self-similarity, our attempts to process the emergent LOVE IS MONEY blend and particularly the concrete/abstract oscillations of 'gold' within it. Lorenzo's fleeting description is striking for its strongly visual schematics, calling on the CMT image schemas of up/down and small/large bodily orientation and then mapping these orientations onto the *values* of dirty/clean to form what Fauconnier and Turner (2002) call a 'double-scope' blend. It is only through such concretized and embodied schematics that we are able to conceptualize – to visualize and thus experience imaginatively – the mysteriousness of eschatology:

> Look how the floor of heaven
> Is thick inlaid with patens of bright gold.
> There's not the smallest orb which thou behold'st
> But in his motion like an angel sings,
> Still choiring to the young-eyed cherubins.
> Such harmony is in immortal souls,
> But whilst this muddy vesture of decay
> Doth grossly close it in, we cannot hear it.
>
> (5.1.58–65)

Such semantic scaling-up first frames and then distributes (on analogy with what cognitive scientists call 'parallel distributed processing') semantic matter into the epistemological dimensions to which this play has repeatedly, insistently returned its modern critics.[31] To cite a particularly salient example for my purposes: Arguing from within the Marxist cultural framework that critics like Woodbridge, Korda, and Bailey are now richly historicizing, Robert Watson links *The Merchant*'s cognitive vituperativeness to its bewildering dialecticalism and finds that it enacts an 'epistemological revolution on behalf of … a traditional Christian order':[32]

Certainly the play provokes a range of anxieties about verbal and monetary systems, with puns functioning as an analogue of usury … But

Shakespeare seems to expose this wound in the epistemological fabric in order to repair it, or at least anesthetize it, and maybe even hallow it. Words and coins produce redemption on earth, as Christ's mediation does in heaven. The *adaptability of language* can be used to correct, or render innocuous, the sinister functions of a contract; the mental universe (as the cognitive advantages of 'fuzzy logic' confirm) is a court of equity, not law. The queasiness of puns yields to the beauties of metaphor.[33]

The 'adaptability of language' that Watson describes here, from its smaller-scale, text-based semantic/poetics framework to its larger-scale dialectical/epistemological frameworks, encompasses myriad 'couplings' (a cognitive science term) of materiality-to-culture and potentially offers us a vocabulary to further clarify such couplings – in designating, for instance, a critical term for how a particular historical performance of *The Merchant of Venice* might have influenced, say, the literary creativity of a spectator-author in the play's original audience, easily one of Shakespeare's many poet-friends. To my knowledge, we do not currently have a term for this particular coupling of material/cultural intersubjectivity (although we might all point to examples of such influence having taken place and left its traces in any number of early modern texts). A bottom-up theory of cognitive/cultural cumulativeness may reveal many such transmedial nodes of intersection that our tendency to privilege the site of the text arbitrarily diminishes.

I would even go so far as to suggest (even more overtly than my earlier hint) that the constant reshuffling and recoupling of semantic features in cognitive blends – such as what we find with the gold ducats and the LOVE IS MONEY reconfigurations in Shakespeare's play – renders effects comparable to those of strange attractors in physical systems.[34] Since all cognitive blends amount to highly compressed containers of system-wide semantic information, their prototypical features may well serve to skew cognitive focus in unpredictable but (understood in retrospect) patterned ways as do the information-rich attractors of other kinds of dynamic systems. And just as the attractors of physical systems distribute deep structure toward increasingly complex scales of physical organization, cognitive blends spread their framing structure throughout the systems of human experience, linking small to large scales of phenomena, patterning in uncannily similar outlines the forms of human thought, ideology, and material and immaterial aspects of human culture.

Epistemology and complexity

Cognitive scientists, in small but increasing numbers, are considering the efficacies of models of complex 'connectionism' for linking the patterns of cognitive processing to abstract domains of human experience. The epistemology of cognitive- or mind-embodiment found in cognitive (and increasingly other schools of) linguistics provides an etiology for how the brains and minds of people – agents – beget the cultural systems within which people inhabit the world familiarly. While minds are capable of intellectualizing cultural systems that even belie those minds' own actual functioning, as I have tried to show through my exploration of the semantic tensions in *The Merchant of Venice*, they are nonetheless shaped and limited by the constraints placed on *how knowledge itself is organized*. This engages a very different epistemology from those subsumed within both traditional and postmodern sets of critical theory, and its recognition, I argue, is essential to our ongoing efforts to grasp the complex material cultures that early modernists routinely puzzle.

But to return, finally, to the ethical imperatives for criticism to which this essay referred briefly in its first section: A key current in early modern materialist studies has been a fundamental shift of focus away from social difference and toward the phenomena and discourses of the 'trans-'. That is, we are moving toward the *crossing, mixing*, and *exchanging* dynamics of processes governed by the rampantly associative. New concepts are being actively generated to fill critical lacunae that were mostly invisible just a few decades ago but that are being revealed, bafflingly and gestalt-like, by Marxist and neo-Marxist historiography. Alongside our more traditional vocabulary of 'transformation', 'transportation', 'transgression', and 'translation', today's critics situate 'transhistory', 'transculture', 'transnationalism', and 'transversality' – to name only a few of the proliferating coinages – and witness how such neologistic inspiration can reawaken metaphoricity in even the most formal and apparently ahistorical apparatuses. This shift represents not a rejection of the discourse of social difference, but a profoundly important extension of its implications, and in my opinion, at least, it is amounting to yet another paradigm shift in critical theory in general.

Critical theorists in the social sciences and humanities have developed tools for identifying and evaluating the categorical effects arising from social difference within cultures, forging descriptive models of culture that serve a range of disciplines from anthropology and economics to literature and the fine arts. Their charges are contextualized by powerful ethical visions that give them strong disciplinary identities, especially in the humanities. However, the imperatives to describe and to delineate ethical prescriptions of and for culture may work against each other to the extent that the first necessitates a view of categorical effects arising overly narrowly from social difference to the exclusion of other kinds of social cognition. As a cognitive phenomenon, the perception of *any* kind of difference – including but not limited to social difference – provides the boundary demarcations that make all semantic categorization possible. To focus overly narrowly on any one kind of difference is simply to undertheorize both cognition and culture. This is because all human cultures are also by definition functional alliances based at various dimensions on human commonality. Members of cultures coexist and participate in complex acts of intersubjectivity to produce material artifacts, social contracts, innovations, adaptations, geographical, political, and intellectual expansions, and expressions of their experiences both within and beyond their immediate environments. Given this tension, it would seem that social difference must itself be contextualized within frameworks of broader constraints. Such a framework is not an idealist fantasy but a structural necessity, the mechanism that enables the dialectic between forces of change and stability inherent in every society and that makes visible, if only incompletely and in retrospect, the traces of culture's complex patterns.

This paradigm shift we are witnessing of late represents a suppleness and willingness within critical theory to (1) recognize the powerful constructive forces that constitute 'culture', (2) develop a vocabulary describing culture's systemic dynamics, (3) study the patterns of complexity within those systems, (4) extend materialist methodologies to those systems' conceptualization, and (5) cultivate both descriptive narratives about and ethically inflected responses to the resulting models. In early modern studies, our models must accommodate human subjectivity and agency in such ways as to embrace the effects of human individuality and idiosyncrasy that poststructuralism has been largely unsuccessful at effacing and that are especially instructive about the phenomena of our period: e.g., intentionality, creativity, authorship, and other effects of the materiality of minds. My instinct here is that complexity *accumulates* and that, if so, to understand cultures, we must understand the information that has accumulated within them over time. Small-scale effects at local sites give rise to large-scale effects at cultural dimensions.

This vision is basically harmonious with and admiring of the science, but I hope it is also suggestive of something that offers a comprehensive disciplinary overview: Far from colluding in perpetuating the unhelpful 'two campuses', 'two cultures' divides, I embrace the humanities and the sciences as complementary approaches to knowledge-gathering and – importantly – as potentially more fully benefiting each other in recognizing how their purviews represent differences in scope. What aspects of knowledge ought *not* to be interesting and useful to culture critics, whose training is by far the widest-spanning – and which quality makes them the most specially licensed for speculative inquiry and thus the most generous in their views – of anybody in the academy today?

–Dedicated to the prescience of Rosalie Colie.

Notes

The author extends deep gratitude to two fellow readers at the Folger Shakespeare Library, spring 2009: Nicholas Moschovakis and Joshua Calhoun, who read an earlier version of this essay and offered generous commentary and support.

1. See F. Elizabeth Hart, 'Matter, System, and Early Modern Studies: Outlines for a Materialist Linguistics', *Configurations* 6 (1998), 311–343, for notes pertaining to the relationship between varieties of Marxism and neo-Marxist – particularly Foucauldian – materialist historical methodologies.
2. One of many recent examples is Alison Games' *The Web of Empire: English Cosmopolitans in an Age of Expansion, 1560–1660* (New York: Oxford University Press, 2008), whose description of the impact of travel technologies on early modern conceptual 'frameworks' is worth quoting at length: 'Global processes knit the early modern world together, enabling people to perceive in its entirety a world once experienced only in fragments. Maritime and navigational advances accelerated the pace of connection and the rate of communication among distant continents. Places long familiar became more quickly reached, their goods more readily and cheaply exchanged, their oddities more rapidly assimilated. In their sea voyages, Europeans also came upon entirely unknown lands in the western hemisphere, places populated by such a bewildering array of unfamiliar people, languages, cultures, plants, animals, and commodities that old intellectual frameworks were irrevocably challenged' (6).
3. The groundbreaking work of Stephen Greenblatt, whose rhetoric of 'circulatory' poetics would become the basis for New Historicism, provides an early example that has been reiterated by others too numerous to list here. See Stephen Greenblatt, *Shakespearean Negotiations: The Circulation of Social Energy in Renaissance England* (Berkeley: University of California Press, 1988). Henry Turner, 'Travel, Trade, and "Translation" in the Work of Richard Hakluyt', a talk given at the New England Renaissance Conference, Wesleyan University, 2008, provides perhaps the most self-conscious illustration of an early modernist's adoption of a scientific rhetoric of non-linear dynamic systems in his reading of Shakespeare's *A Midsummer Night's Dream*. At a particularly instructive meeting of the New England Renaissance

Conference (Wesleyan University, October 2008), Turner and Jacques Lezra deployed similar imagery of 'networks', 'circuitry', 'chemical bonds', 'logical fluidity', and 'conceptual economies' in their respective talks on the convergences between early modern trade and translation. Overheard at this event were observations by respondents about the inadequacies – as currently theorized – of such rhetoric to capture the genealogy of authority within such systems and the difficulties of distinguishing 'facts' from the 'fantasies' of unwieldy 'grand narratives' (author's notes). This essay will seek to address such problems by formulating theory for a bottom-up etiology of concept-building, an approach made inevitable, I believe, by materialist historical methodologies.

4. See especially Mary Thomas Crane, *Shakespeare's Brain: Reading with Cognitive Theory* (Princeton: Princeton University Press, 2001); Ellen Spolsky, *Satisfying Skepticism: Embodied Knowledge in the Early Modern World* (Aldershot and Burlington: Ashgate Publishing, 2001); 'Darwin and Derrida: Cognitive Literary Theory as a Species of Post-Structuralism', *Poetics Today* 23:1 (2002), 43–62; *Word vs. Image: Cognitive Hunger in Shakespeare's England* (Basingstoke and New York: Palgrave Macmillan, 2007); and Bruce McConachie, *Engaging Audiences: A Cognitive Approach to Spectating in the Theatre* (Basingstoke and New York: Palgrave Macmillan, 2008).

5. See Evan Thompson, *Mind in Life: Biology, Phenomenology, and the Sciences of Mind* (Cambridge, MA: Harvard University Press, 2007), which also helpfully summarizes the history of cognitive science in its development toward an embrace of dynamic systems theory: 'Three major approaches to the study of the mind can be distinguished within cognitive science – cognitivism, connectionism, and embodied dynamicism. Each approach has its preferred theoretical metaphor for understanding the mind. For cognitivism, the metaphor is the mind as a digital computer; for connectionism, it is the mind as neural network; for embodied dynamicism, it is the mind as an embodied dynamic system. Cognitivism dominated the field from the 1950s to the 1970s. In the 1980s, connectionism began to challenge the cognitivist orthodoxy, followed in the 1990s by embodied dynamicism. In contemporary research, all three approaches coexist, both separately and in various hybrid forms' (4).

6. I am influenced in this model-building particularly by the work of the cognitive-evolutionary psychologist Merlin Donald, who argues that 'culture' is what happens when collections of individual human minds come together and, as such, must be treated as an aggregate phenomenon. See Merlin Donald, *Origins of the Human Mind: Three Stages in the Evolution of Culture and Cognition* (Cambridge, MA: Harvard University Press, 1991). Incidentally, the tissue connecting minds to cultures in Donald's model is memory – the range in types of memory from an individual's 'biological storage' to the expansive 'external memory systems', or writing technologies, of which the Internet is only the latest example in an impressively brief 7,000-year history.

7. My approach resembles Crane's (*Shakespeare's Brain*) but for one key difference: While she recognizes and deploys the bottom-up dynamics that have always intrigued me in developing such theory, as a literary critic she maintains the top-down stance of the humanities in general, moving from cultural contexts to texts and then to the brains and minds of human agents,

specifically of early modern authors and scientists. Crane's approach points to the literary text as a critical intermediary between culture and agency, context and subjectivity; whereas, in my model the literary text is just one of many types of cultural products that scholars must consider within a mosaic of interdisciplinarity. The text, in other words, is an available but not a necessary repository of the complexities that human cognition stamps onto the world; and in my opinion its relations within the linked media of cognition and culture have yet to be fully theorized (i.e., the question of 'What is literariness?' is still open and vital).

8. See F. Elizabeth Hart, 'The Epistemology of Cognitive Literary Studies', *Philosophy and Literature* 25 (2001), 314–334.
9. See Spolsky, *Satisfying Skepticism*, as well as her 'Darwin and Derrida', on 'good enough' epistemology; on literary universals, see Patrick Colm Hogan, 'Literary Universals', *Poetics Today* 18:2 (1997), 223–249.
10. See George Lakoff and Mark Johnson, *Metaphors We Live By* (Chicago: University of Chicago Press, 1980).
11. See Hart, 'Matter, System, and Early Modern Studies'.
12. Ivo Kamps, 'Materialist Shakespeare: An Introduction', in Kamps, ed., *Materialist Shakespeare: A History* (New York: Verso, 1995), 4.
13. See Hart, 'Matter, System, and Early Modern Studies', for elaborations of these points.
14. See Lakoff and Johnson, *Metaphors We Live By*, and George Lakoff and Mark Johnson, *Philosophy in the Flesh: The Embodied Mind and its Challenge to Western Thought* (New York: Basic Books, 1999).
15. See Gilles Fauconnier and Mark Turner, *The Way We Think: Conceptual Blending and the Mind's Hidden Complexities* (New York: Basic Books, 2002).
16. This field of linguistic research has burgeoned during the past ten years, far beyond my capacity to eavesdrop on all the conversations. A recent study that was brought to my attention too late for me to incorporate it into this essay is Jerome A. Feldman, *From Molecule to Metaphor: A Neural Theory of Language* (Cambridge, MA: MIT Press, 2006). See also Lakoff and Johnson, *Metaphors We Live By* and *Philosophy in the Flesh*; and Ronald Langacker, *Foundations of Cognitive Grammar*, Vol. 1: *Theoretical Prerequisites*, Vol. 2: *Descriptive Application* (Stanford: Stanford University Press, 1987, 1991), to cite only a few of the pioneering works.
17. Kamps, 'Materialist Shakespeare: An Introduction', 7.
18. Louis Montrose, 'Professing the Renaissance: The Poetics and Politics of Culture', in H. Aram Veeser, ed., *The New Historicism* (New York and London: Routledge, 1989), 15–36 (21, emphasis in original).
19. See Crane, *Shakespeare's Brain*, 10–13, for a similar discussion of cognitive categorization tailored specifically for early modern literary and cultural theory. Crane's sources include John R. Taylor, *Linguistic Categorization: Prototypes in Linguistic Theory*, 2nd edition (Oxford: Clarendon Press, 1995), and George Lakoff, *Women, Fire, and Dangerous Things: What Categories Reveal about the Mind* (Chicago: University of Chicago Press, 1987); and supporting them are Eleanor Rosch and B. B. Lloyd, eds, *Cognition and Categorization* (Hillsdale, NJ: Lawrence Erlbaum Associates, 1978). See also Feldman, *From Molecule to Metaphor*.

20. The foregoing description of blending has been adapted from F. Elizabeth Hart, 'The View of Where We've Been and Where We'd Like to Go', *College Literature* 33:1 (2006), 225–237.
21. Gail Kern Paster, *Humoring the Body: Emotions and the Shakespearean Stage* (Chicago and London: University of Chicago Press, 2004); Robert N. Watson, *Back to Nature: The Green and the Real in the Late Renaissance* (Philadelphia, PA: University of Pennsylvania Press, 2006); Bruce R. Smith, *The Key of Green: Passion and Perception in Renaissance Culture* (Chicago: University of Chicago Press, 2009).
22. To avoid confusion, I present the fuller citations here rather than in the text. In roughly chronological order: Donald C. Freeman, '"According to My Bond": *King Lear* and Re-Cognition', *Language and Literature* 2:1 (1993), 1–18; '"Catch[ing] the Nearest Way": *Macbeth* and Cognitive Metaphor', *Journal of Pragmatics* 24 (1995), 689–708; '"The Rack Dislimns": Schema and Metaphorical Patterning in *Antony and Cleopatra*', *Poetics Today* 20:3 (1999), 443–460; 'Othello and the "Ocular Proof"', in Graham Bradshaw, Tom Bishop, and Mark Turner, eds, *The Shakespearean International Yearbook: 4. Shakespeare Studies Today* (Aldershot/Burlington: Ashgate Publishing, 2004), 56–71; F. Elizabeth Hart, 'Cognitive Linguistics: The Experiential Dynamics of Metaphor', *Mosaic* 28:1 (1995), 1–23; 'Matter, System, and Early Modern Studies'; 'Embodied Literature: A Cognitive-Poststructuralist Approach to Genre', in Alan Richardson and Ellen Spolsky, eds, *The Work of Fiction: Cognition, Culture, and Complexity* (Aldershot/Burlington: Ashgate Publishing, 2004), 85–106; Mary Thomas Crane, *Shakespeare's Brain*; 'The Physics of *King Lear*: Cognition in a Void', in Bradshaw, Bishop, and Turner, eds, *The Shakespearean International Yearbook: 4. Shakespeare Studies Today*, 3–23; 'Roman World, Egyptian Earth: Cognitive Difference and Empire in Shakespeare's *Antony and Cleopatra*', *Comparative Drama* 43:1 (2009), 1–17; Ellen Spolsky, *Satisfying Skepticism*; 'Women's Work Is Chastity: Lucretia, *Cymbeline*, and Cognitive Impenetrability', in Richardson and Spolsky, eds, *The Work of Fiction*, 51–83; *Word vs. Image*; Evelyn B. Tribble, 'Distributing Cognition in the Globe', *Shakespeare Quarterly* 56:2 (2005), 135–155; '"The Dark Backward and Abysm of Time": *The Tempest* and Memory', *College Literature* 33:1 (2006), 151–168; Lalita Pandit, 'Emotion, Perception and Anagnorisis in *The Comedy of Errors*: A Cognitive Perspective', *College Literature* 33:1 (2006), 94–126; and Amy Cook, 'Staging Nothing: *Hamlet* and Cognitive Science', *SubStance* 35:2 (2006), 83–99.
23. See Henry S. Turner, *Shakespeare's Double Helix* (London and New York: Continuum, 2007), and Alison Games, *The Web of Empire*. Notice the 'web' of Games' title and her repetition throughout of terms like 'emergence' and 'circulation' as critical tropes to account for shifts emanating from smaller-scale human interactions toward larger-scale but retentively patterned cultural phenomena. For example, Games writes: 'The repeated migration of the people at the heart of English expansion reveals the *webs of connection* that first linked England to a wider world and then, through multiple voyages, tightened the *web*, and embedded England in a world of uncertain and tantalizing opportunity' (15, emphasis added).
24. Valerie Traub, 'Mapping the Global Body', in Peter Erickson and Clark Hulse, eds, *Early Modern Visual Culture: Representation, Race, and Empire in*

Renaissance England (Philadelphia, PA: University of Pennsylvania Press, 2000), 44–97.

25. To borrow a term used by Henry Turner and Jacques Lezra in their 2008 conference talks.

26. See Linda Woodbridge, 'Payback Time: On the Economic Rhetoric of Revenge in *The Merchant of Venice*', in Paul Yachnin and Patricia Badir, eds, *Shakespeare and the Cultures of Performance* (Burlington/Aldershot: Ashgate Publishing, 2008), 29–44; Natasha Korda, 'Dame Usury: Gender, Credit, and (Ac)counting in the Sonnets and *The Merchant of Venice*', *Shakespeare Quarterly* 60:3 (2009), 129–153; Amanda Bailey, *Of Bondage: Debt and Dramatic Economies in Early Modern England* (in progress).

27. See Ralph Berry, *Shakespeare's Comedies: Explorations in Form* (Princeton: Princeton University Press, 1972).

28. Citations to the play are from David Bevington's *The Complete Works of Shakespeare*, 6th edition (Harlow: Longman, 2009).

29. See Crane, *Shakespeare's Brain* (13) for an overview of prototype category organization (after Rosch and Lloyd, *Cognition and Categorization*).

30. The BALANCE schema and its semantic entailments are discussed in detail by Mark Johnson, *The Body in the Mind: The Bodily Basis of Meaning, Imagination, and Reason* (Chicago: University of Chicago Press, 1987), 74–98.

31. Starting (at least) with Berry, *Shakespeare's Comedies*, and continuing through the readings of Norman Rabkin, *Shakespeare and the Problem of Meaning* (Chicago: University of Chicago Press, 1981), Marc Shell, *Money, Language, and Thought: Literary and Philosophical Economies from the Medieval to the Modern Era* (Berkeley: University of California Press, 1982), Walter Cohen, 'The Merchant of Venice and the Possibilities of Historical Criticism', *English Literary History* 49 (1982), 765–789, and most recently Robert N. Watson, *Back to Nature*. In a comment similar to one cited by Watson, *Back to Nature*, 284, Rabkin writes: '*The Merchant of Venice* is a model of our experience, showing us that we need to live as if life has meaning and rules, yet insisting that the meaning is ultimately ineffable and the rules are provisional. The experience of the play, like the experience of a sonata – or of life itself – is one of process, and involves not just a final cadence or even the recapitulation of some main themes, but a whole sequence of contrasted but related developments' (*Shakespeare and the Problem of Meaning*, 31).

32. Watson, *Back to Nature*, 288.

33. Ibid., 289 (emphasis added).

34. As does a strange attractor in 'phase space', a prototype category feature generates organized but irregularly shaped (radial- and gradient-shaped) clusters of other types of features around it. Such ordering, while not symmetrical in the classical sense, nonetheless shows iterative patterning and may adhere to the geometry of self-similarity at different scales of semantic abstraction (as do dynamic systems). See Rosch and Lloyd, *Cognition and Categorization*, and Lakoff, *Women, Fire, and Dangerous Things*.

2
If: Lear's Feather and the Staging of Science

Amy Cook

> 'If that her breath will mist or stain the stone, / Why, then she lives.'
>
> *– King Lear*

> 'If his occulted guilt / Do not itself unkennel in one speech, / It is a damned ghost that we have seen'
>
> *– Hamlet*

> '… much virtue in If.
>
> *– As You Like It*

In 'if' we have possibilities. In 'if' we have hypotheses. In 'if' we have play. For literary theorists such as Bryan Reynolds, 'if' opens a subjunctive space of transversal movement – the possibility of both/and. For the cognitive linguistic theory of conceptual blending, 'if' is a space builder that prompts the listener to frame the information that follows as a blend of similar yet disanalagous identities in a counterfactual scenario – If he had been his teacher he would have hated himself.[1] For Antonio Damasio, the 'as-if' body loop is the simulation of the emotional and physical experience of a loved one such that our body actually experiences what we witness.[2] For Stanislavsky, the magic 'if' was the launching point for character identification – If I were in the situation of my character, what would I do?[3] For science and literature, there is much virtue in if.

With new theories of cognition, embodiment, and meaning-making come new arguments for the purpose and importance of theater as an instrument with which to alter and expand our relationship to another and our environment. Despite some resistance on both sides, the engagement between the disciplines has already provided exciting work and promises to continue to reshape scholarship in the academy. Along this interdisciplinary coastline there are different research agendas and questions. Some come from literature or theater and look to linguistics or the cognitive sciences to address questions of reception,[4] some start with text but want to understand it as embodied, imagined, and/or projected sound,[5] and still others view the historical products as cognitive instruments, requiring science to recreate the work they did.[6] The questions being asked at the intersection of literature/theater and cognitive science seem to be: How does a new concept of how language and thinking work alter our understanding of art[7] and/or the creation of art?[8] And what can a study of linguistic processing tell us about a historical period or the brain of the person who wrote the text? Rather than describe all of the exciting work being done – or interrogate the field critically in order to suggest ways this interplay between the sciences and humanities can be stronger, a worthy project to be sure – I will focus on two spots or moments in what Reynolds might call 'Shakespace', asking how performance (particularly stagings of Shakespeare) can operate both as theater and as a cognitive instrument. I want to know not just what cognitive science says about Shakespeare, but what the performance and rehearsal of Shakespeare says about science.

Inconceivable

In *The Princess Bride* (1987), Vizzini (Wallace Shawn) responds to each unexpected plot turn by exclaiming: 'inconceivable!' Of course, that which he finds 'inconceivable' is occurring right in front of him, so it is less 'inconceivable' than it is 'unfortunate' or 'unlikely'. Indeed, Fezzik (André the Giant) finally responds: 'I think you should look that word up. I don't think it means what you think it means.' Shawn/Vizzini[9] fashions himself an intellectual – even a scientist/philosopher – so if something happens which he did not expect, it is impossible. But it is inconceivable only because he had not predicted it. It was not part of the imagined possibilities. This calls attention to the mind/body that needs to conceive of the possible in order to run through the thought experiment necessary to assess the likelihood of such an outcome. If Shawn/Vizzini did look up 'conceivable' in the *OED*, he would find the following definition, first used in 1646: 'That can be conceived, imagined, or thought of; imaginable, supposable.' If Shawn/Vizzini had looked up 'conceivable' in the *OED*, he would also notice that prior to 1646 'conceivable' meant: 'That can be received or taken in', as in: 'That ... we might finde therein apt and conceiveable foode' (from Thomas Bowes' *De La Primaudaye's French academie*, 1586). Of course, he would be referencing the first definition, but the physical taking in, suggested by the earlier, and the cognitive taking in, communicated in the latter, are related: both cognition and digestion require penetration from outside the body/mind's container.

Cognitive linguists are now almost unanimous in understanding that thinking and speaking are creative and metaphoric. We do not use language as a code with which we translate what is out there, we organize what is out there around metaphors, image schemas, and mental spaces that come from embodied and embedded experience in the world.[10] Most of what we think of as literal and not creative is in fact understood metaphorically. An abstract idea such as knowledge is understood via a concrete experience of sight, for example, such that we speak of 'seeing' someone's point. The verb 'to know' does not correlate with an action we can perform or contain in a category of experience unless we understand it through our understanding of the visual system. George Lakoff and others posit that we organize our experience through idealized cognitive models (ICMs), compact models of how certain things work when imported to understand a given sentence. Lakoff outlines the ICM for 'seeing', for example, as: '1. You see things as they are. 2. You are aware of what you see. 3. You see what's in front of your eyes.'[11] If we say that we 'see' something, it is generally assumed that we accurately perceive what is in front of us. It is idealized to create efficiency even if it is not always accurate; we might say that we see a cat behind a fence, even though what we actually see is individual strips of fur. I have a physical and primary experience with taking in stimulation visually and then 'seeing' them; I use this process to understand the cognitive process. *I see what you mean. I can see you've given this a lot of thought. I'm not blind to your infidelity.* The cognitive process wherein TO SEE IS TO KNOW is right before our eyes and yet it can be hard to make out the depth of its impact.

We think the way we speak and we speak how we think. In *Women, Fire, and Dangerous Things*, Lakoff outlines the ways in which a new understanding of categories shapes how cognitive linguists think about the brain and language. The traditional view of categorization argues that we categorize things by virtue of common traits shared by the members; if it coheres with a set of traits that define the category, then it belongs in the category. Lakoff traces the development of a new theory of categorization that understands categories in terms of prototypes and basic-level categories. Categories have 'cognitive reference points' and 'prototypes' which organize the category, but do not define the category. This is an important distinction in that it speaks to the discourse around language within the humanities: language can constrain thought without controlling it. Lakoff and Johnson argue that metaphors define what can be viewed as truth: 'In a culture where the myth of objectivism is very much alive and truth is always absolute truth, the people who get to impose their metaphors on the culture get to define what we consider to be true – absolutely and objectively true.'[12] Seeing might be a very good way of conceiving of knowing, but other basic metaphors used to understand the abstract through reference to the concrete (ARGUMENT IS WAR, TIME IS MONEY) might be ideologically, as well as cognitively, constraining.

To test a hypothesis requires a performance of a particular script, a set of assumptions, a cast of characters. In *Making Truth: Metaphor in Science,* Theodore Brown argues that scientific thought is inseparable from the metaphors used to model and talk about the science. He talks about models as metaphors and how they are a mapping of information from a verbal expression of an idea to a 3D representation of that idea. The model then is used in conducting future experiments, motivating thought experiments, and envisioning future elaborations. If atoms are depicted as orbiting balls, it may be difficult to discover that they can be waves. Metaphor theory helps to see that the similarities *exposed* through metaphor can also be similarities *created* by metaphor. Brown gives the example of protein folding:

> Under appropriate conditions most proteins that are active in bio-logical systems coil up and rearrange lengths of the chains so as to assume a characteristic shape. This process was called 'folding' because an analogy was seen between the change the protein under-goes and the folding of objects in the macroscopic everyday world, such as napkins or card table chairs. … As a metaphorical expression it invites us to probe the cross-domain mapping between the literal, everyday act of folding and the changes that occur in a protein as it undergoes the transition we call folding. Thus, the act of naming the process 'folding' *creates* similarities.[13]

Language, made up of poetry, metaphor, and narrative – those things traditionally understood by literary critics – can be a tool to imagine, learn, and probe conceptions and assumptions in other areas; cognitive linguistics have given literary critics the tools by which to 'deconstruct' the dramaturgy of science.

Categories are constructed; there is no 'dog' in the world, there is only 'Fido' and 'Fluffy Wuffy', real panting, furry things that we decide to group together in a category we call 'dog'. Categorizing makes us efficient and offloads cognitive work: we do not need to waste often-precious time figuring out how to behave with this panting, furry thing – Do we shoot it? Talk to it? Pet it? We call it dog and project the information we have from past experiences we have about dogs onto this dog, until we run into an animal that challenges this category (a wolf, a Doberman, a cat) and we must decide to recategorize or alter our category. Whether discussing science, theater, or politics, the language we use – both the metaphors and the categories we use to speak and think – should be probed for its entailments. As Lakoff pointed out: 'Since we understand the world not only in terms of individual things but also in terms of categories of things, we tend to attribute a real exist-ence to those categories.'[14] If my category of 'marriage' stipulates love and commitment, and yours stipulates love and commitment between a man and a woman, we will not be able to agree on how to legislate marriage. If you define the earth as the center of the universe, it might require some demonstration of the truth to alter your conception. If revenge erases the past wrong, perhaps staging revenge can rewrite a future.

Of course, important literary and cultural theorists have explored many of these issues before and cognitive literary and performance the-orists do well to engage with, as well as question, this important work. One does not need to supplant the other. Literary theorists have had hundreds of years to make incredibly astute analyses of Shakespeare's poetry. New historicists insist on the situatedness of the plays and a cognitive reading should consider the context of the minds/bodies of the writer, performers, and audience.[15] Bryan Reynolds argues that the theater of the early modern period was a medium of possession and that through this transversal movement – the becoming other that occurs in performance – the ideological state apparatus can be challenged. Transversal theory posits a way of understanding the cultural and cogni-tive instrument that is and was the theater. His intervention is to see the way ideological change is facilitated by a re-imagining possible through staging and performing what is not.[16] In 1935, Caroline Spurgeon argued for the importance of the disease metaphor in *Hamlet*. It is not my intention to disagree, but finding the prevalence and significance of disease in *Hamlet* would be where a cognitive linguistic reading begins, not ends. She does not question the metaphoricity of disease, or unpack the web of mental spaces necessary to understand 'rank' offenses or something 'rotten'. None of these are referring to a literal thing, but rather a metaphoric conception of illness that relies on an understanding of the body as container, illness as war, infection as invasion, seeing as knowing, etc. With the disease metaphor broken down further, it is possible to see a link between the disease metaphor and Hamlet's obsession with not seeming or with his need to write down that a man may smile and still be a villain: if the body is a container, how does the outside reveal information about the inside? By applying this linguistic method to poetry, the end game is not an explication of the image, but an interrogation of the integration network necessary to understand the image and the performance thereof necessary to compose that image and therefore the spaces, connections, and images recruited for this comprehension.

The polysemy and word play in Shakespeare can create a further linguistic richness, since in one word he can pack two ideas. Bruce Bartlett argues that Shakespeare uses 'waight' at the top of the Lady Anne wooing scene in *Richard III* to refer to the weight of the coffin and to Richard's wait to ascend the throne: 'The manifestly visible waight may be Henry VI's coffin, but a subtly abstract waight is also present: the plotted ambition of Richard's short-term designs on the crown (present even back in *3H6*).'[17] He then turns to the 'waight' in *Othello* that occurs midway through the play in the pivotal scene where Desdemona pleads with Othello to speak with Cassio soon and then to hear her 'suite', that is 'full of poize and difficult waight' (3.3.83). Bartlett argues that since poize is glossed by most editors as meaning 'balance or weight', 'waight' is redundant if it does not also convey the hand-wringing that Desdemona is likely engaged in, impatient with her husband's delay. It is just this delay that allows Iago to convince Othello to wait for the encounter with Cassio and that provides time for the handkerchief dropping. Finally, Bartlett examines the 'waight' at the end of *King Lear*. Arguing that the lines should be given to Edgar (even though some editors give them to Albany), Bartlett connects the 'waight of this sad time we must obey' (5.3.299) of the final moment of the play with Edgar's earlier assertion that 'Men must endure / Their going hence even as their coming hither' (5.2.9–10). Here, the 'miracle' of life referred to in Act 4 becomes 'a two-fold waight to be obeyed'.[18] The 'waight' of life in *King Lear* creates in one word the network of meanings evoked by both words.[19]

Bartlett interrupts his text-based analysis at the end to suggest 'Performances should give us a pregnant pause of a couple of seconds after Edgar's *waight*.'[20] Though he does not extrapolate, presumably this pause is meant to help the audience hear both meanings, in part because it allows time for the word(s) to reverberate (or 'land' in acting parlance) and in part because the pause becomes the wait that helps evoke the 'wait' in 'waight' – particularly if the actor actually paused for 'a couple of seconds' which is a long pause on stage. Bartlett assumes that since there is no aural difference between 'wait' and 'weight' the performance must help convey the textual analysis he has unearthed. Yet Shakespeare encoded the double meaning in the sentence, which prompts for the multiple meanings in 'waight' however the actor says the line:

> The waight of this sad time we must obey,
> Speake what we feele, not what we ought to say:
> The oldest hath borne most; we that are young
> Shall never see so much, nor live so long.

(5.3.299–302)

The start of the sentence – 'the waight' – makes us think of weight, since 'the weight' is a noun rather than a verb and is more often delimited in the precise measurements that *the* weight suggests. The sentence quickly prompts the other meaning, since time rarely has weight but often has wait. Through this, Shakespeare gives matter to time while at the same time turning the 'life's miracle' (4.5.55) into a specifically allocated 'wait'.[21] Shakespeare used language to perform this double meaning; he did not need to rely on the actor's 'pregnant pause'.

In an argument based solely on the texts of the plays – and even more specifically two words that in performance sound the same – Bartlett's penultimate sentence shifts all the responsibility for conveying this meaning to the actor. While I argue that multiple meanings are present in the text, Bartlett is not wrong to remind us that these words are always prompts for performative meaning. In the context of an old king carrying the body of his dead daughter, weight/wait is not an abstract concept; watching King Lear hold Cordelia, an audience can feel what happens when wait becomes weight. We can hear this polysemy in the words, we can see it as we read through the text, but it is only when one actor carries another actor (an older man holding a grown woman) that the audience knows it. This stunning moving Pietà, the father cradling a grown woman the way he may have cradled the baby she once was, breaks our heart because we know it, we do not just perceive it. What we see, hear, and imagine still must be staged and embodied in order to be known.

Patrick Colm Hogan uses blending theory to unpack Cordelia's explaining the 'nothing' she gives her father by saying that, 'I cannot heave / My heart into my mouth' (1.1.96–97). For Hogan, the blend is organized structurally by the human body, such that, 'we have a blended space in which Cordelia's mouth as a physical space that can contain something solid is identified with her mouth as an organ of speech and her physical heart is identified with her feelings, all through the organizing structure of Cordelia's body.'[22] He goes on to suggest that the blend creates emergent meaning in which such an action (heaving one's heart into one's mouth) leads to death because 'Cordelia does ultimately die by choking and suffocation (from hanging).'[23] This is emergent meaning that

> conceptual blending theory may help us understand, not only particular metaphors, but many levels of complexity in literary works – and in our response to those works. Indeed, with regard to the latter, one of the most remarkable things about blending theory and other cognitive approaches is that they reveal to us the subtlety of our ordinary reactions to literature and the arts. ... For, in the end, our experience of the play is necessarily the experience of a complex, blended space that we have made, cognitively.[24]

I would add, however, that a study of blends in performance shows that linguistic structures – coming from embodied experience of the world (as Lakoff, Johnson, Turner, and others have shown) – return to the body in a staging of the spaces and the counterfactual spaces primed to understand textual theatrics.

Lear's feather

When Lear enters in Act 5 carrying Cordelia, he spends the majority of his dying breath trying to conceive of the death of his daughter. He first declares that she is dead but then doubts his own perception; he asks for a mirror to make sure: 'I know when one is dead, and when one lives; / She's dead as earth. Lend me a looking glass, / If that her breath will mist or stain the stone, / Why, then she lives' (5.3.261–264). When the mirror fails to bring the sign of life for which he is so desperate, Lear repeats his experiment with a feather hoping for different results: 'This feather stirs, she lives! If it be so, / It is a chance which does redeem all sorrows / That ever I have felt' (5.3.266–268). What he 'knows' has broken down and so he must perform an experiment. If x happens, then y will be true. Moreover, these are experiments staged for the members of the court still living: they are his spectators. They can validate his experiment through witnessing it. Though his words fail him, he turns to scientific observation and the performance of the experiment to understand what his mind cannot conceive.

But why does this work? Why assume that the chain of events are linked? A feather stirring is not the same as a daughter's life. But connected through breath, the two spaces are related. Without mapping the movement of the feather to what it represents, no sorrows will be redeemed. How do we know what we think we know? In the century following *King Lear*, philosophers/scientists were asking and answering new questions. Through conducting experiments and then writing about their experiments, Sir Francis Bacon, Galileo Galilei, and others changed how data were gathered and examined; they understood knowledge as mediated and therefore changed the tools used for seeing and the performance of their results.

In 1611, Galileo asked spectators to look at the heavens through his telescope and see the moons of Jupiter and the mountains on the moon. Despite the ocular 'proof', the viewers were unconvinced; such visions were inconceivable. They were, however, impressed with the telescope's ability to render readable text on a far-away building.[25] Others, however, would not be able to look through his telescope; an instrument so rare and expensive could not provide ocular evidence for all. Galileo produced his accounts not so that they could be verified but, knowing that they could not be verifiable immediately, he used his writings and his drawings to persuade. Integrating the visual distortion of the telescope into the text, Galileo makes *The Starry Messenger* into a visual instrument for readers to see what he has seen. Elizabeth Spiller points to the drawings in Galileo's *The Starry Messenger*, images created not to duplicate what he saw in his telescope but as 'viewing aids' to give his readers an experience of observing: 'Galileo does not want to reproduce the image he sees in the telescope; rather he makes *The Starry Messenger* into a kind of textual telescope for his readers so that his readers will experience this new way of observing as a new way of reading.'[26]

He does this not through perfect reproduction, but through distortion. Galileo must make what he has seen conceivable: 'with the glass you will detect ... such a crowd of others [stars] that escape natural sight that it is hardly believable. ... But in order that you may see one or two illustrations of the almost inconceivable crowd of them, and from their example form a judgment about the rest of them, I decided to reproduce two star groups.'[27] Galileo's illustrations make a distinction between the stars that are visible to the naked eye – with double outlines – and the ones that are only visible with a telescope. For Spiller, this itself is invention:

> What separates Galileo's work, however, is a new definition of the invisible. Where earlier illustrators had shown parts of the cosmos that were invisible to human sight, Galileo now depicts that which is invisible without a telescope. In doing so, Galileo adapts the familiar as a visual template within which new stars can be recognized. Such schematizations allow readers literally to 'see' new formation inside an identifiable framework.[28]

Galileo needed more than the telescope to produce new knowledge: he needed to find a 'framework' within which such new data can make sense to his witnesses: this is what you see with your eye and this is what you would see if you had a telescope. What creates knowledge is the relationship between the experiment of looking and the story told to make sense of what is seen.

Lear turns to a mirror and a feather to provide information that he may not be able to see with his naked eye. He observes, hypothesizes, and tests. As Francis Bacon suggests in *The Advancement of Learning*, perception alone does not lead to knowledge:

> For the mind of man is far from the nature of a clear and equal glass, wherein the beams of things should reflect according to their true incidence; nay it is rather like an enchanted glass, full of superstition and imposture, if it be not delivered and reduced.[29]

Bacon does not suggest that delivery or reduction requires fiction, however. In fact, in his *Novum Organon* he warns against several things that can distort perception: (1) Idols of the cave, (2) Idols of the market place, (3) Idols of the theater. The first causes the individual to bias perception according to limited perspective and emotional valence. The second is a kind of group-think: we are biased by the 'received wisdom' of our culture. The third reflects the power of the narrative – the story, the metaphor, the model – to bias perception. These Idols are important to acknowledge and work against because, as the quote above says, the mind does not take in information equally. For Bacon, and others of the scientific revolution, what is needed are rigorous observation, performance, and reception.

Bacon believed 'Man, being the servant and interpreter of Nature, can do and understand so much and so much only as he has observed in fact or in thought of the course of nature: beyond this he neither knows anything nor can do anything.'[30] Man had to witness in order to conceive of the new. Important as the tool is, it is also the performance of the tool in action that creates and distributes knowledge and, despite Bacon's concern about the Idol of the theater, stories told about investigation, about the observed, are required for conception. Lear uses the feather and the mirror to see that which he cannot see without a mediating device yet his final words are commands to the spectators: 'Do you see this? Look on her! Look her lips, / Look there, look there!' (5.3.311). If they see it, it will be so. Lear holds the mirror up to Cordelia's lips hoping that the invisible will be visible on the glass; the breath of life will manifest in tiny droplets on the surface of the glass. If he does not see the invisible, his daughter is dead.

We use fiction to find the truth. In *The Fate of Place*, Edward Casey argues that the discussions of space accustomed the Renaissance thinkers to think in terms of space without end (in counterdistinction to finite space, or: place). The understandings of Thomas Aquinas and Newton, for example, may differ, but

> both tendencies share one important thing in common: they were both conceived as ways in which infinite space can be *imagined* … what the world and the universe would be like if God were to choose to alter things as they are radically … When one begins to think this 'otherwise,' one is approaching things *secundum imaginationem*, 'according to imagination' – not according to how things in fact are, have been, or will presumably be. … infinite space is a matter of what can be imagined, of what *could be*; finite space is a matter of what *is* the case.[31]

Thinking *secundum imaginationem* explores what *is* by thinking about what *is not*. This counterfactual space[32] allows for, as Reynolds shows, subjunctive movement, defined as 'the hypothetical dimension of "what-ifs" and "as-ifs" that unsettles certain authoritative processes in critical practice'.[33]

There is no breath on the glass and the feather does not stir. He can see that she is dead. He has demonstrated that she is dead to the witnesses of the court. He refuses to know that she is dead. In the line before pleading with the court to see movement on her lips that is not there, Lear engages in a profoundly painful *secundum imaginationem* thinking: 'Thou'lt come no more, / Never, never, never, never, never!' (5.3.308–9) Shakespeare upends the iambic rhythm with five trochees in a row: we are stopped short by the sound of this line as compared to the relatively regular iambic flow of the line before. In counterdistinction to the iambic pentameter we expect, these 'nevers' are an alternate and jarring world. Never is the answer to one question – when will Cordelia come again? – and yet he gives it five times. Each time he says never there is a corresponding time where she does come; never, like her death, is only comprehensible in relationship to the counterfactual space that is not present. 'Never' occurs twenty-seven times in *King Lear* and each time it is used, the sentence focuses more on the time against which the 'never' operates: 'Sure I shall never marry like my sisters, / To love my father all', 'And from her derogate body never spring / A babe to honour her!' and 'The oldest hath borne most; we that are young / Shall never see so much, nor live so long'. These three examples use an absent future to vividly paint the present present. Never can only be grasped by repeated calls to those times when she is there. He tells himself a story where she is missing then and then and then and then and then in order to know that she is dead. Lear uses the feather and mirror to enlarge what he can see and a story to improve what he can imagine. Lear – and Shakespeare – then performs the experiment for an audience, creating knowledge where there was perception and imagination.

To represent the previously invisible, to perform the seemingly impossible, is vitally important to creating the visible and the possible. Drama and performance offer a new staging of an old story that then begins a reconception of the questions and answers we are interested in. Scientists can see the same things under a microscope in two different centuries, but the stories they are told by artists and writers about who they are can impact what they see on the slide.[34] Light has always operated the same way and yet in one century it is seen as a particle and the next it is a particle and a wave. Since, as Lakoff and Johnson argue, the metaphors we use constrain what we can think of as 'true', the speaking and performing of a new metaphor can scaffold new conceptual structures.

'Denmark is a prison'

Theater stages ideas of self and other. Bruce McConachie, Tobin Nellhaus, and others have argued that moments in theater history can be understood as informed by – shaped by – image schemas. McConachie sees the theater of Cold War America as structured around the image schema of containment: a contained self in a bunkered family within the nation.[35] He finds the dramas of Tennessee Williams, Arthur Miller, and others as centering on the activities of 'empty boys' (e.g., Brick) and 'fragmented heroes' (e.g., Willy Loman) within a home – scenically designed to frame the interior rooms within the house and within the outside world visible through the inside and all within a proscenium arch (as in Jo Mielziner's set design for *Death of a Salesman*). Acting training at the time produced actors that generated the appearance of outer shells that belied a 'true', but often hidden, interior. McConachie suggests that the dramaturgy and performance conventions of the time both structured and were structured by an experience of containment that made inside real and safe and outside threatening and foreign. Nellhaus connects the shifts in performance strategies with shifts in communication methods and notes the relationship between the sentimental dramaturgy of the early eighteenth century and Locke's philosophy of the self as a text written on a 'blank slate' and that, therefore, understanding the self and the other required reading the outside for information about the inside.[36] For Nellhaus, the communication framework of a particular age (the print and newspaper culture of the eighteenth century or the visual/allegorical system of communication of the medieval period) is a cognitive instrument that shapes the theater of the age, which, in turn, provides the models by which we organize our ideas of self: '… changes in communication frameworks not only introduce new metaphors but also reorganize the roles played by older and, in some cases, quite basic metaphors'.[37] Since, as cognitive linguists now almost uniformly agree, thinking is visible through speaking and speaking shapes the thoughts available to us, then how we talk about theater, science, and self enables and constrains what we see as the power and possibility within them.

Different performers and performance conventions evidence different conceptions of inside and outside, self and other. Ira Glass, the host of *This American Life*, introduces a segment on *Hamlet* as performed by prisoners by asking what any of us can relate to in *Hamlet*: who among us has a father killed, a mother stained, or is plagued by a revenge-obsessed ghost? Reporter Jack Hitt suggests that there is much to learn about *Hamlet*, a play about a man contemplating a violent crime and its consequences, from people who have committed such a crime and are currently living with the consequences. His 'Act V' story follows the rehearsals and production of *Hamlet*'s Act 5 within a maximum-security prison in Missouri. With very little education and no acting training, these inmates, according to Hitt, make sense of Shakespeare's play because of where they have been and what they have seen. The hour-long piece tracks Hitt's experience re-reading the play through the actors in it and re-reading the inmates through the conflict between their history out of prison and their present in their characters. Hitt argues that their individual perspectives, their crimes and victims, were powerfully present in their performance. Perhaps they were – without having been there, we must take Hitt's word for it – but what strikes me most about his piece is the way in which the container metaphor structures his understanding of acting in *Hamlet* but does not necessarily reflect the language used by the prisoners to describe their experience of being Horatio, the Ghost, or Laertes.

Most Method actors refer to being 'in' character or getting 'out' of themselves: their bodies are the containers that temporarily allow a character inside, assuming their actor training can help them get out of the way. Derek 'Big Hutch' Hutchison, the 'killer whale' in a cast mostly made up of prisoners on the 'minnow' end of the prison hierarchy, does not seem to talk about his character this way; Horatio, for him, is a 'chump', and Hutch seems to feel no need to 'get in' to the character of Horatio. His performance is one that Hitt singles out as successful, not because Hutch has gotten into character, but because 'Horatio has Hutch under control and the audience in his hand.'[38] In analyzing the performance, Hitt speaks of the character having the inmate under control and yet throughout the rehearsal process, Hitt talks of acting as being a relationship between an inside and an outside: 'One of the problems of doing any play in prison is that being a good actor is the exact emotional opposite of what it takes to be a good inmate. Rather than close off all feeling and look tough, you have to open your vulnerable self up and withstand the, often cruel, laughter as you try to find some authentic emotion within you.'[39] For these actors, it seems to be not an act of finding something in them but rather finding a world, a possibility outside of them, outside of the prison of their past and present.

One of the actors speaks of the life-changing power of being cast in the play: 'the day opened up a whole world for me … it was like the day my daughter was born' and another spoke of how being in the play allowed him to 'get into something else. That did open my eyes. To get into reading Sylvia Plath, and Frost, and Wordsworth.'[40] These actors are not opening themselves up to let out some emotion or let in another character; things are opening up for them. According to Hitt, James Word (the actor who takes over the role of Laertes from the Amish pedophile who is transferred to another prison) 'channels Laertes' character in a way that should make any Method actor cringe with jealousy';[41]

a Method actor, he assumes, aims at possession, aims at being an empty conduit for a character. Hitt asks Word: 'Do you feel like you can be Laertes because so much of Laertes is in James Word?' There is a long pause and Word responds: 'I am Laertes. I am. I am.' Hitt wants – and assumes Word wants – Word to be possessed by Laertes, to have someone else inside him. Word responds that he has escaped Word and has entered Laertes. Onstage, Word is not *in* something else and nothing is *in* him; in *Hamlet*, Word *is* something else.

Being someone else, someone other than the man who shot two men in the head and left them for dead, for example, can reformulate the future. One actor admits to Hitt: 'I took a man's life. Do I deserve to be out there? I cannot say'; finding an affirmative answer to that question is key to staying 'out there'. It is by hearing themselves speak words meant for another, words that put air – four acts of air – between murder and revenge, that they can stage and embody another option for their present and their future. Edgar, a former post office employee, explains what the purpose of playing has been for him:

> She [the director, Agnes] makes us feel human, man. She really does. When I go in there and I have to take my clothes off and get butt naked and bend over and spread my cheeks so some man can look up my butt, you know all the humiliating things that they do to us in here, and she comes in and does what she does for that minute, for that two and a half hours – all these guys with PhDs and can be doing other things, they come in and … I at least can feel human. In here.

The rehearsal room is the container for them that creates a possible future.

After the ten minute 'cast party', Hitt tells us: 'Manuel is leaving for a halfway house in 48 hours. He could have been out weeks before but chose to stay in prison so he could finish the play. [...] a few others have parole board meetings in a couple of weeks, to decide whether they have changed enough to mingle with us on the outside. To that extent, this whole night, including the cast party, is just another rehearsal.' Hitt wants to tell the story that most NPR listeners want to hear: like Baby Einstein turning infants to geniuses, Shakespeare can turn murderers into citizens. Rather than writing it off as a fairy tale, though, I would like to suggest that the distinction between how these actors speak of acting and how Hitt and Method actors refer to acting is important: a container metaphor for the body suggests that the men can be possessed by characters, but that these men are not being changed by possession but rehearsing escape. The staging of *Hamlet* allows them to form a hypothesis about how the past does, and does not, inform the future.

Telling and understanding stories – embodied and perspectivized language – teaches us to see and thus to conceive and then to test, and finally, perhaps, to know. King Lear performs an experiment to show himself and the court what he cannot bear to know: Cordelia is dead. *Hamlet* provides escape through embodiment – and not because the bodies of the actors make room for the characters, but because the body does not need to be a container that holds a past and an identity that forms the future. If we are interested in knowing, in finding our humanity by bridging the gap between what we think is and what we think could be, then we need stories and we need science. Science uses stories and theater is a hypothesis, tested and staged every night: 'If that her breath will mist or stain the stone, / Why, then she lives'.

An interplay between cognitive science and theater or literary studies has the potential to provide new methods of inquiry and new insights into the questions that engage us. F. Elizabeth Hart and McConachie have argued that much of the theory that has dominated literary studies for the last thirty years is based on a theory of language which conflicts with current research within cognitive linguistics,[42] and, while I do not think that scholarship needs to be empirical, I do believe that scholars should attend to the research in the sciences that challenges the 'master theorists' of the home discipline. When cognitive linguists insist that their research examine the complexity of 'now is the winter of our discontent' and not restrict themselves to 'the cat is on the mat' simply because the latter is easier to study, then literary theorists should avail themselves of the information gained through that work. When cognitive scientists study how an emotional body engages with and is formed by narrative, then those of us interested in narrative should pay attention – if for no other reason than to make sure they are accurately representing the nuances of narrative. Such work should involve a rigorous investigation of the science it applies while providing new answers and questions to our home discipline; it should not merely be a parlor trick done for humanities viewers at home.

Cognitive science offers literary and performance theorists new ways to ask and answer questions about the object of their study. I want to know how we know what Shakespeare means and why that meaning continues to be important to us. I want to know why being and seeing bodies speak onstage is powerful – how it manages to change minds. Since the seemingly simple ability to watch, understand, appreciate, and be moved by a theatrical production is, in fact, an extraordinary cognitive and biological feat, my investigation has profited by research in science. I deploy the sciences not because it is more 'objective' or true than previous theoretical movements in the humanities but because the interests and findings within that field shed light on my field. When researchers within the cognitive sciences fail to provide creative methods of approaching exciting questions, then I will turn my attention elsewhere.

Notes

1. On conceptual blending theory, see Gilles Fauconnier and Mark Turner, *The Way We Think: Conceptual Blending and the Mind's Hidden Complexities* (New York: Basic Books, 2002).
2. See Antonio Damasio, *Descartes' Error: Emotion, Reason, and the Human Brain* (New York: Avon, 1994) and Damasio, *The Feeling of What Happens: Body and Emotion in the Making of Consciousness* (San Diego, CA: Harcourt, 1999). In *The Actor, Image, and Action: Acting and Cognitive Neuroscience* (New York: Routledge, 2008), Rhonda Blair discusses the possibility for actor training of the 'as if' body loop and in 'Wrinkles, Wormholes, and *Hamlet*: The Wooster Group's *Hamlet* as a Challenge to Periodicity' (*TDR* 53:4 [2009], 92–103), I suggest that such research could have a profound impact on theories of performance. Also, in this volume, Jen Boyle engages with Damasio's 'as if' body loop and its possible relationship with subjunctive theory.
3. See Konstantin Stanislavsky, *An Actor Prepares* (New York: Routledge, 1964). Stanislavsky tells actors that 'with my magic *if* I shall put myself on the plane of make-believe, by changing circumstances only: the hour of the day. [...] Out of that simple circumstance there follows a whole series of consequences' (65).
4. Some examples of this would be: Barbara Dancygier and Eve Sweetser, *Mental Spaces in Grammar: Conditional Constructions*, Cambridge Studies in Linguistics 108 (Cambridge: Cambridge University Press, 2005); Donald C. Freeman, '"Catch[Ing] the Nearest Way": *Macbeth* and Cognitive Metaphor', *Journal of Pragmatics* 24 (1995), 689–708; Freeman, '*Othello* and the "Ocular Proof"', in Graham Bradshaw, Tom Bishop, and Mark Turner, eds, *The Shakespearean International Yearbook* (Aldershot: Ashgate, 2004), 56–71; F. Elizabeth Hart, 'Matter, System, and Early Modern Studies: Outlines for a Materialist Linguistics', *Configurations* 6:3 (1998), 311–343, and Chapter 1 in this volume; Amy Cook, 'Staging Nothing: *Hamlet* and Cognitive Science', *SubStance* 35:2, Issue 110 (2006), 83–99; Lisa Zunshine, *Strange Concepts and the Stories They Make Possible* (Baltimore: Johns Hopkins University Press, 2008).
5. Some examples of this would be: Bruce R. Smith, *The Acoustic World of Early Modern England: Attending to the O-Factor* (Chicago: University of Chicago Press, 1999); Smith, 'Hearing Green', in Gail Kern Paster, Katherine Rowe, and Mary Floyd-Wilson, eds, *Reading the Early Modern Passions* (Philadelphia, PA: University of Pennsylvania Press, 2004), 147–168; Mark Turner, *The Literary Mind: The Origins of Thought and Language* (Oxford: Oxford University Press, 1996); Amy Cook, 'Interplay: The Method and

Potential of a Cognitive Approach to Theatre', *Theatre Journal* 59:4 (2007), 579–594; Bruce McConachie, *American Theater in the Culture of the Cold War: Producing and Contesting Containment, 1947–1962* (Iowa City: University of Iowa Press, 2003); and McConachie, *Engaging Audiences: A Cognitive Approach to Spectating in the Theatre* (New York: Palgrave Macmillan, 2008).

6. Some examples of this would be: Mary Thomas Crane, *Shakespeare's Brain: Reading with Cognitive Theory* (Princeton: Princeton University Press, 2001); Crane, 'The Physics of *King Lear*: Cognition in a Void', in Bradshaw, Bishop, and Turner, eds, *Shakespearean International Yearbook*, 72–97; Philip Davis, *Shakespeare Thinking* (London: Continuum, 2007); Ellen Spolsky, *Gaps in Nature: Literary Interpretation and the Modular Mind* (Albany: State University of New York Press, 1993); Spolsky, *Satisfying Skepticism: Embodied Knowledge in the Early Modern World* (Aldershot: Ashgate, 2001); Spolsky, *Word vs Image: Cognitive Hunger in Shakespeare's England* (Basingstoke: Palgrave Macmillan, 2007); and Adam Max Cohen, *Shakespeare and Technology: Dramatizing Early Modern Technological Revolutions* (New York: Palgrave Macmillan, 2006).

7. Neuroscientist V. S. Ramachandran's essay (co-written with William Hirstein: 'The Science of Art: A Neurological Theory of Aesthetic Experience', *Journal of Consciousness Studies* 6:6–7 [1999], 15–51) audaciously explores how and why the brain responds to art. They articulate 'Eight laws of artistic experience' used by artists to 'titillate the visual areas of the brain' (17). They posit that visual metaphors in art are powerful because the act of connecting the analogized image with the referent stimulus is pleasurable and because our brains reward the establishment and cross-referencing of categories as it helps it store and code information more economically. The evocation of the limbic response often happens prior to the full comprehension of the metaphor, suggesting that metaphor comprehension might be something the brain rewards for its 'economy of coding' (31). While this and work like it is appealing and exciting, it is important to note that it is speculative; no more so, perhaps, then literary theory, but no less so just because it was written by a neuroscientist. Further, one hopes that an interplay between the humanities and the sciences would provide scientists with the scholarship on art and literature necessary to make their work rigorous.

8. Both John Lutterbie and Rhonda Blair are providing interesting insights into this. See Lutterbie, 'Neuroscience and Creativity in the Rehearsal Process', in Bruce McConachie and F. Elizabeth Hart, eds, *Performance and Cognition: Theatre Studies and the Cognitive Turn* (New York: Routledge, 2006) and Blair, *The Actor, Image, and Action*.

9. McConachie argues that since the character requires the actor to exist, how we refer to the embodied character should always reflect this inseparability. See *Engaging Audiences*.

10. I provide a longer articulation of the potential for re-reading Shakespeare on the page and on the stage in *Shakespearean Neuroplay: Reinvigorating the Study of Dramatic Texts and Performance through Cognitive Science* (Basingstoke: Palgrave Macmillan, 2010) and in 'Interplay: The Method and Potential of a Cognitive Approach to Theatre', *Theatre Journal* 59:4 (2007), 579–594. In this volume, Hart provides a clear and compelling case for how recent work in cognitive science challenges work being done in materialist criticism specifically but also in the humanities more generally. In 'Matter, System, and Early

Modern Studies', she offers a 'science based critique of Derridean language theory, focusing on the formalism latent in Jacques Derrida's narrative of différance' (314).

11. George Lakoff, *Women, Fire, and Dangerous Things: What Categories Reveal About the Mind* (Chicago: University of Chicago Press, 1987), 128.

12. George Lakoff and Mark Johnson, *Metaphors We Live By* (Chicago and London: University of Chicago Press, 1980), 160.

13. Theodore Brown, *Making Truth: Metaphor in Science* (Urbana and Chicago: University of Illinois Press, 2003), 25.

14. Ibid., 9. Moreover, there is evidence that we use stories, religion, and myth to generate categorical variants to reinforce the category. Pascal Boyer finds that, across cultures and times, religious figures or stories tend to contain an element that *'contradicts information provided by the ontological category'; 'religious concepts invariably include information that is counterintuitive* relative to the category activated', *Religion Explained: The Evolutionary Origins of Religious Thought* (New York: Basic Books, 2001), 64–65. This religious 'other' helps clarify, then, what we know as normal.

15. This is not to suggest that we can know what their brain looks like – or that it would necessarily be interesting if we could – only that their different world and experience might impact what they thought. Hart insists on the relevance of the language explosion in understanding Shakespeare's use of language and his many neologisms and Crane explores the use of language and imagery in Shakespeare's plays to imagine the way his mind processed his world.

16. His theory, for example, can fit with a cognitive linguistic reading that insists on the power of counterfactual space in speech and thought. One cannot refer to a missing chair without imagining a chair that could be or was in the space now being held by a 'missing chair'. Something that is not (a chair that is not there) can become a thing (missing chair) to be used conceptually and semantically as a noun. See Amy Cook and Bryan Reynolds, 'Comedic Law: Projective Transversality, Deceit Conceits, and the Conjuring of *Macbeth* and *Doctor Faustus* in Jonson's *The Devil is an Ass*', in Bryan Reynolds, ed., *Transversal Enterprises in the Drama of Shakespeare and His Contemporaries: Fugitive Explorations* (Basingstoke: Palgrave Macmillan, 2005), 85–111. The inconceivable and the conceivable are mutually dependent; one is the *tool* with which to create the other.

17. Bruce R. Bartlett, 'Bearing the "Waight": Double-Entendre in *Richard III, Othello,* and *King Lear*', *The Shakespeare Newsletter* 49:1.240 (1999), 7–8 (7).

18. Ibid., 8.

19. Crane, 'The Physics of *King Lear*'. Mary Crane hears in the play echoes of debates at the time about divisibility and argues that *King Lear* exists in light of an 'epistemological rupture' where the breakdown of Aristotelian physics also calls into question basic mental concepts of weight, space, divisibility, and existence. In addition to turning to the history of science for circulating debates at the time, Crane suggests that Shakespeare's basic mental model of weight and wait, 'woven into the fabric of poetry', provides a 'shaping presence' or 'frame that supports plot, characterization and theme' (20). See also Henry Turner, '*King Lear* Without: The Heath', *Renaissance Drama* 28 (1997), 161–193.

20. Bartlett, 'Bearing the "Waight"', 8, an interesting choice of words for an essay focusing on the polysemy of words. Bartlett does not explain what the pause is pregnant with or what the gestation period is for pauses. Nor does he stipulate what makes one pause pregnant and another barren. Presumably this has to do with the performance energy of the actor, but he does not give any more acting notes on how to achieve the pregnancy of the pause.

21. Annabel Patterson has found a similar case of meaning in juxtaposition in her analysis of the abbreviated answer to why the Players have been away ('their inhibition comes by meanes of the late innovation') in the second Quarto of *Hamlet*: 'Here I want simply to argue that what Hamlet's "innovation" means must be affected by its semantic proximity to "inhibition", that concreteness comes by the vibrations between these two Latinate abstractions' (*Shakespeare and the Popular Voice* [Cambridge, MA: Blackwell, 1989], 27).

22. Patrick Colm Hogan, *Cognitive Science, Literature, and the Arts: A Guide for Humanists* (New York: Routledge, 2003), 113.

23. Ibid., 114.

24. Ibid.

25. For more on Galileo's experiments and his attempts to make better observers out of his contemporaries, see Elizabeth A. Spiller, 'Reading through Galileo's Telescope: Margaret Cavendish and the Experience of Reading', *Renaissance Quarterly* 53 (2000), 192–221.

26. Ibid., 200.

27. Cited in ibid., 202.

28. Ibid., 202.

29. Francis Bacon, *The Advancement of Learning* (Rockville, MD: Serenity Publishers, 2008), 119.

30. From Bacon's *Novum Organum* (1620), cited in *The Portable Enlightenment Reader*, ed. Isaac Kramnick (New York: Penguin Books, 1995), 39.

31. Edward S. Casey, *The Fate of Place: A Philosophical History* (Berkeley, CA: University of California Press, 1997), 110.

32. See also Spolsky's brilliant cognitive historical study of early modern art and theater, *Satisfying Skepticism*.

33. Bryan Reynolds, *Becoming Criminal: Transversal Enterprises in the Drama of Shakespeare and His Contemporaries* (New York: Palgrave Macmillan, 2006), 284. Reynolds and I argue that the comedy of Ben Jonson, for example, works to move spectators into an 'as-if' space through its comedy: 'Comedy can confuse and create doubt. It can reorder the existing logic and expose the possibility of alternate ways of thinking. These potential alternatives might not fit the simple, ordered world of the dominant paradigm and may allow us to imagine new worlds.' See Cook and Reynolds, 'Comedic Law', 92.

34. Steven Shapin, 'Pump and Circumstance: Robert Boyle's Literary Technology', *Social Studies of Science* 14:4 (1984), 481–520. Laura Otis has shown that germ theory did not take hold until the mid-nineteenth century, even though the microscope technology had been around since 1670, and this is because what was seen under the microscope was culturally defined – 'Cell theory relies on the ability to perceive borders, for to see a structure under a

microscope means to visualize a membrane that distinguishes it from its surroundings' – and that the conception of the self as a semi-permeable container was the result of the literature and art of the period. Laura Otis, *Membranes: Metaphors of Invasion in Nineteenth-Century Literature, Science, and Politics* (Baltimore: Johns Hopkins University Press, 1999), 4.

35. McConachie, *American Theater in the Culture of the Cold War*, 10.
36. Tobin Nellhaus, 'Performance Strategies, Image Schemas, and Communication Frameworks', in McConachie and Hart, eds, *Performance and Cognition*, 76–94.
37. Ibid., 91.
38. Jack Hitt, 'Act V', in *This American Life*, edited by Ira Glass, 1 hour: NPR, 2007.
39. Ibid.
40. Ibid.
41. Ibid.
42. See, for example, McConachie's 'Falsifiable Theories for Theatre and Performance Studies', *Theatre Journal* 59:4 (2007), 553–577, and Hart's 'Matter, System, and Early Modern Studies'.

3
Ghosting the Subjunctive: Perceptual Technics in Daniel Defoe's *A Journal of the Plague Year* and Transversal (New) Media

Jen Boyle

In Robert Lazzarini's new media installation, *Skulls*, participants enter a room with three-dimensional skulls displayed on the surrounding walls. What initially appears to be an exhibit of minimalist art slowly materializes as a macabre yet playful exploration of perceptual and cognitive suspense: the skulls slowly become skewed, changing shape in relation to the participant's location in the room. These objects, though affixed to the wall, seem to warp and transform (Figures 1 and 2). *Skulls* combines the perceptual affects of digital media and physical anamorphosis, a technique that flourished in early modern Europe,[1] to create a physically stationary object that appears to be static in one instance, transforming in the next. Lazzarini begins by digitally scanning images of anamorphic skulls, thereby preserving the skewed representation of the image, and then reconstitutes the images as three-dimensional sculptures of resin and bone. The material, sensorial, and somatic confusion evoked by Lazzarini's installation serves as an experiential critique of mediated affect and perception.

Skulls is an interactive performance that also embodies what Bryan Reynolds has called 'subjunctive space', a 'multidimensional', hypothetical timespace of 'as if' and 'what if'.[2] This imaginative and active space, a version of 'transversal movement', allows one to 'hypothesize scenarios and experiences'.[3] Lazzarini's exhibit creates a space for embodied interaction with the unstable temporality of the subjunctive, creating an experience where both the imagination and the perceptual sensorium of the participant are engaged with the present and future tense. That is, *Skulls* produces a mode of perceptual suspense that invokes affective states of the present and future tense simultaneously. A first glance reveals a stable object on the wall, while a second look is

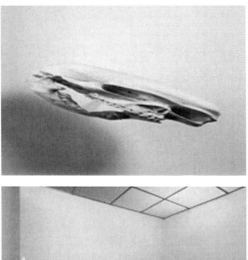

Figures 1 and 2   Robert Lazzarini, *Skulls*, 2000 (14" × 3" × 8", resin, bone, pigment). © Whitney Museum. Courtesy of Robert Lazzarini

'tricked' into seeing the object as a virtual image, seeming to move and change form. In a description resonant with Reynolds' characterization of the subjunctive, the *Bitstream* exhibition narrative describes *Skulls* as a 'paradoxical space', a 'place of intensity' made possible by the fact that the skulls 'linger somewhere in between an object and image'.[4] This 'somewhere' is an experience and emplacement that emerges, literally, in time and *in situ*.

My focus in what follows is on how the kind of 'intensity' evoked by Lazzarini's exhibit (which would seem to imply, literally, an arching inward) elicits a subjective experience at once localized with immediacy in and on the body and yet productive of a proleptic image of 'what's to come'. This experience can be described as a mode of subjunctivity that is perceptual and autonomic, as well as cognitively imagistic. I explore this figure of the 'perceptual subjunctive' as a model for affective mediation in Daniel Defoe's *A Journal of the Plague Year* (1722) and within the contemporary science of mirror neurons to reconsider the relationship between the mediated image, temporality, and subjunctive consciousness. Our digital moment offers up new forms of the perceptual subjunctive. Indeed, we are able to 'see' for the first time the interactive relationship between subjunctive space, somatic and cognitive response, and perceptual technics through the use of advanced fMRI (functional Magnetic Resonance Imaging) scans: images that track changes in the brain's neural activity in response to perceptual and cognitive activities. In what follows, I blend contemporary neuron science, the transversal subjunctive, and the perceptual and autonomic explorations in Daniel Defoe's *A Journal of the Plague Year* to experiment with a theory of mediated perception that negotiates the space between cultural artifacts and 'aggregates of mind'.[5] My theoretical explorations here play with the potential of the subjunctive for reimagining mediated perception across the literature and science divide, and within an early modern context that challenges our assumptions about the synchrony of embodied temporality and the diachronic pulse of historical time. Our images of mediated perception in the contemporary moment are specters of past theories of mediation, apparitions that have been cast as 'new', but which are uncanny returns of the past (in the present and future tense).

Empathic technoscience

Empathy is chemistry – in part. New developments in neurobiology (discoveries made possible by brain imaging media) demonstrate that we are coded chemically and biologically for subjunctive feeling.[6] Recent research on 'mirror neurons', neurons that fire in the brain both when an action is performed and when that action is observed, are challenging long-held beliefs about how a theory of mind develops in animals and humans, as well as how feeling, language, and imitation are connected to our biological substrate.[7] The discovery of mirror neurons generates a bioscientific narrative for how and why we are able to 'feel with' others. Our perceptions function not as affirmations or contestations of empirically derived 'truths', but as formations of consciousness that emanate out of the flows between the 'states' of subjects and objects, and as events of becoming that operate, as Barbara Stafford has phrased it, as affective 'echoes' rather than mimetic forms.[8] That is, mirror neurons offer evidence of how consciousness is based literally in the mediated exchange of images – it is not material information that our perceptions represent back to us in constructing consciousness, but the traces of the very process of the exchange of these time-images. Thus mirror neurons represent how consciousness emerges out of the procedures of mediation. As such, mirror neurons serve as biological reflections of the way in which physiological feelings and image-feelings can be virtually indistinguishable. In this sense, we could argue, becoming conscious is becoming-media.[9] Perhaps even more significantly, these findings in neuroscience further validate the role of subjunctivity in how shifts in consciousness take place.

In *Proust Was a Neuroscientist*, Jonah Lehrer summarizes Antonio Damasio's insights on the immediacy of the interactive relationship between body and mind. By examining a range of literary genres in light of recent neuroscience, Lehrer shows how the transfer of information between the mind and body operates on the principle of image transmission: 'In fact, even when the body does not literally change, the mind creates a feeling by *hallucinating* a bodily change.'[10] As Lehrer highlights, Damasio refers to this affective image (a virtual 'hallucination' that impacts consciousness and physiology) as the 'as-if body loop, since the brain acts as if the body were experiencing a real physical event'.[11] The implications of this are striking: our body's physical and cognitive response network functions not like a machine or circuit but as an image theater and subjunctive time-space.

This essay revisits the emergence of science, mediated images, and affective space in early modern England to explore how seventeenth-century science and literature formed its own conjunction in response to new technologies of perception amid theories of body and mind. This commerce between technology and literature made possible by explorations into the nature of the mediated image inflects the discourse of contemporary neuroscience. This conjunction also challenges the still majoritarian view that early modern technoscience signaled the ascent of a world picture based in a version of empiricism that saw subjective space and time as singularities and unmediated states.[12]

Daniel Defoe's *A Journal of the Plague Year* is a complex critique of the majority views surrounding early modern empiricism. Defoe's text explores the intersection of images and perception as subjunctive space, as events of becoming-media that refigure notions of space, and, most significantly, of subjective temporality. Combining the influences of early modern views on the techno-mediated image with the poetics and science of Lucretius' *De rerum natura* ('On the Nature of Things'), an early Roman work of philosophical poetry tremendously popular throughout the mid-1600s in England, *A Journal of the Plague Year* sets forth a theory of embodied consciousness as becoming-media. The early modern fascination with Lucretian philosophy provides us with a space for reimagining a genealogy to mediated perception. The second half of this essay, then, argues for a theory of mediated perception conversant with the transversal subjunctive.

Within transversal theory, subjunctive space is conjured out of the virtual and the actual, and within doubled 'time-images' that promote hypothetical and empathic responses formed at the intersection of 'what if' and 'as if'. I focus here on the temporal dimensions of subjunctivity and its evocation of the creative instability of mediated perception. Samuel Weber has explored media and mediated events as turning points between consciousness, history, and theory. Weber's *mediaura*, a reworking of Walter Benjamin's lost aura in the age of mechanical reproduction, invokes the figure of the *une passante*, a profoundly moving but flowing, fleeting brush with a passer-by.[13] In my reading, the transversal subjunctive extends the potential of such figures as temporal disjunctions to the immediacy of cognition and the mediated image's ghostly 'echo' across time. Transversal media unfold a creative power in reading Lucretian image-flows, mirror neurons, and Defoe's 'speaking sights' as fellow ghosts of the subjunctive – past and present-future time-images we feel with and through.

Early modern mediation: Lucretius and perception

Lucretian Epicureanism has been read as a precursor to the mechanical and materialist philosophies that would come to inform empirical science.[14] As a result, on the one hand, Restoration critics have tended to focus almost exclusively on Lucretius' concern with atomism and how such discourse influenced and anticipated the development of early modern empirical science. On the other, the linkages developed between atomism and Enlightenment liberalism have absorbed Lucretius' influence into political theories of emergent individualism. As a result, much of Lucretius' theories and poetics of perception located between more familiar concepts of temporality, spatiality, and subjectivity have been obscured or neglected. It is in Book iv, a part of *De rerum natura* that has received less attention than the earlier books, where Lucretius maps out perception as the main impetus for the 'eternall motion' of 'nature'.[15] Lucretius' particular version of Epicurean doctrine argues for a perceptual heterodoxy to the world, with perception varying in degree and intensity relative to the affects of *simulacra* (Lucy Hutchinson translates this word as 'images', 'species', or 'bodies', interchangeably). Lucretius views all perception as a result of the movement of *simulacra* through the air and between animate and inanimate objects. A radical nominalism is implied in Lucretius' theory of the *simulacra*, in that perception of the world is highly contingent upon images reworking their affect on bodies – images are never universal in their essence. The discrete 'bodies' that constitute a specific image cause variations in the simulacra affect. That is, perception is an affect of *simulacra*, images that are continually regenerated as different forms by the *elementa*, or 'seeds', that make up these images. 'Seeds' are the fundamental unit to Lucretius' atomism, but their discrete affects are determined by the 'flowes' of images and the continually shifting nature of such 'visible skins' as they move between bodies 'cast[ing] off which their owne semblance beare'.[16]

The reproductive image evoked by the casting off of 'semblance' is made even more explicit in Lucretius' first extended metaphor in Book iv for the character of images, when he likens the progress and affect of *simulacra* to 'bodies' that 'send much out' like the 'skins of new fallne calves'.[17] This birthing metaphor used to explain the nature of mediated images returns throughout *De rerum natura* and is the primary figure for ascribing a kind of willful autonomy to images; a 'desire' they possess both to give and receive 'impressions' from the bodies they interact with. The 'desiring' nature of the *simulacra* creates sympathetic perception between inanimate and animate agents: even 'stones' when they send forth their images produce 'strong impressions' of a 'sence' of 'one whole body' shared between object and observer.[18]

As Lucretius progresses through Book iv, he considers how the procreative power of sense images and sensory impressions are byproducts of the 'comingl'ng' of virtual and material forms, as 'every kind of nature doth containe a mixed seed'.[19] Such mixing becomes the basis to a theory of 'phantom' images that come to us in dreams and fantasies:

> Whence Centaurs, Scylla, threefaced Cerberus,
> Are oftentimes presented unto us,
> Together with the images of those
> Dead men, whose drie bones sepulchers enclose;
> Because the images are carried every where,
> Part, of their owne accord, formd in the ayre,
> Part falling off, from severall kinds of things.[20]

At this point in Lucretius' reflections on the affected/affective observer, there is a swerve toward the plenitude and desire inherent to both bodies and atoms. His vision of plenitude connotes a kind of baroque sensibility of the virtual. That is, *simulacra* and *elementa* are unrestricted in their desire and power to bring together the 'monsters' from the imagination with sensory impressions of the 'real': images spatially 'conjoyne' things made of both 'natural subtiltie' and 'thinn webs' that in 'Nature never [were]'. While these baroque surfaces, or 'webs', imply a multi-dimensional experience to simulation, they also weave an intricate 'texture' of desire in constructing real objects around us. Indeed, corporeal desire as such becomes a matter of 'conjoining' the 'agitated flowes' of 'desiring *simulacra*'. Lucretius thus finds a remainder to the mediated image that disrupts the fixity of objective and subjective states. Image-bodies – both as *simulacra* and as 'conjoyn'd' male and female forms – continually produce new forms and flows. This trans-reproductive energy, understood as a process that occurs between images and objects, is envisioned as a form of collective memory as well: these 'lost objects' that exist between forms are described as memories of forgotten desires and their former or potential corporeal states.[21]

Lucretius' 'visible skins' are floating images that make material form a consequence of temporal disjunction – literally it is the 'casting off' that becomes the determinate action in how images affect us, and thus, the mediating power of images is a temporal power of the virtual, situated between the casting off and bearing forth of 'semblance'. We find echoes of this form of temporality and the mediated image in contemporary mirror neurons. Mirror neurons, like Lucretius' image 'skins', are affective surfaces that require temporal disjunction and dissolution in order to re-form the material of cognitive and physiological response. Damasio's 'as-if body loop' argues for a model of our biological substrate as a corporeal image in a state of perpetual becoming. Physiological change and cognitive impressions are never in stasis. The body and mind operate together as a theater of proleptic images that are never separate from the emergent material form and function of physiology and cognition. Moreover, the exchange of 'cast off' images that are also 'hallucinated' as consciousness and somatic experience posit a temporality that is in between past-present-future tense. This model of subjunctive temporality has tremendous implications for a theory of media and cultural memory, past and present. Moving across scientific and literary expressions of mediated perception illuminates traces of the interactive commerce between the macro-structures of cultural meaning-making and the micro-structures of affect and perception. This play with mediated images as phantasms that 're-tissue' both historical time and embodied temporality skirts both the over-determinedness of the conceptual machinery of history and material culture, on the one hand, and the fixity and isolation of subjective interiorities, on the other.

Defoe's visible skins and speaking sights

Perception is the world in Daniel Defoe's *A Journal of the Plague Year*. At once aspiring to be a chronicle of the actual plague of London in 1665 and a spectacular novel of choice and action in the face of disaster, the *Journal* is in many senses an attempt at understanding a catastrophic historical event as a flow of mediated and perceived images: the images of diseased bodies; the images of death and loss calibrated graphically, taxonomically, and narratively; and the hallucinated images of those mentally and physically affected. Like many of Defoe's works, the *Journal* also slyly enacts an ironic critique of the social, religious, and philosophical machinery of the time.

   Published in 1722, a moment when the scientific and philosophical institutions of Enlightenment empirical science were becoming well instantiated, Defoe's fictional chronicle invokes a historical context fifty-seven years prior. The Restoration period that Defoe's text describes is a London still at the threshold of the emergence of modern technoscience. Defoe had already explicitly expressed his worry over the emergence of a more mechanized 'Projecting Age', replete with 'inventions' and new instruments for the 'art of war'.[22] However, the *Journal* creatively sets forth an alternative to the mechanization of mind and body that still today is made synonymous with early Western theories of embodiment. Defoe's interest in Lucretius goes well beyond the latter's influence on the development of the verse essay that scholars have noted.[23] Indeed, throughout the *Journal* the narrator, H.F., recounts multiple encounters with the perceptions of those afflicted by the plague disease (physical, mental, social). The perceptions of the people H.F. encounters or reports on are as significant, if not more so, than the 'facts' offered up concerning the manifestation of the disease. In one of the many passages where the narrator reanimates how perceptions become both a sign of those infected by the physical disease and those suffering the plague's virtual social contagion of 'terror', he talks of how 'air' and 'vapor' become the principal mediums for the most affecting images:

> Some heard Voices ... Others saw Apparitions in the Air ... but the Imagination of the People was really turn'd wayward and possess'd: And no Wonder, if they, who were poreing continually at the Clouds

saw Shapes and Figure, Representations and Appearances, which had nothing in them but Air and Vapour. Here they told us, they saw a Flaming-Sword held in a Hand, coming out of a Cloud ... There they saw Herses, and Coffins in the Air ... Heaps of dead Bodies lying unburied, and the like; just as the Imagination of the poor terrify'd People furnish'd them with Matter to work upon.[24]

We can think of this as a kind of image double for Lucretius' observations on how mythical figures join up with images of dead men in our dreams. Like Lucretius' images that are 'carried every where', both of 'their owne accord' and 'formd in the ayre', the perceptual images of those affected by the plague are a confused mix of the 'actual' and the 'virtual' ('Heaps of dead Bodies lying unburied' and a 'Flaming-Sword held in a Hand'). The emphasis here is not on discerning those images that represent the empirical reality of the plague's affects, but on the *movement* between images that represent the physical realities of the plague, and images that surface as 'airy' emblems and allegories, reanimated as a virtual collective memory (floating coffins carrying the elect heavenward; or an emblem for the end of days seen in a flaming sword and hand).

Moreover, H.F. offers us a kind of empathic rationale for this 'conjoyning' of virtual and actual images in referring to the 'matter' that emerges out of them. The 'matter' of these images is their real affect on the constitution of collective and individual experience of the plague, but it is also the media/medium through which conjurers, soothsayers, wardens, and scoffers 'worked' the perceptions and responsiveness of the 'poor people'. That is, the matter that emerges from these perceptual images *is* the mediation and re-mediation of social and individual consciousness and their potential literally to be re-formed over and again. Defoe follows this account by quoting himself via a bit of philosophical poetics on perception that seems almost a direct imitation of *De rerum natura*:

> *So Hypocondriac Fancy's represent*
> *Skips, Armies, Battles, in the Firmament;*
> *Till steady Eyes, the Exhalations solve,*
> *And all to its first Matter, Cloud, resolve.*[25]

It is not clear what constitutes such 'hypochondria' in the *Journal*. In contemporary terms, a hypochondrium is a 'somatoform' disease; that is, a disease that mimics physiological response through images conjured out of fear of disease. However, fear of disease in this context can be understood simply as fear of the material body itself – a terror of the body as a medium. In this sense, the 'matter' of the poem is both the proliferation of multiple perceptions and affects of the plague and its signs and meanings, but also the 'first matter' that is mediation itself: the matter that is 'worked upon' by and through embodied and imagined actualizations of the plague's affects; as well as the 'first matter' of airy exhalations that allow for such actualizations to take form, to appear or disappear. Defoe reproduces here a Lucretian theory of mediated perception where images materialize in 'part falling off, from severall kinds of things', and 'part, of their owne accord, formd in the ayre'.

Samuel Weber might detect in Defoe the anticipation of a Hegelian critique of mediation, articulated by Weber as 'an infinite process of becoming other in order to become the same'. This model of mediality is a procedure invested in the 'safeguarding' of 'finitude from an alterity'.[26] That is, it could be argued that Defoe's version of becoming-media relies on an appeal to a 'first matter' that offers up neither a distinct 'medium' nor the proliferation of real difference. This 'first matter', in light of Hegel's notion of 'medium qua mediation' – a 'media theology' that aspires to a kind of *creatio ex nihilo* – privileges the idea of a substantive matter out of which real differentiation occurs, but which in the end only reifies the idea of universal 'oneness'.[27] Weber elaborates on this dialectical process by pointing out that despite its implied context for the potential of mediation and the emergence of new forms, the temporality inherent to this model of mediation is 'something that will always already have taken place, in the future perfect of the concept'.[28] Yet, as I'll explore in the second half of this essay, there is something else at play here in Defoe's approach to the plague's images in terms of Lucretian perception and mediation. The temporality of mediation in Defoe's *Journal* is neither dialectical future perfect nor (historically or subjectively) present, but subjunctive. Defoe's subjunctive tense for perceptual images, like the mirror neurons that now haunt our understanding of how images work the matter of the body, operates at the threshold of the *as* of 'as if' and the *what* of 'what if'.

Ghosts and skins: seeing 'as if'/'as if' seeing

Ghosts and skins appear, disappear, and reappear in the *Journal*. Skin is discussed by H.F. as the single most legible sign of the plague. The only reliable indicator of who has been fully infected are the 'boils' and marks that cover the bodies of those afflicted. Such signs are needed, as other signs of the disease, such as fits, agues, or delusions, are attributable to the contagion of fear and distress. The skins and ghosts in Defoe's text take on a kind of metonymic function, standing in, respectively, for the material surface of the plague, and for the virtual images of the plague. When describing the horrible devastation wrought by contagion, H.F. focuses on how the disease progresses to the surface of the body. In one instance a woman is 'frighted' to death by the discovery of signs of the plague on her daughter's skin: 'She looking upon her body with a candle, immediately discovered the fatal tokens on the Inside of her Thighs.'[29] The discovery of these 'fatal tokens' launches the woman into such a terror that the 'Fright … seiz'd her Spirits' and she fainted, dying soon after. These 'fatal tokens' or 'spots' that appear on the surface of the skin become the principal image for the materiality of the outbreak. In one sense, then, the image of spots on the skin operates as a kind of actualization and grounding for the virtual spread of the plague's terrors (via images) and its various contagions.

Yet, such actualized images are never far away from the virtual in the *Journal*, and, indeed, even seem to call forth virtual representations. This is no more apparent than in the recurring references to ghosts. As often as we see attempts to describe the 'realness' of the plague's affects in terms of 'fatal tokens' on the skin, we see the narrator's invocation of the disease's virtual powers in terms of the appearance and disappearance of ghostly apparitions. These ghosts take up some of the most detailed scenes in the *Journal*, and are almost always associated with both material form, as in the case of the 'fatal tokens', and a motion that defies stasis:

> seeing a Crowd of People in the Street, I join'd with them to satisfy my Curiosity, and found them all staring up into the Air, to see what a Woman told them appeared plain to her, which was an Angel cloth'd in white, with a fiery Sword in his Hand, waving it, or brandishing it over his Head. She described every Part of the Figure to the Life; shew'd them the Motion and the Form; and the poor People came into it so eagerly, and with so much Readiness; *YES, I see it all plainly*, says one. *There's the Sword as plain as can be.* Another saw the Angel.[30]

What the crowd is seeing here is certainly 'real' in the sense that they are affected and moved; but these images take on their real power as virtual emergences that manifest both 'Motion' and 'Form'.[31] The emphasis on both registers is strikingly consistent throughout the *Journal*. In another instance, H.F. describes a man drawing a crowd to him as he stares into the 'burying place', 'pointing' and 'affirming' a ghost walking:

> he described the Shape, Posture, and the Movement of it so exactly … On a sudden he would cry, *There it is: Now it comes this Way:* Then, *'Tis turn'd back;* till at length he persuaded the People into so firm a Belief of it, that one fancied he saw it, and another fancied he saw it.[32]

In every case where H.F. narrates the appearance of ghostly images he stresses the motion, form, and affective power of such images on the crowd. In many senses these virtual scenes come to stand in for a vaguely familiar notion of 'media' in the contemporary sense, as events that emerge as virtual images but with very discrete and persuasive actualizations. The moving images conjure up a crowd that is 'moved'. The 'air' that produces these scenes serves as a medium, an 'element "in" and "through" which the data of sense pass on the way to their addresses', and which is seemingly caught up in a 'telos of virtuality as actualization'.[33]

Yet, the actual and virtual appear less as two models of meaning-making in H.F.'s account than they do mirrors of one another in every possible sense of the word. The 'fatal tokens' that appear on the skins of plague sufferers are the surface material that argues for the 'real' affects of the disease. Yet the images of ghosts are 'skins' as well, both in terms of their function as doppelgangers for the physical skins that show their 'marks' and then fade from sight, and in terms of the apparitions' affects on the collective body of the crowd. Indeed, this body becomes that 'matter' that is 'worked' and remade throughout the text. We have in this case something again closer to a Lucretian theory of mediated perception, where the images of the plague viewed collectively come into view *in between* the 'casting off' and 'bearing forth' of 'semblance'. Of importance here is the fact that this model is less spatial than it is temporal. The images appear and disappear as a product of the in-between transport of the 'casting off' and 'bearing forth' of sense data – as a remainder to both the actualized 'tokens' on the skin, and the virtual 'form and motion' of images in air.

The emergence of ghostly apparitions in particular signals a material that is both skin and not-skin, an image that simultaneously evokes that which 'came before' and that which 'came after'. Thus, the plague's ghosts are a medium in every sense possible and every possible sense.

As such, the *Journal's* images are traces of a different incarnation of the virtual in time. This re-conjured relationship between temporality and materiality points to a subjunctive subjectivity, and is wholly dependent upon the notion that images and actions are (can be), in the most literal sense, the same thing. However, it would be a mistake here to argue for the actual and the virtual as precisely one and the same. Indeed, what distinguishes subjunctive subjectivity in relation to mediation in this instance is the way in which approximation, understood in the temporal sense, becomes decisive in determining perceived reality. Approximation is the only way to describe an affective image that is caught in suspense between states of 'as if' ('casting off' semblance) and 'what if' ('bearing forth'). Two crucial aspects to images formed as such is their ability to remain temporally open and yet to manifest on/ as 'skins'. Mirror neurons, which I return to below, are contemporary ghosts of this form of Lucretian perception.

A 'visible call': mediated images as echoes

Mirror neurons, it turns out, are not all the same. While it is true that brain images point to a direct relationship between the images produced when witnessing an action and the affects of such actions, such affects cannot be mapped as a simple cause and effect relationship. More specifically, research indicates that there are competing scenarios for how mirror neurons function. In a minority of cases, these mirror neurons are activated by an entirely mimetic relationship between actions seen and actions performed. That is, in a smaller number of cases, the action seen is exactly similar to the action performed by the observer. In a greater proportion of the instances in which the same neurons fire when observing or performing an act, however, the acts in question are only approximate to one another.[34] For example, it might be the case that observing someone reaching out to grab someone by the arms fires the same brain neuron system that is activated when the observer reaches out to embrace someone. In this case, there is an approximation between the somatic and cognitive feelings generated and internal hypotheses or understandings about intentionality and outcome. The somatic and cognitive mapping of the images of actions observed and performed actions creates a physiological and cognitive sense of the relatedness between the actions of others and our own. Yet, this mapping also creates an opening around or ambiguity about the intentions or outcomes of such actions. The implications of this point to how we are not just learning through 'feeling with' others around mirrored actions, but how we are encountering potential for shifts in the conceptual, affective, and cognitive meanings of such actions. Moreover, the images of such actions are caught up simultaneously in the immediacy of things felt or experienced 'in the moment', and the proleptic potential for new meanings and affective states to emerge from these image feelings. Mirror neurons imply that images are both virtual and actual, at once 'here and now' in the autonomic sense, and in the process of becoming something else entirely.

The most arresting image of Lucretian perception in Defoe's *Journal* is the burial pit at the center of the city. The description of this 'dismal object' and its emergent role in representing a model of mediated perception has much in common with contemporary theories of mirror neuron systems. The pit becomes a kind of nucleus of Defoe's text. H.F'.s narratives seem to orbit repeatedly around this space and the bodies contained within it. The burial pit becomes an exploration of how mediated images transform not only subjective and collective perception, but also the physical and temporal spaces of the city. Plague victims who have been retrieved on 'dead-carts' from homes and churches by day are transported at night to the large burial pit at the center of London. H.F. exhibits an almost obsessive fascination with the physical dimensions of the pit: '40 feet in length and about 15 or 16 Foot abroad; the first time I looked at it, about nine feet deep'.[35] He initially focuses his observations on the shape and form of the pit, but then moves into a descriptive pathos of how the image of the pit affects him. He is particularly stunned at one point by the visible grief and trauma of a man who follows to the 'gulph' of the pit the dead-cart carrying the bodies of his wife and children.[36] H.F's fascination with seeing into the pit is tempered by a warning from a pit-yard guard who warns H.F. that the pit is a 'speaking sight', one that 'has a voice with it, and a loud one'.[37]

The 'speaking sight' of the pit is the most vocal and visible image in the *Journal*. It is a space, as Carol Houlihan Flynn has observed, that is 'always changing shape' yet a 'matter that is impossible to transcend, impossible to ignore'.[38] Yet the pit is not a stable image in either the actual or virtual sense. The man who follows his family to the burial pit watches the bodies be interred and then, 'looked into the pit again as he went away ... [finds] the buriers had covered them so immediately with throwing in earth ... nothing could be seen'.[39] Like the ghostly apparitions described earlier, this burial scene is referred to as a 'visible call', an image-sound that calls forth recognition of both the permanence and changeability of images of such terror. The 'call' that is sent out by the pit would seem to imply a kind of interpellation.[40] That is, the pit appears to stand in as a combined image for all the institutional, religious, and social ideologies that define the plague as an event. These forces might seem to coalesce in the image of the pit, to produce a Foucauldian subject that is over-determined by the discipline and terror that the plague mediates. Yet, from the standpoint of both Lucretian perception and mirror neurons, this site is both actualized as an image of corporeal stasis and virtually reanimated as a different kind of 'call': an echo image that sounds off across the present and future tense. The bodies go into the pit, becoming discrete images of death, much like the 'fatal tokens' that appear on the skins of the afflicted. But once in the pit, with seeming immediacy, bodies are transformed into an image of disappearance, as vanishing apparitions.

This 'pit', where things appear with material and figural immediacy while simultaneously becoming an image for alternative future 'states', surfaces again in contemporary representations of mirror neurons in brain science. The pit in Defoe's chronicle becomes a center to the text, a media object that recasts a theory of mediation itself. In the *Journal*, the pit is a figure for Lucretian perception. The 'evidence' of mirror neuron theories is of course an image construct as well. Though they speak to a form of meaning-making that creatively and transversally dislodges certain structural assumptions derived from the humanities about images as signs and symbols, mirror neurons are no less circumscribed by representation. Images generated by functional Magnetic Resonance Imaging (fMRI) and popular reproductions of these scans render forth an image of the mirror neuron that seems an uncanny reappearance of Defoe's pit (Figures 3 and 4).[41] Figure 3 seems an almost 'historical' version of the mirror neuron, as an antiquated drawing that fuses together scientifically realistic detail with an anticipation of what's to come (represented with the dark center, but evocative of the 'black holes' that signify potential transformation, transience, and hypothetical suspension across the discourses of science and science fiction). Figure 4 is representative of a futuristic art rendering of the mirror neuron image, and within a future tense, it too highlights the real and phantasmic registers of mirror neuron science.

What is so striking about both Defoe's Epicurean images and the oft-produced contemporary images of mirror neurons is how they inhabit a space/time of 'experimental' indeterminacy and openness. These images are evoked not as evidentiary logics that reveal a contact between the distinct epistemologies of science and literature, but as figures that experiment with the approximate limits of each – not a performance of science imitating art (or vice versa) but a critique of imitation itself

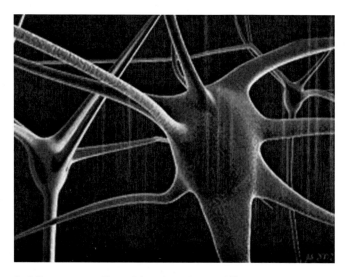

Figure 3 Mirror neuron (from 'Blog about Science')[42]

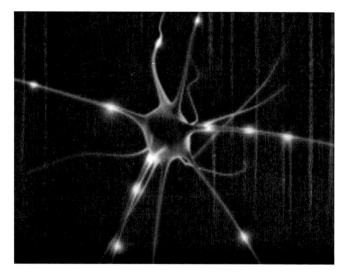

Figure 4 Mirror neuron image (from *Slog*, 'Visual Art: The Future of Criticism')[43]

across science, literature, and across the distinctions between 'virtual' and 'actual' materiality.[44]

At the center of the mirror neuron is a 'hole' where nothing can be seen. In part, as I have said, this representation speaks to an exciting and productive confusion surrounding mirror neuron science at the moment. Recently, the discourse of mirror neurons has been embraced in popular and professional contexts. The fascination with mirror neuron bioscience extends across literature, science, art, primatology, anthropology, childhood studies, and performance theory. Moreover, the frames and context for mirror neuron research often invoke a kind of futuristic nostalgia for the power of science to offer us 'evidence' of our sentimental and empathic tendencies.[45]

To this end, one of the most popular demonstrations of the function of mirror neuron bioscience is the example of primates learning to understand the motives of one another. This discourse is caught up not just in a privileging of the logos of scientific explanation for questions conventionally regarded as humanistic in nature, it is also entangled in the desire for empirical procedures that define and conceptualize the increasing confusion of the heterological boundaries of what 'makes us human'.[46] Thus the romance with the narrative of brain chemicals and pulses that is popularized as a device to allow us to 'mirror' one another has obvious appeal in the contemporary moment. However, it is this popular idea of a natural empathy within us that, to my mind, has obscured the more challenging possibilities associated with the actual science of mirror neurons. Indeed, mirror neurons argue less for how we all have the capacity to 'just get along' if we could let 'nature' prevail, than they do for a new way of thinking about images, bodies, and mediated consciousness. The hole at the center of both Defoe's text and the mirror neuron (like Lazzarini's skulls described in the opening to this essay) re-forms the relationship between temporality and consciousness. Rather than sites that point to the limits of the virtual and actual as absences, as a dialectic of lack, the mediated images within Defoe's Epicureanism and aspects of mirror neuron science evoke thresholds of rematerialization. Such space-time images rely on temporal difference – images actualizing and 'casting off' in the present tense, while virtually 'bearing forth' other possible future states, meanings, and material forms. We are only at the edges of a conversation between the humanities, science, and technology that would argue for a present that *actually* and *virtually* informs the past and the future.

The transversal subjunctive pushes us to the edge of such a conversation. The transversal subjunctive is an experiment in the mediating affects of images at several levels simultaneously. That is, rather than crossing *over* to see what the empirical 'facts' of bioscience can tell us about our fictional and phantasmic 'imitations' within the humanities and arts, the transversal subjunctive is a crossing *through* the givenness of such categories in the premodern, modern, and postmodern con-texts. This is at once an historical problem, in that a text like Defoe's does not fully recognize the contemporary demarcation of literature and science; and it is also a theoretical intervention to the extent that transversality takes up the challenge of hypothetical suspension ('as if' and 'what if') as critique – moving across the sacral limits of hard/soft (science and fiction) and virtual/actual (metaphysical and material).

Modernity (again): 'casting off' or 'setting off'?

Michelle Brandwein has argued that Defoe's *Journal* is literally a 'formative' work to the extent that it proves a new model of social and individual consciousness coming into relief. Brandwein wishes to demonstrate how the *Journal* reenacts the way in which 'a human being turns him- or herself into a subject' (becoming then an object) amid the seventeenth and eighteenth centuries' 'veritable technological take-off in the productivity of power'.[47] To this end, the conscious and uncon-scious images contained within the narrative are examples of a conflict between a time when individualism was 'pre-emergent' (1665) and individualism was 'emergent' (1722). Brandwein attempts to capture in this a sense of subjective temporality that is still traceable in the text, a condition of becoming caught between the representation of subjective time and historical time. Yet, we are still left in the end with a model for how Defoe's text represents a kind of pre- and over-determined history, albeit as it 'is happening' in the context of the moment.

Samuel Weber's *mediaura*, a reworking of Benjamin's lost aura in the age of mechanical reproduction, invokes the figure of the *une passante*, a profoundly moving but flowing, fleeting brush with a passer-by as a figure for mediation and modern subjectivity.[48] This figure of the *une passante* – an allegorical image drawn from a poem by Charles Baudelaire – becomes in Weber's reading an image of the modern urban 'mass'. Modern subjectivity, Weber argues, takes the form of an interior consciousness that is itself a product of mediation. In this reading, the observer imagines a brief encounter with a passer-by as a moment out of which consciousness of the self surfaces as a mediated experience with a 'ghostly crowd' – 'ghostly because, like the apparition of the *passante* in Baudelaire's poem, they only *come to be* in *passing away* ... What *is* comes to *pass* as *nothing* ... but a certain aura'.[49] Weber expands here on how this allegory of mediation can be read as a construction of the self as a 'setting itself *off*' from the crowd that simultaneously reveals its affinity 'with everything pedestrian': 'It is the affinity of an *apparition*. The *passante appears* only to *disappear*, almost instantaneously.'[50]

   What Weber is in part after here in his re-reading of Benjamin's aura is a theory for how we can reanimate the concept of mediation as an allegory for consciousness itself. Significantly, this recasting of mediation requires an understanding of the movement between language and images as performativity. In other words, as Weber discusses, we have slipped into complacency about subjects and objects (again). 'New media' and 'the media' have become signposts for a brave new world – digitized, phantasmic, and emergent. In the end, however, we find floating in the digital ether some familiar apparitions: subjects that are 'set off' from the historical and ideological conditions that define them, on the one hand; and 'self-contained and detached "objects" of study', on the other. Despite its 'calls' to newness, 'digital culture' often employs 'new media' to re-conjure some old Aristotelian and Platonic ghosts. I quote at length here Weber's recounting of how reimagining Benjamin's aura offers a different version of mediation as *movement between*:

> Benjamin's concept of history knows neither goal nor 'global integration' but at best, an 'end.' But this end does not come 'at the end,' but rather is always actual, always now. The actuality of this 'now,' however, is never self-contained, integrated, simply present: rather it is a divider, a dividing-line or point, producing a 'cut' that is never in-between but always outside of that it divides. History emerges, for Benjamin, only insofar as the 'here and now' imparts itself as a 'there and then,' encountering its innermost division 'outside of itself': 'before' itself, in a past that opens-imparts itself – to the future.[51]

Subjunctive subjectivity is perceptual approximation, an approximation that is neither wholly *actual* nor *virtual*. This perceptual approximation is not just an abstract phantasm of emergent potential, but a time-image that requires us to rethink the affect of temporality vis-à-vis mediation. Daniel Defoe's *A Journal of the Plague Year* is transversally subjunctive in that it resists the casting of time as a 'setting off' of a historically contained individualism that Brandwein sees as central to the text. Defoe employs Lucretian perception to perform mediation as an 'open-impart[ing]' that is a 'casting off' leaning toward a 'bearing forth'.

   Transversal theory has talked about the idea of 'paused consciousness' as an opening onto the powerfully 'affective presences' that can emerge from transversal readings and performances – forces that can 'inspire emotional, conceptual, and/or material deviations from the established norms for any variables, whether individuated or forming a group'.[52] Paused consciousness is a 'cut' with the subjectively familiar and opens up the potential for new movements and conceptual or emotional formations. Significantly, this method requires that we look past the 'quagmires' of progressive and conservative ideologies in investigating new ways to read, perform, and create.[53] To this end, I want to suggest that bringing mirror neurons into dialogue with Defoe's early modern chronicle of disaster and terror is not just an exercise that forces science into the discourse of literature and vice versa. Indeed, a careful treatment of both discourses reveals a potential for finding new models for not just how we 'do history', but how we think of the affective presence of time and the materiality of mediation (new and old).

In his discussion of the spectacle of war in the wake of 9/11, Samuel Weber echoes the frustrations at present with the seeming over-determined nature of mediated images. As he laments, '*theatricalization* seems to constitute one of the essential components of war'.[54] The successful spectacle of war and terror, Weber argues, is conducted through 'exacerbate[d] anxieties of all sorts by providing images to which they can be attached, ostensibly comprehended, and, above all, *removed*'. The image of death and threat is 'schematically' refigured as both a material image; but also as a kind of procedure that reaffirms the 'setting off' (to use an earlier term) of the self: the 'viewer is encouraged to "move forward" and simultaneously to *forget the past*'.[55]

The reading I have offered here of Defoe and neuron science points to another way of thinking about the *affective* temporality of the 'past' and the 'future' in terms of the mediated image. Reimagining Defoe's mediated images in the *Journal* as artifacts of Lucretian perception challenges our understanding of a crucial moment in what Weber calls the 'Western dream of self identity'.[56] Defoe's images of terror are projections that point to the fixity and corporeality of the mediated image, and forward to the potential for new time-images to emerge, creating shifts in collective and individual consciousness. This model of mediation and perception is finding new life in neuron science as well. Reading transversally, and thus reading Defoe in light of mirror neuron science, reveals how attempting to 'move forward', in Weber's words, requires us to be in the past. The ghosts and 'visible skins' of Defoe's early modern chronicle and mirror neurons are apparitions that are casting off and bearing forth at this moment:

> *A dreadful Plague* in London *was,*
> *In the year sixty five,*

> *Which swept an Hundred Thousand Soul*
> *Away; yet I alive!*

All these ghosts: fellow fugitives. *'There it is: Now it comes this Way ...'*[57]

Notes

1. Anamorphosis is a perceptual affect that relies on the juxtaposition of two perspective planes within one viewing space. The impact of anamorphosis depends on the inscription of two images or portraits within a single viewing area. Typically, a viewer would be required to shift their position physically in order to see an alternate image within the portrait or scene, usually rendered along a second perspective geometrical plane. The perceptual doubling of anamorphosis produces a rupture in the viewer's gaze and the stability of the object under view. For some further examples of anamorphosis as a literary and figural device throughout early modern literature, see Lyle Massey, 'Anamorphosis Through Descartes or Anamorphosis Gone Awry', *Renaissance Quarterly* 50:4 (1997), 1148–1189, and Alison Thorne, *Vision and Rhetoric in Shakespeare* (New York: Palgrave Macmillan, 2000).
2. Bryan Reynolds, *Transversal Enterprises in the Drama of Shakespeare and His Contemporaries* (New York: Palgrave Macmillan, 2006), 16–17.
3. Transversal space is described by Reynolds as a 'spacetime encompassing, among other known and unknown qualities, the nonsubjectified regions of individuals' conceptual-emotional range'. *Transversal Enterprises*, 16–17.
4. 'Skulls Installation at the Whitney Museum of American Art', http://www.pierogi2000.com/flatfile/lazzarins.html (accessed 8 November 2008).
5. Though I will be focusing in this essay on theorizing affect, perception, and (new) media, my 'blending' of the images of brain science with literary representations of virtual and actualized images shares approaches with the contributions in this volume by F. Elizabeth Hart and Amy Cook. I will be making references throughout this essay to connections with Hart's and Cook's more thoroughgoing treatments of cognitive literary studies. See in this volume, F. Elizabeth Hart, 'A Paltry "Hoop of Gold": Semantics and Systematicity in Early Modern Studies' for her thoughts on systems, 'aggregates of mind', and culture.
6. As I will discuss below, one of the unfortunate consequences of the recent interest in mirror neurons in disciplines outside of the sciences is the way in which the specifics of how mirror neurons actually perform are reduced to a simple cause and effect relationship. The actual research surrounding mirror neurons points to a more complex and diverse set of conditions for their activation. I hope to make productive use of these distinctions later in this essay.
7. For a survey of the important developments in mirror neuron research see Maskin I. Stamenov and Vittorio Gallese, eds, *Mirror Neurons and the Evolution of Brain and Language* (Amsterdam: John Benjamins, 2002); Stein Braten, ed., *On Being Moved: From Mirror Neurons to Empathy* (Amsterdam: John Benjamins, 2007); and V. S. Ramachandran and Sandra Blakeslee, *Phantoms in the Brain: Probing the Mysteries of the Human Mind* (New York: Quill, 1998).
8. Barbara Stafford, *Echo Objects: The Cognitive Work of Images* (Chicago: University of Chicago Press, 2007).

9. My coinage of this term is a play on the multiple performative and material 'becomings' of Reynolds' transversal poetics and the further adaptations of the two models of 'becoming' in Gilles Deleuze and Félix Guattari, *A Thousand Plateaus: Capitalism and Schizophrenia*, trans. Brian Massumi (St Paul: University of Minnesota Press, 1987). Becoming-media is also an idea in conversation with F. Elizabeth Hart's 'bottom-up dynamics' for thinking the 'linked media of cognition and culture'. A recent project that recontextualizes philosophy via systems theory and affectivity is John Protevi, *Political Affect: Connecting the Social and the Somatic* (St Paul: University of Minnesota Press, 2009).

10. Jonah Lehrer, *Proust Was a Neuroscientist* (New York: Houghton-Mifflin, 2007), 20.

11. Ibid.

12. There have been some richly provocative challenges to this view of late, particularly in studies of mapping and/or theatrical space in the early modern period. To name just a few: Henry S. Turner, *Shakespeare's Double Helix* (London and New York: Continuum Press, 2008); Mary Baine Campbell, *Wonder and Science: Imagining Worlds in Early Modern Europe* (New York: Cornell University Press, 1999); Jonathan Gill Harris, 'Untimely Mediations', *Early Modern Culture: An Electronic Seminar* 6 (2007); Julian Yates, *Error, Misuse, Failure: Object Lessons from the English Renaissance* (Minneapolis: University of Minnesota Press, 2003); and Carla Freccero, *Queer/Early/Modern* (Chapel Hill, NC: Duke University Press, 2006).

13. Samuel Weber, *Mediauras: Form, Technics, Media*, ed. Alan Cholodenko (Stanford, CA: Stanford University Press, 1996), 94.

14. See Richard Kargon, *Atomism in England from Hariot to Newton* (London: Clarendon Press, 1966), and Richard W. F. Kroll, *The Material Word: Literate Culture in the Restoration and Early Eighteenth Century* (Baltimore: Johns Hopkins University Press, 1991). For a recent and powerful reimagining of Lucretius in both contemporary theory and Renaissance culture see Jonathan Goldberg, *The Seeds of Things: Theorizing Sexuality and Materiality in Renaissance Representations* (New York: Fordham University Press, 2009).

15. As I am interested in the appropriation of Lucretian ideas in a seventeenth-century context, the majority of the passages quoted throughout this section are taken from Lucy Hutchinson's manuscript translation of *De rerum natura*: *Lucy Hutchinson's Translation of Lucretius: De rerum natura [c. 1656]*, ed. Hugh de Quehen (London: Duckworth, 1996). I have also consulted and refer to terms from the Latin version of *De rerum natura* (Titus Carus Lucretius, *T. Lucreti Cari Poetae Philosophici antiquissimi De rerum natura natura liber primus [-sextus]* (Impressum Venetis, 1495); Hutchinson, iv, 79.

16. Ibid., iv, 49.

17. Ibid., 55–61.

18. Ibid., 269–274.

19. Ibid., 670.

20. Ibid., 767–773.

21. I draw attention here to Hart's contribution in this volume and her discussion of the work of Merlin Donald. Donald views the entire range of memory types (from 'biological storage' to the media of 'external memory systems') as the 'tissue' connecting an aggregate of minds to culture.

22. Daniel Defoe, *An Essay Upon Projects* (London: Tho. Cockerill, 1697), 1.
23. Paula R. Backscheider, 'The Verse Essay, John Locke, and Defoe's Jure Divino', *ELH* 55:1 (Spring, 1988), 99–124.
24. Daniel Defoe, *A Journal of the Plague Year*, ed. Louis Landa, with a new introduction by David Roberts, 2nd edition (Oxford: Oxford University Press, 1990), 22.
25. Ibid., 23.
26. Samuel Weber, 'The Virtuality of the Media', *Sites: Journal of the Twentieth-Century/Contemporary French Studies* 4:2 (2000), 303.
27. Ibid., 302.
28. Ibid.
29. Defoe, *Journal*, 50.
30. Ibid., 23.
31. That is, the synchronic ('form') and diachronic ('motion') registers of the image, while not identical to one another, are certainly inseparable to the extent that the affective power of the image requires the interplay between materialized form and the alteration of form in time.
32. Defoe, *Journal*, 24.
33. Weber, 'The Virtuality of the Media', 302.
34. Leonardo Fogassi and Vittorio Gallese, 'The Neural Correlates of Action Understanding in Non-Human Primates', in Stamenov and Gallese, eds, *Mirror Neurons and the Evolution of Brain and Language*.
35. Defoe, *Journal*, 80.
36. Ibid.
37. Ibid.
38. Carol Houlihan Flynn, *The Body in Swift and Defoe* (Cambridge: Cambridge University Press, 1990), 11–12.
39. Defoe, *Journal*, 81.
40. Interpellation refers to the Althusserian concept of a subject being 'called' into recognition. Individuals operating with the language codes and spaces of a particular ideology are produced (called) into socially legitimized identities, gestures, and spatial practices. See Louis Althusser, 'Ideology and Ideological State Apparatuses' [1970], in *Lenin and Philosophy and other Essays*, trans. Ben Brewster (New York and London: Monthly Review Press, 2001), 121–176.
41. The popular images of mirror neurons that now flourish in general readership non-fiction, blogs, and documentaries are predominantly variations on the neuron with the absent center or hole. While fMRI scan images are still reproduced in scholarly treatments of mirror neurons, in the popular imagination, the representation of the mirror neuron as a figure that embodies both the persuasiveness of bioscience and the mystery or open indetermination of the speculative aspect of such figures seems to prevail as a dominant image.
42. http://jmgs.wordpress.com/2007/05/06/mirror-neurons/ (accessed 23 October 2008). This figure – a stock image that is reproduced on multiple general audience science blogs and websites – imagines the power of the mirror neuron as a 'hole'; but it invokes not absence but the power of becoming: the 'black hole' as a site of future and past potential.
43. http://slog.thestranger.com/2008/07/the_future_of_criticism (accessed 7 November 2008). This image is featured at a popular news and art blog:

'representation is as real as reality'. The small lights that balance the density of the center appear as sparks or codes of new information headed to or moving from the center.

44. 'Experiment' here harks back to an early modern context that, as Henry Turner reminds us, conjures the 'fictive', the 'factual', and the 'imaginary' 'within the larger problem of the relationship between "art and nature," which itself formed the discursive domain for many arguments that we would today describe as "scientific" or "technological"' ('Life Science: Rude Mechanicals, Human Mortals, Posthuman Shakespeare', *South Central Review* 26:1 & 2 (Winter & Spring, 2009), 197–217. A further play on the term, however, also crosses into Foucault's notion of an 'experimental critique'; a mode of critique that puts to the test the extreme limits of accepted categories and concepts; Michel Foucault, 'What is Enlightenment?' in Paul Rabinow, ed., *The Foucault Reader* (New York: Pantheon, 1984).

45. For a thoughtful reconsideration of this connection between empathy and mirror neurons, see Evan Thompson, *Mind in Life: Biology, Phenomenology, and the Sciences of the Mind* (Cambridge, MA: Harvard University Press, 2007).

46. It is important to note here as well that neuron science itself has become a site of such confusion and contestation. Debates have ensued over the very existence of mirror neurons in humans versus primates. See Gregory Hickok, 'Eight Problems for the Mirror Neuron Theory of Action Understanding in Monkeys and Humans', *Journal of Cognitive Neuroscience* 21:7 (2008) and Christian Keysers, 'Mirror Neurons', *Current Biology* 19:21 (2009), R971–R973.

47. Michelle Brandwein, 'Formation, Process, and Transition in *A Journal of the Plague Year*', in Paula Backscheider, ed., *A Journal of the Plague Year* (New York and London: Norton, 1992), 349.

48. Weber, *Mediauras*, 94.

49. Ibid. (emphasis in the original).

50. Ibid.

51. Weber, 'The Virtuality of the Media', 310.

52. Reynolds, *Transversal Enterprises*, 2.

53. Ibid., 18.

54. Samuel Weber, 'War, Terrorism, and Spectacle, or: On Towers and Caves', *Grey Room* (2002), 14–23, 15.

55. Ibid., 21.

56. Weber, *Mediauras*, 4.

57. Defoe, *Journal*, 24.

4
What was Pastoral (Again)?
More Versions

Julian Yates

> My favorite trope for dog tales is 'metaplasm.'
> Metaplasm means a change in a word, for example
> adding, omitting, inverting, or transposing its let-
> ters, syllables, or sounds. The term is from the Greek
> *metaplasmos*, meaning remodeling or remolding. I use
> metaplasm to mean the remodeling of dog and human
> flesh, remolding the codes of life, in the history of
> companion species relating ... Metaplasm can signify
> a mistake, a stumbling, a troping that makes a fleshly
> difference ... Woof!
>
> Donna Haraway, *The Companion
> Species Manifesto*[1]

This essay is eco-friendly but trades in echoes. It reprises Paul Alpers' iconic *What is Pastoral?* (1996) via the figure of a parenthetical repetition that demands to be reminded, one more time, 'What was Pastoral (Again)?' My answer or simply what follows will be 'More Versions', an addendum, if you like, to William Empson's *Some Versions of Pastoral* (1935).

My aim is not to recover some idealist definition of pastoral that will see off competitors or settle the matter once and for all.[2] On the contrary, by troping Alpers' and Empson's titles I aim to reinhabit old critical skins too quickly sloughed, and to fill them out in a way that, as Donna Haraway might say, makes a 'fleshly difference'. For me, this difference entails taking up the burden that there exists a history of technology, of the machine, the plant, and the animal, that is simultaneously and necessarily also a history of human life, and so of approaching pastoral as what, a little while ago now, in *Of Grammatology*, Jacques Derrida would have called a key element in the 'arche', 'general, or generative text', that set of routines by whose repetition the world is successively rewritten.[3] It is worth recalling here that Derrida's staging of 'the history of life ... [or] difference' as the 'history of the *grammè*' aims to make visible modes of cognition, historical consciousness, and forms of personhood that do not respect the ratio of the line or the linearization of the world that occurs in a phonetic writing system. The story, as you remember, begins with the observation lethal to any metaphysics of presence that 'life' begins with the writing event of 'genetic inscription' and 'short programmatic chains regulating the behavior of the amoeba or the annelid up to the passage beyond alphabetic writing to the orders of the *logos* and of a certain *homo sapiens*'.[4] Writing manifests as a question of coding. In what follows I am going to model pastoral as a relay in the 'general text' in this sense – as an ongoing folding together of places, times, phrases, and things, which replay or refold a set of routines, programs, or tropes for making up persons and worlds.

The essay falls into three parts. The first sponsors a partial, necessarily impressionistic, redefinition of pastoral in relation to a long history of *otium*, and so to a script that codes labor and leisure, writing the world via a constitutive 'anthropogenesis'.[5] The second poses the question of definition, redescribing the constitutive tropes of pastoral with an eye to the way the discourse produces shepherds, sheep, and dogs, gentlemen, city-dwellers, and countryside, according to the paradigm of the 'companion-species' or 'multi-species' as developed by Haraway.[6] In the third section, I recover some of the latencies, resistances, tariffs, failures, or errancies within the discourse, from which it may be possible to launch scripts that process humans, animals, machines, and world differently, with an ear to what Derrida calls the 'pluri-dimensionality' of other 'level[s] of historical experience', figuring pastoral now as an archive or contact zone with occluded or botched ways of being.[7]

What animal *otium*?

Like many Westerners of a certain age my first encounter with pastoral was in front of the television. I don't mean to say that as a child I was deprived or didn't get outside very much, but rather that my world was punctuated by the times at which certain programs began and ended, by what, in *Television* (1974/5) Raymond Williams names 'the mobile concept of "flow"' – the experience of temporality as mediated by the technologies of broadcast television.[8] Growing up in the UK before the advent of the remote control, the VCR, the DVR, or a multiplicity of channels, most evenings I got to luxuriate in the 'flow' programmed for us down in London that materialized up North by the turn of a knob on the 'set' in the living room. In the evenings 7.15 p.m. heralded the end of the national news and assorted local news magazines and the airing of this or that rerun of *The Rockford Files*, *Star Trek*, or any number of newly minted sitcoms. So when, now, I invent my childhood all over again, I can pinpoint with seeming clockwork accuracy the arbitrarily decided moment at which, with the *negotium* of schoolwork handily put away, time refolded itself into the collective *otium*, leisure, down time, or release that the TV afforded.

'It's good to put your mind in neutral, to idle', was the moral my home-grown humanist father deduced from this activity – speaking a democratized version of the script that Quentin Skinner finds rehearsed in Book 1 of Thomas More's *Utopia*, where the Platonically inclined Raphael Hythlodaeus holds forth while the more sanguine, Ciceronian *Morus* listens, contemplating the pros and cons of Platonic withdrawal versus a life of public service.[9] Depending on whether or not you agree with Skinner, the Ciceronian script More lifts from *De Officiis* wins out, and Book 2 imagines a world not of self-cancellation or human annulment so much as of humanist self-actualization, in which *otium honestum* or *negotiosum*, good, which is to say, useful *otium* is available to all, and mere idleness a structural impossibility.[10] *Utopia*, claims Erasmus in a letter, was written *'per otium'*.[11] And *'otium'* (*'sed res occium poscit'*) is what Raphael claims his description of the Utopian *res* will take.[12] If, in other words, *Utopia* imagines a humanist collective or habitat, then this habitat is premised on the figure of the neutral, the idle, of leisure or the space that comes between, winking in and out of being, ratified by its proleptic tasking with the good of the commonwealth, work, and world. Cradled by the guarantee that 'tomorrow' meant a return to the worlds of school and work, we idled in the tele-topical pastoral idyll afforded by the programmed flow that issued from the 'box.'

I have begun with this self-indulgent (idle) snapshot of a fictive (idyllic) childhood in order to make the point that the story of pastoral is keyed to a long history of *otium* and the humanist calculus that programs a set of relations between home and work, leisure and labor. Indeed, as Anthony Grafton and Lisa Jardine have argued, scripts such as that Skinner finds retooled in More's *Utopia* remain operative, even today, in the educational protocols and prerecorded ideologies of the humanities.[13] Obviously, by their translation to new media platforms and different demographics, those scripts change – the technical humanist term *'otium'*, for example, ceases to be keyed to the production of knowledge in the form of what Timothy Reiss calls a 'passage technique', and instead morphs into a notion of *tempus* merely ('free time'; down time; the time that comes between).[14] Distributed more freely through the collective, *otium* become *tempus* recasts Platonic withdrawal or idleness now as a weekly or daily technique, by whose observation human subjects are enlisted in maintaining their stability and happiness, and so also the stability of a labor force. The allied discourse of 'wellness' similarly replays the age-old question of philosophy, 'how to live well', as a question of optimization merely, of good somatic and psychic hygiene.

If my staging of pastoral *otium* as a calculus of labor and leisure keyed to its role in making human persons seems slightly esoteric, it is because I seem to be demoting key elements that are frequently taken to define the activity – shepherds, their 'natural' milieu, the 'natural world' itself – to the role of scenery. Venture to the mid-seventeenth century, for example, when, as Keith Thomas tells us, 'sophisticated city-dwellers, like Queen Henrietta Maria ... dallied at Wellingborough because she liked the countryside',[15] and pastoral seems to have materialized as an orientation to some 'thing', the countryside specifically, for which 'sentimental longing' and 'idealization' was encouraged. Thomas goes on to cite Samuel Pepys' 1667 encounter with 'an authentic country shepherd and his son on the downs near Epsom' as a case in point. Yet, to read Pepys' description of his encounter is to follow him through the overlay of time frames and places, each with its own codes, each inflected by the presence of the other, that shapes the way he takes his Sunday rest by going mobile.

We wake with him on 14 July, the 'Lord's Day', 'a little before 4', to feel his 'vexation' with his wife who delays their departure, and get under way a little 'past 5 a-clock' as they board the coach whose windows frame the 'very fine day', and 'the country very fine; only, the way, very dusty'.[16] They get to Epsom by '8 a-clock'. Pepys drinks 'four pints' (338) of water 'and ha[s] some very good stools by it'. They stop at a tavern; visit friends nearby; take the 'ayre' by coach, 'there being a fine breeze abroad', and then take a walk. Pepys gets them lost in the woods, 'sprains [his] right foot', manages to walk it off. They go up to the downs where the sheep are and are greeted by

> the most pleasant and innocent sight that ever I saw in my life; we find a shepheard and his little boy reading, far from any houses or sight of people, the Bible to him. So I made the boy read to me, which he did with the forced tone that children do usually read, that

was mighty pretty; and then I did give him something and went to the father and talked with him; and I find he had been a servant in my Cosin Pepys's house ... He did content himself mightily in my liking his boy's reading and did bless God for him, the most like one of the old Patriarchs ever I saw in my life, and it brought thoughts of the old age of the world in my mind for two or three days after. (338–339)

Pepys notices the father's 'woolen knit stockings' and 'shoes shod with iron ... both at the toe and heels, and with nails in the soules of his feet', which he pronounces 'mighty pretty' (339). The father explains that the heavy boots are necessary because '"the Downes, you see, are full of stones, and we are fain to shoe ourselves thus; and these ... make the stones fly till they sing before me"'. Pepys gives the 'poor man something' and observes that he 'values his dog mightily ... and that there was about 18 scoare sheep in his flock, and that he hath 4 s[hillings] a week the year round for keeping them'. The conversation over, back we head to town, stopping for milk from a milkmaid 'better than any creame' to be had in tavern or town. As the coach flies, 'it being about 7 at night', we see 'people walking with their wifes and children to take the ayre'. 'The sun by and by going down', Pepys tells Mrs. Turner, one of their companions, 'never to keep a country-house, but to keep a coach and with my wife on Sunday and to go sometimes for a day to this place and then quite another place; and there is more variety, and as little charge and no trouble, as there is in a country-house' (339–340). The glow-worms appear. Pepys finds them 'mighty pretty'. But his foot still hurts – and even Mrs. Turner warming it gently does no good. He needs help down the lane to his house – and has to spend the following day in bed.

While the countryside presences here, it serves as an effect to be generated and also as a vector or anchoring point for a carefully reconstructed account of an economical day out – saving the expense of a country house via the maximized convenience of a coach and horses. Indeed, the week is punctuated, so he says, by these Sunday jaunts, and framed by the double invocation of the calendar, solar and liturgical, Pepys' reconstruction is mediated by the technical apparatus that programmed the flow of his day: the pocket watch; the coach; the roads. It is haunted by the residues of the Lord's Day. The shepherd and his son, literal or figurative, their Bible between them, locate in the heart of the downs a scene of innocent instruction that transports Pepys back into the world of the Hebrew Bible as the pastoral scene becomes suffused with the residues of the religious experience Pepys is not having in London. The sheep, whose appearance overwhelms, serve as a summary synecdoche of the scene. They constitute its material occasion, for it is by their minding at four shillings a week that father and son share this life and have the leisure that they put to such good or innocent use.

But, if Pepys pronounces the scene 'pretty', then it is in part also because at every point he pays his way – the reckoning at the tavern, the successive 'somethings' he gives to the son and his father, whom he learns was once in service to his cousin, who owned the country house that Pepys claims he's better off without. As removed as the world of the country seems from the city – its water serving as a corrective enema to Pepys's dodgy bowels; its milk better than the cream in the tavern; its air cleaner than the road the coach's passage renders dusty – its 'prettiness' is funded by a series of monetary exchanges or 'reckonings' that record the linkage of places and the interpenetration of country and city and points between. Of course, as Pepys' rendering of the day makes clear, he does not belong on the downs. He gets everyone lost; sprains his ankle because he wears the wrong shoes; trips on stones that the shepherds send flying – the skipping of the stones at the nails on their toes their song, the biblical text the boy reads, their lyric, their walking a kind of writing that only Pepys can recognize, read, record. And yet it is this lack of belonging (his having the wrong shoes; writing with a quill and not the nails on his shoes) that permits his sense of having reached an outside that is nevertheless familiar, pictured in miniature, as a series of stations or topoi, the coach annihilating the space that comes between, and all funded and maintained by the reassuring relays of, once upon a time, the ties of service and now by modest financial 'somethings'.

The diary records the rites of access to the 'country', the technical modification of psyche and soma the routines effect, and the efficacy of Pepys' weekly punctuation by the countryside. But as a rewriting or revision, the diary marks also a further miniaturization or framing of what was already framed by the windows of the coach: an encounter with the countryside 'in small'. No wonder then that Pepys ventures out every Lord's Day he can, religiously we might say, taking such steps he must to invest in the production of this idyll ('the word idyll derives', as Terry Gifford reminds us, 'from the Greek *eidyllion*, meaning a small picture … [or] short poem of idealized description').[17] We could say then that Pepys' autobiographical urge, just like my own earlier fictive turn, further miniaturizes the experience, replaying the sound of the stones flying before him, recording and revising the scene in his diary and so reconstituting day as kind of auto- or automatic archive that installs an authentic Sunday experience among the literal flocks whose shepherds' Bible study enables him to render them figural. Reread the diary (which is already a rerunning of the day) and the idyll materializes all over again – a tele-topical Sunday outing for those weeks when work or gout precludes getting away.

If, for Williams, television serves as a key 'actor' or 'actant' in the most recent chapters to the story of pastoral, it is because, in the early twentieth century, 'there is', as he writes, 'an operative relationship between a new kind of expanded, mobile and complex society and the development of a modern communications technology',[18] replacing Pepys' pocket watch, coach, and diary with their fossil-fueled inheritors, the train and car, and such other miniaturizing technologies that come to replace the paper, pen, and ink with which Pepys crafts his account. Although at no point does Williams say so explicitly, in *Television* he revises and expands the analysis of pastoral forms as they relate to the organization of material life that he undertakes in *The Country and the City* (1973), but now transformed by the tele-topical figure of the broadcast. Television figures as a material, semiotic, and rhetorical event. It has a technical and political history all its own – but its effects, as in the passage from letter and table to pocket watch and coach, and beyond, are essentially to amp up the already installed humanist script for making persons and constituting collectives.

Welcome, we might say, to the writing machine. Welcome, Giorgio Agamben might say, to the alliance between writing systems and the 'anthropological machine', ancient and modern, and so to the routinized procedures by which the 'human' is produced as a 'space of exception' (38) as distinct, that is, from any other form of life.

Crucial to the functioning of this machine in its modern iteration is a generalized experience of what Martin Heidegger calls 'profound boredom' (63) ('*tiefe Langweile*'), which prompts a 'being left empty' or 'abandonment in emptiness' (63) ('*Leergelassenheit*') that, as Agamben notes, constitutes 'nothing less than an anthropogenesis, the becoming *Da-sein* of living man' (68). As Agamben notes, for Heidegger the *locus classicus* of this experience was the '"tasteless station of some lonely minor railway"' (63) where the world of disposable objects (train timetables, magazines, chocolate bars, cigarette stubs, fellow passengers – if there are any) fails to captivate. And this waking to an overwhelming emptiness, this sensory detachment, voids all sense of an interior or essence. For Heidegger, then, as Agamben observes, the conception of *Dasein* (being there) 'is simply an animal that has learned to become bored; it has awakened *from* its own captivation *to* its own captivation' (70). 'This awakening of the living being to its own being-captivated', he continues, 'this anxious and resolute opening to a not-open is the human' (63). Thus 'being there', becoming 'human', designates merely that animal which has learned to enter into a becoming idle. No *negotium* here, then, no return to the world of work or civics, idleness merely, an excess of undirected *otium*.

Ever eager to burst the Heideggerian idyll, Agamben prefaces his account of Heidegger on boredom with the story of a very famous tick. For coincidentally, while Heidegger was metaphorically cooling his heels at that lonely country station, just a few hundred miles away, in a laboratory in Rostock, zoologist Baron Jakob von Uexküll had managed to keep a tick 'alive for 18 years without nourishment ... in a condition of absolute isolation from its environment' (47). As Agamben notes wryly, under precise conditions, this tick had 'effectively suspended its immediate relationship with its environment, without, however, ceasing to be an animal or becoming human' (47) and he wonders whether, if Heidegger had known of said tick, this knowledge would have made a difference in his modeling of *dasein*. 'Perhaps the tick in the Rostock laboratory', concludes Agamben, 'guards a mystery of the "simply living being," which neither Uexküll nor Heidegger was prepared to confront' (70). For Agamben, the prospect of animal *otium*, an *otium* that is decoupled from the mechanisms of the anthropological machine, might constitute an opening on to another order of propositions about life (*bios/zoos*) than that found in Heidegger.

Agamben ends his analysis of the 'anthropological machine' by asking whether or not we have it in our power to render this machine 'less lethal and bloody' (38). At the end of the book, he raises the ante still further by stating that 'to render inoperative the machine that governs our conception of man will therefore mean no longer to seek new – more effective or more authentic – articulations, but rather to show the central emptiness, the hiatus that – within man – separates man and animal, and to risk ourselves in this emptiness: the suspension of the suspension, Shabbat of both animal and man' (92). And yet, what seems like a manifesto or slogan in favor of a relentless pursuit of the negative is almost immediately put on hold, left to idle, as Agamben wonders whether there might already exist attempts to render the 'simply living being' that preserve the possibility that 'there is still a way in which living beings can sit at the messianic banquet of the righteous without taking on a historical task and without setting the anthropological machine in motion' (92). Just as Agamben embarks on what appears to be a systematic destruction of all attempts to positivize the 'human', he introduces a further figure of suspension, and so raises the possibility that within the archive of textual traces from which various 'pasts' are summoned, there may exist latencies, neglected routines, from which the 'pluri-dimensionality of historical experience' might be summoned in order to imagine alternate scripts.[19]

Drawn back to that tick (and emphatically not to the likes of Pepys' 'innocent' and 'pretty' sheep) – a tick, it must be said, who fails to take note that it inhabits a pastoral enclave, who fails, as it were, to take notice of pastoral – Agamben worries the figure of the idle/idyll, trawling through the prospects of pastoral to see if something subsists there, something that may somehow constitute a 'fleshly difference'. Quoting from Uexküll, who writes that 'every country dweller who frequently roams the woods and bush with his dog, has surely made the acquaintance of a tiny insect', Agamben notes that 'the opening has the tones of an idyll' (45) – 'but here the idyll is already over, because the tick perceives nothing of it' (46): the 'eyeless' tick, this 'blind and deaf bandit' comes to know of the 'approach of her prey ... only through her sense of smell'. Maddeningly, Uexküll 'gives no explanation' of how he was able to keep this tick alive for eighteen years and merely supposes that during this period 'the tick lies in "a sleep-like state similar to the one we experience every night"' (47). Very understandably, this ellipsis prompts Agamben to ask 'what becomes of the tick and its world in this state of suspension that lasts eighteen years? How is it possible for a living being that consists entirely in its relationship with its environment to survive in absolute deprivation of its environment?' This figural tick, whose perspective Uexküll and Agamben after him solicit or seek to inhabit via an inconclusive or botched *prosopopeia*, stands as an encrypted archive, as the sign of an event not yet judged to have taken place. For by what order of metaphysics might we account for the condition of *this* tick? Was it profoundly bored? Or was it, as a Renaissance humanist might have framed the question, 'at leisure' (*per otium*), idling, free of everyday tasks and distractions, its system in neutral? Is there something then that subsists within the tropes of pastoral and its various ways of factoring *otium* that speaks to a desire to deactivate the anthropological machine? Agamben appears to think so – but that 'something' seems to be boxed up within the tick's failure to take notice of the idyll, that is, with the failure of a certain zoographic or deployment of the non-human in the rhetorical crafting of a world in words or paint – an idling of *prosopopeia*, a failure or delay in summoning a face or voice. For by that failure what issues forth is a figure of the figure, and perhaps with that the possibility of opening a space within discourse for an unanticipated set of responses – the 'human' reduced now to a waiting on the tick, a waiting for or to see whether the tick will respond.[20]

What was pastoral (again)?

Given the stakes, as Agamben defines them, the question sounds faintly ridiculous. We know what pastoral is. Not quite a genre, but sometimes a mode, activity, vehicle, or poetic device – idealizing or ironic – pastoral treats high themes in low costume – enabling its producers to explore issues of patronage, poetic servitude, etc., set against a world that appears static, innocent, or, as Pepys might put it, 'pretty'. As Robert N. Watson writes in *Back to Nature*, 'pastoral is thus another cultural phenomenon explicable as a response – often simultaneously as an inscribed banner of protest and a blank flag of surrender – to the burdensome knowledge of mediation'[21] – in other words, the knowledge that there will be no unmediated access to some 'thing' we call 'nature' or the 'natural'. In his hands, the Renaissance vogue for pastoral, which dominates the mechanisms by which they attempted to 'dream away the time', becomes a vision of a society backing into a Nature that everywhere was felt to be receding from them.

Plagued by a problem of definition, pastoral is reduced then either to a set of tropes subordinate to its historical deployment – to what, as Louis Montrose puts it, pastoral might be said to 'do' in its successive historical iterations – or, perhaps worse, it becomes a cipher for a fairly lumbering conception of ideology as mystification. Either way, pastoral is, apparently, bad news – something we might hope that the recent fluency of literary and cultural studies in ecology and ecological modes of description might enable us to pass beyond, do without, or supersede. For, in essence, pastoral (and its persistence) has become something used to frighten eco-critics who offer up what they consider to be friendlier distributions of knowledge or more capacious modes of description than those previously on offer. 'Don't you think that your more holistic than thou vocabulary of networks, ecologies, webs or collectives is a just a wee bit too holistic, nostalgic, romantic, or, pastoral?' 'How do you really feel about slime?' – that is to say about all those forms of life that lack charisma, which are lethal, or so unphotogenic that it's not possible to grant them the status of 'honorary primates' and so process them as partial, potential, citizen-subjects.[22]

Invoking Leo Marx's dictum, for example, 'No shepherd, no pastoral',[23] or citing Paul Alpers' very sensible reminder that 'the central fiction of pastoral – in [Kenneth] Burke's terminology, its representative anecdote – is not the Golden Age or idyllic landscapes, but herdsmen and their lives',[24] offers little shelter from such charges. For, whatever its origins, and whatever its minimal units, the pastoral metaphor spins off images of worlds that are taken for 'nature', if only in retrospect, if only when successively rerun on whichever media platforms program the flow of this or that historical moment. The metaphor effects too durable a transport – we are, after, or before all, according to Heidegger, 'shepherds of Being'.[25] Famously, for Raymond Williams, pastoral and its kitschy, downgraded, 'equivalents' from the sixteenth century onwards are understood as a perpetual stasis machine that sustains a murderous environmental and social fiction. An 'escalator' Williams calls it, 'moving without pause' forever backwards into an ever-receding image of the Golden Age *du jour*.[26]

Rhetorically, it may seem that I should disagree with Williams, but it's not quite that simple. One thing I have learned is that tactically it is best to agree with everything everyone has ever said about pastoral. This is not because I am some happy shepherd advocating an idyllic, pastoral pluralism but rather because everyone, in his or her own way, tends to get his or her particular section of the rhizome correct. The extended or general text of pastoral, pastoral as archive, manifests in different times and places folded in strange or seemingly contradictory ways, as its core tropes are successively performed.[27]

Williams flags this point early on in *The Country and the City*: 'the initial problem', he writes, 'is one of perspective' (9) – where you find yourself distributed, that is, in relation to pastoral. He begins the book with a cannily lyrical reconstruction of the two 'networks' he has inhabited, or fallen between, familial and academic, which parse out loosely into 'country' and 'city' – but which already, by his description, and by the transport technologies (but not yet the television broadcasts) that interpenetrate, fail to sit happily in either category. His rendering of the history of pastoral as the gradual localizing of 'a dream and then … into a description and thence an idealisation of actual English country life and its social and economic relations' (26) takes aim at the process whereby 'the charity of consumption' (a feast) occludes and defers the possibility of a community which may have existed but which remains still to come: 'a charity of production – of loving relations between men [and women]' (30–31). In *Television* this 'charity of production' finds expression in Williams' valorizing of those unscripted programs which 'open themselves towards people not assumed in advance already to be represented' – programs which run live and which therefore risk the possibility of 'dead' or dull 'air'.[28] The issue here is not Williams' desire for a utopian figure of unalienated labor – though that is present – but instead the way this occlusion requires that all analysis of social relations occurs belatedly, '*post festum*', to use Karl Marx's figure from *Capital*. For Williams then, deeply concerned also by the scandalous ways in which generations of readers misconstrue literary references for historical reality, the story of pastoral yields a theory of ideology – what Paul de Man would call 'a confusion of linguistic with natural reality, of reference with phenomenalism'.[29]

Read this way, Williams' figure of the 'escalator' seems much less hostile to the likes of a William Empson and his *Some Versions of Pastoral* (1935) than Alpers at least implies.[30] Sometimes a 'trick',[31] but, by turns, a 'myth', a 'process' (22–23), a 'machine' (30), a 'double plot', an 'organism' (144–145), or a 'tap root' (261), 'pastoral is a queer business ... permanent ... and not dependent on a system of class exploitation'. It works by what sounds a lot like taxidermy – 'putting the complex into the simple' (22–23) so as to 'imply a beautiful relation between rich and poor' (11), 'giv[ing] the impression of dealing with life completely' (29). 'In pastoral', writes Empson, 'you take a limited life and pretend it is the full and normal one' – it's a metaplasmic refolding of surfaces that creates the appearance of depth, of 'life' or 'liveliness', not so much by what it adds as by what it leaves out or subtracts. No wonder pastoral endures – 'takes refuge' in a 'larval Alice' (269) is how Empson puts it – the activity comes to stand for a 'predatory' or 'pro-active' sense of mimesis, not exactly the 'ambient' or ambulatory quality of nature writing that Timothy Morton has recently named 'ecomimesis'[32] and which informs Pepys', Williams', and my own autobiographical self-staging – but a more fundamental, un-Romantic 'stuffing'. For Empson, then, pastoral names a metaplasmic operation by which writing touches itself, and by that touching which is in truth a folding, traces the skin of the world, producing a set of effects that may be taken for phenomena, spun off as so many 'worlds' or 'pasts'. Indeed, in Empson's hands, pastoral takes on a privileged relationship to Agamben's 'anthropological machine', producing the 'human' as a positive category only by the subtraction of a range of beings it spins off to lead simpler or less lively lives than those of their readers.

Track back in the company of David Halperin's *Before Pastoral* to the literary world of Theocritus' Idylls, and it is possible to isolate the 'trick' of pastoral, as Empson called it, still further. For what we find is not a writer of what Virgil will codify as 'pastoral', but instead of a 'bucolic poetry' defined in terms of its formal features as opposed to its thematic concerns. As Halperin observes 'at the time Theocritus was composing the Idylls, the principle of classifying poetic genres according to meter and the doctrine of fixity and separateness of the poetic genres had long been powerfully established'.[33] Moreover, in describing the apparent disunity of subject matter in the Idylls, Halperin argues that this thematic breadth is indexed to a programmatic re-reading of Homer and the function of epic in Theocritus' present. 'An examination of Theocritus' poetic technique', he writes,

> as well as his treatment of themes reveals a pattern of contrasts or oppositions between bucolic *epos* and heroic *epos* – or perhaps between the heroic and non-heroic registers *within* the tradition of *epos*. Theocritus does not in general introduce totally alien material into the epic genre; rather, he elaborates the non-heroic alternatives to the traditional heroic mode, which are already present, albeit inchoately in early Greek epic.[34]

Halperin finds this recoding or resignifying of epic enacted in the form of an extended examination of the workings of the *ekphrastic* figure of the ivy-cup which is given a famously lengthy description in the First Idyll and which stands in relation to the singing competition that the description defers.

Tracing the philological origins of this wooden milking bowl or *kissybion*, and comparing it to the way similar words are used in the *Odyssey*, Halperin concludes that it 'seems to be a large bowl, used by rustics – in short, a humble implement belonging to a primitive economy'.[35] But whereas in the *Odyssey* such bowls are made from metal and are used for drinking wine, Theocritus has transferred the decoration of such bowls to an object of much more humble origins (172–173). Halperin reads this transfer or translation of decorative motifs to material of more humble origins as a statement which makes readable Theocritus' aim both to outdo his material and also to 'provide ... a picture of the life and feelings of little people, portrayed in situations that are not earthshaking' (177). Reading the successive survivals of the First Idyll in papyrus as evidence of its programmatic nature, he agrees with those critics who read the Idyll as providing '"the poet's *sphragis*"', or signature.[36] For Halperin, then, what the question of origins yields is a moment of reading or re-reading where Theocritus' 'static vignettes [idylls or] *eidyllia*' (literally paintings with captions) (186) constitute themselves as a rerunning or relayering of epic themes and techniques – a way of reckoning or coming to terms with the latent elements of epic by extending epic's techniques (both comically and seriously) to broach subjects of everyday concern. Their root in *ekphrasis* serves to emphasize the way the idylls operate as a doubling or refolding of origins, creating tiny, augmented worlds or pictures in words.

What was pastoral (again)? Ideology? Mimesis? *Ekphrasis*? What about all those shepherds? 'No shepherds, no pastoral' – to which we must add, 'sheep'. Sheep are the given, the stock/capital, so the story goes, that shepherds 'own' in order to maintain the fiction that they are not hirelings but free men. This is one of the points that John Fletcher is keen to make in the preface to *The Faithful Shepherdess* (1608), when he instructs his readers 'you are ever to remember [that] Shepherds [are] ... owners of flockes and not hyerlings' – which is to insist, in a sense, that they are not themselves 'sheep'.[37] Summarizing numerous readers of pastoral, Montrose observes that the genealogy of the poet-shepherd that Fletcher invokes here stems from the way 'literary celebrations of pastoral *otium* conventionalize the relative ease of the shepherd's labors'.[38] He credits George Puttenham's *The Arte of English Poesie* (1589) with a strategic 'conflat[ion] ... of the attributes of the pastoral world with those of the Golden Age', but also with a careful rearrangement of labor practices such that the '"ease" and "idleness" typical of the Golden Age' derives 'from the organization of pastoral society itself, rather than from the spontaneous fertility of nature'.[39] In Puttenham's hands, then, pastoral becomes a way of placing poetry at the foundation of the world, and of crafting a genealogy of different poetic forms keyed to the progress of the social order. In the beginning there is pastoral, sheep, and shepherds. Then follow castles, towns, kings, citizens, tragedy, and comedy. And, for Montrose, this genealogy is allied to the crafting of gentlemen-poet-shepherd-lovers as a script for pursuing an ennobled poetic-politic career. Contrast Sir Walter Ralegh's worldly anti-pastoral self who refuses an office he deems beneath him with the words, '"I would disdayne it as miche to keap sheepe"', with his pastoral poetic willingness when it comes to Queen Elizabeth I to cast himself as both shepherd and '"gentill Lamm"'.[40]

As Ralegh knows, and Fletcher feels mightily, there were times when it was important for human 'not-sheep' to insist that they really were, in fact, not sheep – except, of course, when it is good for a person to be a 'lam' or a 'sheep', and you find yourself, along with everyone else, singing Psalm 23, 'lying down in green pastures', 'the Lord [your] Shepherd, [you] his Sheep'. Or, you found yourself signing letters addressed to Queen Elizabeth I, with the epithet: 'your majesty's "sheep", and most bound vassal', as did Lord Chancellor Sir Christopher Hatton in the early 1590s – Elizabeth coming to refer to him affectionately, as Robert Cecil would opine later in a letter, as 'her mutton', fantasizing Hatton's presence with her as she recreated out on the 'Downes covered with sheep'. Sometimes it was good for the human 'not-sheep' to be a sheep – reassuring, or, maybe, even, given the right shepherd, a little sexy.[41]

Obviously I am not the first to remark how the bio-politics of pastoral and the networks of pastoral care with which it is allied trade on a sheepy metaphorics in which human persons oscillate between the roles of shepherds and their four-legged charges.[42] Sheep and 'not-sheep' – the 'human' continuously derived not as a permanent entity so much as an ongoing subtraction. The discourse is as old as the West. It is tempting therefore to deploy an anthropological fix that would posit pastoral, much as did Puttenham, as a discursive exaptation from a primary organization of agricultural production and animal husbandry that defines the geography, history, and metaphors of the built worlds we live. As Fernand Braudel writes, the sheepy metaphors we live by are indexed to an alliance of human, animal, and plant resources or actors, demonstrating 'an ancient choice or priority [in the West for ruminants and wheat] from which everything else descended'.[43] Certainly, pastoral trades on a putative agricultural settlement and mobilizes metaphors that derive from Haraway's figural 'multi-species' (human, dog, sheep, goat) or the mutual domestication of animal and human. Transhumant husbandry is etched into the land. It built the circuits that paper and electronic pastoral will and now travel.[44]

We could even, as Terry Gifford did or claims to have done, in his excellent introduction to pastoral, travel to those places where the Golden Age seemingly endures today, accepting 'an invitation to observe real pastoralism in practice'. Get up early, hop on a plane, and, like some latter-day Pepys, we could pay a visit to real shepherds 'of a village in the mountains of Crete', say, as Gifford did, talk with them, listen to their songs, and time travel back to the world that Puttenham envisaged.[45] Gifford finds shepherds – even shepherds who still talk about Arcadia, 'and still have witty song competitions too, each singer trying to outdo the other'. But they also save their money to buy 'cattle trucks', and brandish Kalashnikovs made in China. Gifford makes the exemplary observation, that here we see 'from the beginning of its long history [that] the pastoral was written for an urban audience' – or, in the case of these Cretan shepherds, we discover the interpenetration of localities with global networks of exchange. But, as Puttenham's already pastoralizing genealogy of poetic forms demonstrates, the larger point might be to observe that all moves to phenomenalize the metaphors we find expressed in 'pastoral' will tend to positivize the 'past' via a retro-projection of those very metaphors. Tempting as it is, therefore, to derive pastoral from pastoralism proper, and so to invoke 'sheep' as a literal, material, basis, inquiries into questions of origin tend simply to transform themselves into this or that anthropology or invocation of a physicalist base.

When, instead, David Halperin tracks the discourse of pastoral back to its putative origins in pastoral societies in the Near East (where human persons, sheep, and cows first domesticated one another) what he discovers is a semi-readable coding of different types of human speech, in which the voices of herdsman, cowherd, and shepherd were heard by and in their relation to an 'inhuman liminality' which derived from their relative remoteness from the city and so proximity to beast and the divine.[46] The story that Halperin discerns of shepherd as 'go-between', as *homo sacer*, the person who marks the connection and boundary between differing species and between human and divine, corresponds more closely to the long history of *'homo sacrilization'* that Agamben excavates than to the conventions of literary pastoral. Indeed, following Halperin, we may say that this history, which culminates in the internalization of the 'bio-political fracture' that writes the 'human', such that 'western politics is a biopolitics from the beginning',[47] relies on sheep as its stock in trade metaphor.

This recasting of pastoral as constitutive metaphor is key, for, as primatologist turned sheep-farmer Thelma Rowell would tell us, the metaphors I have traced thus far that render sheep 'sheepish' (if usefully so) derive not from their essential sheepiness but from their modeling and manipulation as livestock or living capital – their physiology, their 'liveliness', as it were, simply a convenient support or backing to their function as producers of wool, meat, and the very many use values and usable matter that early moderns such as Leonard Mascall and Edward Topsell extol.[48] Selective breeding tames their wildness, renders them 'domestic'. Moreover, as Rowell argues, the traditions of primatology and animal behavior studies have dictated that those animals who lead interesting lives (that is lives deemed interesting to us) have tended to serve as privileged experimental subjects – especially if they may be grouped as among the relatives of a certain *Homo sapiens*.[49] Accordingly animals (and that is 'most animals') who 'spend the majority of their time doing nothing',[50] such as sheep, tend to be neglected or asked only the most boring of questions. 'Sheep behavior studies are mostly to do with what they eat, and sheep are not, generally, permitted to organize themselves', she writes.[51] Rowell's solution is to enable sheep to organize their own social structure and then to observe the results. She decides, in effect to 'watch ... sheep in the same way [she has] ... been watching monkeys'. As sociologist Bruno Latour writes in a commentary on Rowell's experimental practices, 'by importing the notion of intelligent behavior from a "charismatic animal" ... she might modify, subvert, or elicit, in the understanding of sheep behavior, features that were until then invisible because of the prejudices with which "boring sheep" have always been treated'.[52]

If it seems that with Thelma Rowell, I have left the orbit of pastoral, then it might be more accurate to say that I have merely invited you to leave the orbit of a particular way of modeling pastoral in relation to metaphors derived from pastoralism. For whatever it is that Agamben seems to think inheres in pastoral *otium* or in the interrupted archive of an idling tick seems likely to be found only via the deactivation of this sheepy metaphorics within which we find ourselves becoming or made 'human', and sheep 'sheepish'. If there is something on the order of an 'event' that subsists in pastoral then, it must endure less in positive representations of a 'past' human–sheep relation that we may hope to recover, than in the zoographic misfiring or troping of key ovine figures.

Given, then, what I think pastoral is and was it seems much too early to be moving on, much too premature to think that we can move on or beyond its metaphors. Instead, let's trace its latencies, its badly behaved metaphors, treating the general text as the warp and weft of a knitted skin. This, I think, is precisely the Empsonian gesture – what is or what will proletarian literature be is the question with which *Some Versions of Pastoral* begins and which it attempts to adjudicate – will it, or must it be 'pastoral?' (1–27). Likewise, Williams' portrayal of the pastoral escalator runs *in tandem* with his deployment of the terms 'residual', 'dominant', 'emergent' as categories of description merely, which, having no referential basis, nevertheless are used by a critic as a sort of advance GPS system as she tries to work out when or where this or that cultural phenomenon falls in a story of human making and manufacture so mediated by the means of production that it is hard to know which cultural forms are progressive or regressive. If there is, in other words, no exit or end to 'pastoral', it is because the activity it designates describes a process by which human discourses apprehend something called 'world', inclining it so that it comes to bear on and speak of human concerns. All that follows then will be 'more versions' – different versions, perhaps, more capacious versions, we may hope, but haunted always by those beings whose cries for admittance to the collective go unheard.

More versions

From time to time, the supplementary sheep of pastoral flicker in and out of view, summoned for just a moment – as the pastoral metaphors unravel into their metonymic chains and a literal sheep, singular, or sheep, plural, materialize on page, canvas, or screen. These appearances constitute the semiotic fine edge of an alternate archive that registers the co-modeling of sheep and not-sheep 'human' as a constitutive and sometimes corrosive process. Space permits only a survey of such appearances – but a privileged scene occurs, as many readers have noted, in Thomas More's *Utopia* (1516/18) when Raphael Hythlodaeus summons the sheep of England to Cardinal Morton's table in order to speak the truth about human labor relations. In order to rebut the arguments of an objectionable English lawyer who maintains that, the poor '"might maintain themselves"' by farming or other '"manual crafts, if they did not voluntarily prefer to be rascals"', Raphael summons a flock of sheep. '"Your sheep"', he answers, '"which are usually so tame and so cheaply fed, begin now, according to report, to be so greedy and wild that they devour human beings themselves and devastate and depopulate fields, houses, and towns."'[53] Raphael animates the sheep of England as a flock of homicidal, man-eating beasts that literally consume the rightful human inhabitants of the land out of house and home, feeding on people-grass in order to fill their owners' stomach-purses. In a perverse reversal of pastoral (and pastoral care) metaphorical shepherds become wolves, enabling Raphael to make visible the depredations of enclosure to his host.

Raphael's animal summoning calves a predatory 'oveme' that passes into Tudor dialogues, as generations of readers subsequently return to the question of what enclosure was or is 'now'. In *A Discourse of the Commonweal of this Realm of England* (1549), the long-suffering knight complains that he can only make ends meet by keeping sheep – to which the husbandman returns 'sheep, sheep, sheep'.[54] Aphorisms such as 'The more sheep, the fewer eggs a penny' and 'We want foxes to consume our shepe' circulated throughout the 1540s, culminating in actual scenes of mass ovicide in July 1549 as Kett's rebels reversed Raphael's figure by slaughtering an estimated '20,000 sheep' in order to feed themselves.[55] Thus were sheep, extolled for their use values, put to use as a food item. Of course, once beaten, Kett's rebels became mere 'sheep' again, 'when they ran confusedly away' as the same chronicle opines – their quasi-utopian interruption of history merely a momentary fitting of the general text.[56]

And such was the fate of most rebels in the representational afterlife that summoned them back into being in pamphlet and on stage. In 4.10 of William Shakespeare's *2 Henry 6*, for example, we meet an uncertainly bovine or ovine Jack Cade 'ready to famish' (4.10.1–10).[57] He has climbed 'o'er a brick wall ... into this garden to see if [he] can eat grass or pick a sallet', where he is dispatched by the *otium* seeking Alexander Iden. Iden prides himself on his self-imposed rustication from the 'turmoils' of Court to 'this small inheritance' his father has left him. 'I seek not to wax great by others' waning' (4.10.16–22), he remarks, or to 'gather wealth'. 'Sufficeth', he adds, 'that I have maintains my state, / And send ... the poor well pleased from my gate'. Speaking from the same script of Platonic withdrawal that we find articulated by Raphael in Book 1 of *Utopia*, Iden enters with a full tummy, perfectly at ease with his self-regulating body and the well-maintained boundaries of his estate. 'Eden' in Holinshed turned 'Iden' (or self-same) in the play comes upon Jack who snarls at him wolfishly. They draw and Iden kills him. Jack dies with famine on his lips – 'O I am slain! Famine and no other hath slain me' (4.10.59) and 'I, that never feared any, am vanquished by famine' (4.10.74).

Before Iden kills him, he subjects Jack to a negative blazon, at every point magnifying their difference, the difference that is between the leisured and well fed and the poor and hungry, between a recreating gentleman and a famished Cade on his way to becoming 'cattel' – to use the period's word for the range of herbivores it herds. 'Oppose thy steadfast gazing eyes to mine', demands Iden, 'See if thou canst outface me with thy looks. Set limb to limb, and thou art the lesser: / Thy hand is but a finger to my fist. / Thy leg a stick compared with this truncheon' (4.10.44–48). The confrontation between a well-fed and well-read country gentleman and a famished illiterate or differently literate stands as a stark reminder of the causes of popular protest in the period even as it may ostensibly stage the victory of private property as such. But it's Cade who speaks the language of butchery in this scene, asking his sword to cut Iden's 'chines of beef' (4.10.56) but the specter of cannibalism it raises seems inclined to magnify the words we hear most often: 'Famine and no other hath slain me!' (4.10.59); or Cade's last words, 'For I, that never feared any, am vanquished by famine' (4.10.74). The Cade sequence ends, then, as uprisings tended to, with the 'populace' becoming once again the figural 'sheep' of historical record, that is by returning to the matter of subsistence that animated their demands in the first place: food. In Act 4, scene 10, in effect, we will have been watching the retraining of Jack's mouth: no longer the self-predicating

Parliament of the land, the mouth of this wolfish sheep is denied flesh as he is forced to eat grass.

Fast forward to Thomas Shadwell's *The Virtuoso* (1676) that satirizes the exploits of the Royal Society, and Jack's final skin-turning performance, his 'becoming cattel' materializes on stage as a source of comedy. Shadwell is making hay of the Society's recent forays into xeno-transfusions, transfusing the blood of a sheep into a man named Arthur Coga. In 2.2, Sir Nicholas Gimcrack, chief experimenter, proclaims the virtues of this procedure, recounting how 'the patient from being maniacal or raging became wholly ovine or sheepish: he bleated perpetually and chewed the cud; he had wool growing on him in great quantities and a Northamptonshire sheep's tail did soon emerge or arise from his anus or human fundament'.[58] Without missing a b/l/eat, Shadwell has Sir Nicholas effect a further enclosure movement premised on an exchange of humors that renders his patients 'sheepish'. He plans to have his tailor make his suits from the wool of these transformed creatures and boasts that he will 'shortly to have a flock of 'em and make all my clothes of 'em'.[59] The play ends with the arrival of an angry mob bent on revenge as they credit Sir Nicholas with the invention of the mechanical loom.

This gradual literalizing or phenomenalizing of the sheepy metaphorics of pastoral finds its emblem, perhaps, in Charlie Chaplin's use of the figure in the opening moments of *Modern Times* (1936) as a crowd of people hurrying to work dissolves into a flock of sheep and then back into people again, the 'human' now merely a filmic skin, our 'little hero' a 'black sheep'. Enter the actually predatory weir-sheep in the emphatically, ludicrously pastoral New Zealand *mise-en-scène* of the recent *Black Sheep* (2007), whose sheep, supped up by a complete misunderstanding of SNCT (somatic nuclear cell transfer), aka cloning, run amok. Dolly, the cloned sheep, as it were comes home to roost and her weir-sheep cousins do in fact quite literally and very graphically eat men and women – just as Raphael and Jack Cade joked in serious play.[60] As they do so, these sheep-become-wolves' bodies morph, their skin stretches, to accommodate their newly sharpened incisors.

If it seems that I have strayed in the direction of *Utopian* discourses and their afterlives, then it seems worth observing that *Utopia* exists by its appearance as a tariff on pastoral, interrupting its idylls with questions of the collective, by orchestrating a momentary stay in proceedings whereby the collective manages to inquire into the relays by which it makes its world.[61] The hiatus or idling effects of *Utopia*, its coding and opening on to *otium*, cause inversions of pastoral to appear, enabling readers to pose questions of or to scoff at their social betters. We are still idling, however, caught like Agamben's tick, tracing the contours of a double archive, pastoral and *Utopian*, as 'sheep' and 'not-sheep' flicker in and out of view. We still wait for something on the order of a figural 'event' that would signify 'sheep' differently and by that signification transform the 'human' 'not-sheep' also.

It is here that we trace the limits of what I can say about Agamben's deactivated *otium*. For if the issue comes down to a modeling of sheep that then rebounds on humans, then one solution, one way of deactivating or stalling the 'anthropological machine' and the mutual definition of 'sheep' and 'not sheep', would be to refuse to treat sheep as 'sheep' at all. Instead, like primatologist-turned-sheep farmer Thelma Rowell, you treat them as members of a much broader group: 'all gregarious long-lived vertebrates capable of mutual recognition'[62] – you give them the chance to be more interesting – less objectal or 'slimy'. For Rowell and ethologists in general, in a world where there is only mediation – where mediation is recognized simply to be the way tomorrow's world is constituted and what passes as a positivized kernel to the 'human' amounts merely to an Empsonian 'stuffing' of the 'non-human' – all that matters is the rubric by which you model 'things', the way in which your putting to use of the world renders it variously lively.

In Rowell's case, this means attempting to imagine epic for sheep – continuing that is the work of Theocritus in his First Idyll and further revising the world of epic, via a kind of 'bucolic poetry' that sheep (historical sheep, this time, sheep in a particular time and place) might be understood to *write*. In effect, Rowell seeks to suspend the 'companion' or 'multi-species' models we have witnessed thus far, and instead seeks to enable sheep to manifest to humans as complex social animals with their own structures and behaviors and without reference to their processing as livestock. As ethologist Vinciane Despret notes, Rowell's 'observations usually start in the morning, with the same ritual: she takes each of her 22 sheep a bowl of its breakfast. But what puzzles any outside observer is that there are not 22 but 23 bowls, that is, always one too many.' 'Why this extra bowl?' asks Despret. 'Is the researcher practicing a kind of conviviality?'[63] For the reader of pastoral poetry it is tempting to suggest that Rowell is further transforming the *ekphrastic* wager of Theocritus' ivy-cup or *kissybion*, but this time the cup materializes in even more humble garb as a feed bowl for a sheep, indeed as a feed bowl that is not used – somehow allowing these 22 historical sheep to refigure themselves and the *prosopopeia* that once rendered them and us 'sheepish'. The 23rd bowl is, as Despret hints in what seems like misdirection, about politeness – about accepting but also tuning out the presence of the human researcher by offering to 'sheep' the chance to transform the protocols of the observation. The presence of the bowl and so the surplus of food transforms the questions that Rowell poses of her sheep, removing or suspending an automatic question concerning competition. The bowl 'is intended', Despret continues, 'to expand the repertoire of hypotheses and questions proposed to the sheep ... [but] to leave them the choice' of answering other questions than those posed to them. Like Theocritus, Rowell prepares the bowl, but it is her 22 sheep whose actions she records that co-*write* the scene it depicts. For Despret, then, 'this [now] emblematic [idyllic/*eidyllion*] twenty-third bowl' implicitly becomes a way of entering sheep into the writing machine which permits or requires the 'human' now merely to idle, to wait or attend, much like Agamben's tick, waiting to be awakened or addressed by an other who now initiates contact.

Is this pastoral? Must it be pastoral? Or has pastoral now been so irremediably altered that it will be hard to know what a sheep or a human has become? What is the morphology of those figures appearing on the ivy-cup become 23rd bowl, figures upon which that which was 'human' now waits, the human figured now as an idling merely, a category held in abeyance, awaiting who knows what?

Bah! Or perhaps, even, Shab/Bah/t!

Notes

1. Donna Haraway, *The Companion Species Manifesto: Dogs, People, and Significant Otherness* (Chicago: Prickly Paradigm Press, 2003), 20–21.
2. As Louis A. Montrose objects, 'merely to pose the question of "what pastoral really is" is to situate oneself within an idealist aesthetics that represses the historical and material determinations of any written discourse'. Louis A. Montrose, 'Of Gentlemen and Shepherds: The Politics of Elizabethan Pastoral Form', *ELH* 50:3 (Autumn 1983), 416.
3. Jacques Derrida, *Of Grammatology*, trans. Gayatri Chakravorty Spivak (Baltimore: Johns Hopkins University Press, 1974), 84–85. Jonathan Goldberg's reading of this section of *Of Grammatology* remains singularly important; see Jonathan Goldberg, *Writing Matter: From the Hands of the English Renaissance* (Stanford: Stanford University Press, 1990), 16–27.
4. Derrida, *Of Grammatology*, 84.
5. Giorgio Agamben, *The Open: Man and Animal*, trans. Kevin Attell (Stanford: Stanford University Press, 2004), 67–68. Unless otherwise indicated, subsequent references will appear parenthetically in the text.
6. Donna Haraway, *When Species Meet* (Minneapolis: University of Minnesota Press, 2008).
7. Derrida, *Of Grammatology*, 85.

8. Raymond Williams, *Television: Technology and Cultural Form* [1974/5] (Routledge: London and New York, 1990), 78.

9. Quentin Skinner, 'Sir Thomas More's *Utopia* and the Language of Renaissance Humanism', in Anthony Pagden, ed., *The Languages of Political Theory in Early Modern Europe* (Cambridge: Cambridge University Press, 1987), 123–158. This essay is revised and reprinted as 'Thomas More's *Utopia* and the Virtue of True Nobility', in Quentin Skinner, *Visions of Politics*, 3 vols (Cambridge: Cambridge University Press, 2002), II, 213–244.

10. For the hugely influential model of self-cancellation, see Stephen Greenblatt, *Renaissance Self-Fashioning from More to Shakespeare* (Chicago: University of Chicago Press, 1980), 11–73. For an allied if differently staged reading in which *Utopia* produces the desired 'silent body' of the middle-class consumer (input equals output) see Richard Halperin, *The Poetics of Primitive Accumulation: English Renaissance Culture and the Genealogy of Capital* (Ithaca and London: Cornell University Press, 1991), 61–100. For an analysis of the tension between *otium* and *tempus* in More's *Utopia*, see Julian Yates, 'Humanist Habitats; Or, Eating Well with *Utopia*', in Mary Floyd-Wilson and Garrett Sullivan, eds, *Environment and Embodiment in Early Modern England* (Basingstoke: Palgrave Macmillan, 2007), 187–209.

11. *Opus Epostolarum Des Erasmi Roterodami*, ed. P. S. Allen et al. (Oxford: Clarendon Press, 1906–58), II, 339. Quoted in *The Yale Edition of the Complete Works of St. Thomas More*, ed. Edward Surtz S.J. and J. H. Hexter (New Haven and London: Yale University Press, 1965), IV, xv.

12. *The Complete Works of Thomas More*, IV, 108/109. The Yale translation renders '*occium*' as 'time'.

13. Anthony Grafton and Lisa Jardine, *From Humanism to the Humanities: Education and the Liberal Arts in Fifteenth and Sixteenth-Century Europe* (Cambridge, MA: Harvard University Press, 1986).

14. On 'passage techniques' see Timothy J. Reiss, *Mirages of the Selfe: Patterns of Personhood in Ancient and Early Modern Europe* (Stanford: Stanford University Press, 2003), 469–487.

15. Keith Thomas, *Man and the Natural World: Changing Attitudes in England, 1500–1800* (Harmondsworth: Penguin Books, 1984), 250–251.

16. *The Diary of Samuel Pepys*, ed. Robert Latham and William Matthews (Berkeley and Los Angeles: University of California Press), VIII, 336. Unless otherwise indicated, subsequent references appear parenthetically in the text.

17. Terry Gifford, *Pastoral* (New York and London: Routledge, 1999), 16.

18. Williams, *Television*, 13.

19. The figures of 'suspension', 'idling', and 'deactivation' are repeated throughout Agamben's work – notably in *State of Exception*, trans. Kevin Attell (Chicago: University of Chicago Press, 2005), 87, where he speaks of a 'halt[-ing of] the machine' (87) and in his reading of Walter Benjamin's 'Critique of Violence' (61–62).

20. On the capacity for response and collapse of the figure of the 'animal' into particularized beings – this cat, this tick – see Jacques Derrida, *The Animal that Therefore I am*, trans. David Wills (New York: Fordham University Press, 2008), 114–119.

21. Robert N. Watson, *Back to Nature: The Green and the Real in the Late Renaissance* (Philadelphia: University of Pennsylvania Press, 2006), 66. For

a similar reading of pastoral keyed this time to the rise of the machine, see Jonathan Sawday, *Engines of the Imagination: Renaissance Culture and the Rise of the Machine* (New York: Routledge, 2007), 294–309.

22. On the perils of a lack of charisma to those forms of life confined under the name 'animal', and for sheep in particular, see Sarah Franklin, *Dolly Mixtures: The Remaking of Genealogy* (Durham and London: Duke University Press, 2007), and also Timothy Morton, *Ecology without Nature: Rethinking Environmental Aesthetics* (Cambridge, MA: Harvard University Press, 2007), 51–52, 159.

23. Leo Marx, 'Pastoralism in America', in Sacvan Bercovitch and Myra Jehlen, eds, *Ideology and Classic American Literature* (Cambridge: Cambridge University Press, 1986), 45.

24. Paul Alpers, *What is Pastoral?* (Chicago and London: University of Chicago Press, 1996), x.

25. Martin Heidegger, 'Letter on Humanism', in *Basic Writings*, ed. David Farrell Krell (New York: Harper & Row, 1977), 234.

26. Raymond Williams, *The Country and the City* (Oxford and New York: Oxford University Press, 1975), 9–12, 13–45. Unless otherwise indicated, all subsequent references appear parenthetically in the text.

27. For an allied modeling on genre as rhizome, see Wai Chee Dimock, 'Genre as World System: Epic and Novel on Four Continents', *Narrative* 14:1 (January 2006), 85–101.

28. Williams, *Television*, 49–50.

29. Paul de Man, *The Resistance to Theory* (Minneapolis: University of Minnesota Press, 1987), 11.

30. Alpers demotes Williams to a footnote and then an in-text appearance only as a source of a 'mis-application' of Ben Jonson's 'To Penshurst' (*What is Pastoral?* 60) but treats Empson at length (*What is Pastoral?* 37–43). Terry Gifford wonders whether (or not) Empson really is the butt of Williams' annoyance in two references in *The Country and the City*; see Gifford, *Pastoral*, 9–10.

31. William Empson, *Some Versions of Pastoral* (New York: New Directions, 1974), 11. Unless otherwise indicated, subsequent references appear parenthetically in the text.

32. Morton, *Ecology without Nature*, 31 and throughout. The sense of mimesis as predatory and pro-active rather than as copy derives from Michael Taussig's *Mimesis and Alterity: A Particular History of the Senses* (New York and London: Routledge, 1993) and Tom Cohen's *Ideology and Inscription: 'Cultural Studies' after Benjamin, De Man and Bakhtin* (Cambridge: Cambridge University Press, 1998).

33. David M. Halperin, *Before Pastoral: Theocritus and the Ancient Tradition of Bucolic Poetry* (New Haven: Yale University Press, 1983), 211. Thanks to Carolyn Dinshaw for sending me to Halperin.

34. Ibid., 238.

35. Ibid., 169.

36. William Berg, *Early Virgil* (London: University of London, Athlone Press, 1974), 13–14 quoted in Halperin, *Before Pastoral*, 186.

37. John Fletcher, *The Faithful Shepherdess*, ed. Florence Ada Kirk (New York and London: Garland, 1980), 15–16.

38. Montrose, 'Of Gentlemen and Shepherds', 427.
39. Ibid., 434–435.
40. Ibid., 439–442.
41. On the permutations and celebratory connotations of Psalm 23 see Bruce R. Smith, *The Key of Green: Passion and Perception in Renaissance Culture* (Chicago: University of Chicago Press, 2008), 181–190. Hatton's letter is quoted in Agnes Strickland, *The Lives of the Queens of England* (London: George Bell and Sons, 1892), 321–322. See also Ian W. Archer et al., eds, *Religion, Politics, and Society in Sixteenth-Century England* (Cambridge: Cambridge University Press, 2003).
42. See '*Omnes et singulatim*', in Michel Foucault, *Security, Territory, Population: Lectures at the Collège de France, 1977–1978* (London: Picador, 2009).
43. Fernand Braudel, *Capitalism and Material Life, 1400–1800* (New York, San Francisco and London: Harper & Row, 1967), 68. Pastoral exists in non-Western societies – but it is not indexed necessarily to shepherds or to sheep. Haruki Murakami's *Wild Sheep Chase*, trans. Alfred Birnbaum (New York: Vintage Books, 1989) is a case in point. Sheep in Japan are a nineteenth-century Western import allied to nationalism, militarism, self-sufficiency and empire-building.
44. M. L. Ryder, *Sheep and Man* (London: Duckworth, 1983).
45. Gifford cites the work of anthropologist Michael Herzfeld, *The Poetics of Manhood: Contest and Identity in a Cretan Mountain Village* (Princeton: Princeton University Press, 1985) and also conversations with the 'current ethnographer and two young shepherds from the village'. See Terry Gifford, *Pastoral* (New York: Routledge, 1999), 13–15.
46. Halperin, *Before Pastoral*, 85–117, 96–97 (especially).
47. Giorgio Agamben, *Homo Sacer: Sovereign Power and Bare Life*, trans. Daniel Heller-Roazen (Stanford: Stanford University Press, 1998), 177–178 and 181.
48. Mascall writes an encomium of sheep; Leonard Mascall, *The First Booke of Cattell* (1591), O1v. Topsell pronounces sheep the most useful of animals, Edward Topsell, *The Historie of Foure-Footed Beastes* (London, 1607), 626.
49. Thelma Rowell, 'A Few Peculiar Primates', in Shirley C. Strum and Linda Fedigan, eds, *Primate Encounters: Models of Science, Gender, and Society* (Chicago: University of Chicago Press, 2000), 65–66.
50. Ibid., 65.
51. Ibid., 69.
52. Bruno Latour, 'A Well-Articulated Primatology: Reflections of a Fellow-Traveller', in Strum and Fedigan, eds, *Primate Encounters*, 368.
53. More, *Utopia*, 65–67.
54. *A Discourse of the Commonweal of this Realm of England* (attrib. to Sir Thomas Smith), ed. Mary Dewer (Charlottesville: University of Virginia Press, 1969).
55. For similar proverbial statements see 'The Decay of Tudor England only by the Great Multitude of Sheep', cited in R. H. Tawney and Eileen Power, eds, *Tudor Economic Documents* (London, 1953), III, 52. On Kett's rebellion, see Alexander Nevil, *Norfolkes Furies; Or A View of Ket's Campe* (London, 1615), a translation of *De Furoribus Norfolciensium Ketto Duce* (1575), D3r.
56. Ibid., K3r.
57. William Shakespeare, *Henry VI, Part Two*, ed. Roger Warren (Oxford: Oxford University Press, 2002).

58. Thomas Shadwell, *The Virtuoso* (1676), ed. David Stuart Rhodes and Marjorie Hope Nicolson (Lincoln: University of Nebraska Press, 1966), 2.2.190–194.
59. Ibid., 2.2.210.
60. For a profile of *Black Sheep*, see http://www.imdb.com/title/tt0779982/ (accessed 1 November 2009).
61. For this characterization of pastoral as a 'cessation of happening' see Fredric Jameson, 'Of Islands and Trenches: Naturalization and the Production of Utopian Discourse', *Diacritics* 7:2 (Summer 1977), 4. Note: 'Naturalization' is a misprint. The title should read 'Neutralization'. For Jameson's suggestion that Empson's *Some Versions of Pastoral* expresses the exhaustion of the Utopian as opposed to the pastoral genre, see Fredric Jameson, *Archaeologies of the Future: The Desire called Utopia and other Science Fictions* (London and New York: Verso, 2005), 12, n. 4.
62. Rowell, 'A Few Peculiar Primates', 70.
63. Vinciane Despret, 'Sheep Do Have Opinions', in Bruno Latour and Peter Weibel, eds, *Making Things Public: Atmospheres of Democracy* (Cambridge, MA: MIT Press, 2005), 360. See also, Vinciane Despret, *Quand le loup habitera avec l'agneau* (Paris: Seuill, 2002).

Part II
Political Theology and the Religious Turn

5
Invitation to a Totem Meal: Hans Kelsen, Carl Schmitt, and Political Theology

Julia Reinhard Lupton

> Thrift, thrift, Horatio! The funeral bak'd meats
> Did coldly furnish forth the wedding table.[1]
>
> *Hamlet*

The term 'political theology' has come to designate the interests of a range of scholars in early modern studies who aim to pursue critical theory (especially psychoanalysis, later deconstruction, and the Baroque meditations of Walter Benjamin) in a direction defined by secularization, sovereignty, and bio-power in the Renaissance and in contemporary life.[2] In this volume, we see signs of this interest not only in the specifically theological-political reflections mounted by Graham Hammill, Ken Jackson, and Gary Kuchar, but also in Julian Yates' provocative re-zoning of humanist pastoral around the specter of bare life, and in William West's generous yet acute critique of the religious turn in critical theory. In his essay on the Marlovian sublime, Hammill demonstrates how Marlowe rethinks and reshapes the sovereign's capacity to decide, itself linked to the creator's capacity to create, in order to stage new scenes of political making. Both Jackson and Kuchar work with new readings of St. Paul by Giorgio Agamben and Alain Badiou in order to draw out what Kuchar, following Eric Santner, calls the Epistles' 'psychotheology': the psychic conditions implied in religious thought, as well as the political-theological affinities of Freud's own thinking of subjectivity in modernity.[3] Yates looks to Giorgio Agamben not for readings of Paul but for scenes in which the human is constituted by excluding other veins of vitality, instituting the human as a state of exception with regard to other creatures. These and other scholars working in response to the challenge of political theology tend to seek out moments of skepticism, recusancy, terror, or inertia whose vertiginous suspensions of religious routine solicit the poesis of new norms and forms of life. In an essay that puts itself at odds with the so-called 'religious turn', West reads Agamben's account of the profane in order to re-zone the secular as that which 'neither stages nor answers the spectacular call from beyond', revealing a space of the everyday left over by the sacrificial cut of religion but by no means determined fully by sacred rules and rhythms. Acknowledging both the diversity and the subtlety of theoretical work being carried out in the penumbra cast by religious texts and traditions, West poses the question of 'what a secular theory worth the name would look like in this developing system of positions'.

Behind much of the current interest in states of exception and the relation of the human, the inhuman, and the political to them lie the writings of Carl Schmitt, the conservative Catholic jurist who helped conceptualize the state of emergency and the shape of dictatorship during the Weimar Republic, activities that prepared the way for Hitler's rise to power. In addition to exhuming the theological dimensions of sovereignty, Schmitt was also an astute critic of the limits of liberalism and of wars fought in the name of peace. Schmitt wrote on both *Hamlet* and Hobbes, and his postwar book *Nomos of the Earth* studies the transition from medieval to modern and late modern spatial orders, along with the forms of war each order supports. In Italy, the writings of Giorgio Agamben as well as Carlo Galli have placed Schmitt's writings in a bio-political framework associated with Hannah Arendt and Michel Foucault as they pass through the landscape of the distinctively Machiavellian Marxism of Italy.[4] The recent publication of the first complete authorized English edition of Schmitt's *Hamlet or Hecuba: The Intrusion of the Time into the Play*, translated by David Pan and Jennifer Rust with an introduction by Jen Rust and myself, promises to open up to a broader audience of Renaissance critics as well as critical theorists Schmitt's idiosyncratic, highly topical engagement with Shakespeare's most Germanic drama. At stake here are not only the details of Schmitt's *Hamlet* reading, but also the chance for more systematic consideration of Schmitt's *uses of* early modern thought as well as his *uses for* early modern studies in the light of twentieth- and twenty-first-century crises in secularization, sovereignty, and geopolitical ordering.[5]

Although I return to Schmitt in the second half of this essay, I would like to broaden the political-theological conversation by introducing a figure less well known in literary circles, a thinker, however, whose stature and import was both equal and opposite to that of Schmitt – a jurist active in the Weimar years, trained in law and philosophy, yet dedicated to constitutional norms rather than sovereign states of exception, and writing as an assimilated liberal Jew rather than a conservative practicing Catholic. The figure in question is Hans Kelsen, born in Prague in 1881, raised and educated in Vienna, and eventually immigrating to the United States, where he would teach law and political science at Harvard and Berkeley. He helped draft the Austrian Constitution of 1920, a document designed to accommodate the pluralist conditions of a multiethnic state (Kelsen 1942). In 1945, Kelsen became legal adviser to the United Nations War Crimes Commission, where he helped design the Nuremberg trials; much of his legal thought is devoted to the legitimacy of international law and the need for international judicial bodies. Kelsen died in Berkeley in 1973. During his Vienna years, from 1881 to 1930, Kelsen was a friend of Freud, a member of the Vienna Psychoanalytic Society, and an active participant in Freud's 'Wednesday meetings'.[6] At the end of 1921, Kelsen delivered a lecture on the conception of the state and social psychology that would be published in the two major Freudian journals over the next three years, first in German (1922) and then in English (1924). Although, unlike Schmitt, Kelsen wrote no essays on Shakespeare, his very first book reconstructed the political thought of Dante.[7] Kelsen's relationship to Freud has been the subject of extensive teaching and writing by Etienne Balibar; I first encountered Kelsen in a seminar led by Professor Balibar, and my thoughts about Kelsen and political theology are very much shaped by Balibar's work.

In this essay, I outline the argument of Kelsen's Freudian excursus and then turn to Schmitt's *Hamlet or Hecuba* in order to test the parameters of political theology in Renaissance studies today. Kelsen offers important constitutional and liberal correctives to the authoritarian arguments of Schmitt; constitutionalism has been by and large a missing piece in contemporary retoolings of political theology, which have tended to counteract the conservatism of Schmitt and Kantorowicz with post-Marxist strains of thought that accept the bankruptcy of liberalism as their premise.[8] The compelling reading of Renaissance sovereignty put forward by Schmitt, on the other hand, discloses the pathological supplement, the political-theological remainder, that Kelsen's normative theory of law is set up precisely to quarantine, yet to which it remains analytically open. Reading Kelsen and Schmitt together at the totemic table set by Freud may help us develop an approach to political theology for early modern studies that remains faithful to the potentialities and achievements of liberalism.

The case of Kelsen

Is the state a group? This seemingly simple question motivates Kelsen's essay, and touches immediately on the problem of citizenship, that is, of legal and subjective membership in a *politeia* or constitutional body. The first answer to the question is straightforward: of course the state is a group, even the group of groups, the group par excellence of modernity. Thus Kelsen begins his essay: 'Like all other social groups, the state, the most significant of them all, is the specific unity of a multitude of individuals, or at any rate of individual activities, and the inquiry into the nature of the state is fundamentally an inquiry into the nature of this unity' (1).[9] Yet even in this initial formulation, questions arise immediately that will continue to animate the pages that follow. Where Freud uses the word *Masse*, Kelsen opens with the word *Gebilde*, which already suggests a higher degree of structuration than the crowd phenomena that form the focus of Freud's study.[10] Is the state really a 'group like all other social groups', as Kelsen suggests at the outset, or does the form that its unity takes ultimately *distinguish* it from all other groups? Kelsen will ultimately declare that the state *is* different, thanks to the rule and role of law, taken as the binding of coercion and obligation into a *Zwangsordnung*, a compulsory order, a set of norms. While Kelsen ultimately criticizes Freud for failing to account for the specificity of state formation, he uses Freud's theory of totemism in order to map the transition from a variable to a fixed rate of group cohesion, from multitude to *politeia*, and to take note of the troubled boundary between them.

After clearing the ground of several failed sociologies of the state, Kelsen summarizes the import of the Freudian model:

> The nature of the group consists – this is the culminating point of all Freud's investigation – in the specific bond which proves to be a double affective tie of the group members to one another and to the leader. It is precisely upon this psychic character of the bond that the spontaneous ephemeral nature, the varying extent of this manifestation which Freud himself repeatedly stresses, depends. (21)

The group becomes a group precisely through a set of identifications that link the members to one another and to a leader. Freud distinguishes between two forms of relationship to other subjects: other people can become *libidinal objects*, in a relationship of love, or they can be *introjected into the ego*, through narcissistic identification, providing the outlines of the super-ego.[11] In the formation of groups around a charismatic leader, the two processes are dangerously fused: '*A primary group of this kind is a number of individuals who have put one and the same object in the place of their ego ideal and consequently identified themselves with one another in their ego*' (116). The leader is object-like insofar as the attachment to him lifts inhibitions and dissolves individuation, impoverishing the ego in the manner of a love relationship. Yet the leader according to Freud is not by and large the focus of specifically sexual impulses, and his leadership role – prescriptive, demanding, despotic – resembles the super-egoic bond with the father.

Insofar as each member of the group takes the leader as a libidinal object, in place of an ego-ideal or super-ego that would have enforced individuation, the members also identify with each other laterally, becoming a *Masse* or group. Groups formed in this way are intense and volatile, prone to waves of emotion and even violence. They are not, however – and this is the core of Kelsen's argument and his point of contention with Freud – precisely the same as the state.[12] Indeed, the organized structures that form the state prevent the forms of regressive identification exhibited by the *Masse*.[13] Organized or artificial groups – groups that remain the same over time, thanks to structures that Kelsen here calls 'institutional' – do not share the attributes of affect and regression that characterize the social groups emphasized by Freud. Moreover, groups congeal around charismatic *leaders*, whereas states form around symbolic *ideas*: 'The distinction between the primitive variable and the artificial stable groups is that between groups with an actual leader and those where the leader is replaced by an idea, and this idea is in turn embodied in the person of a secondary leader' (23). Groups led by an individual *Führer*, himself the immediate object of identification, are more archaic, labile, and regressive than groups organized around ideas, which imply greater symbolic elaboration and mediation. These 'artificial groups' also have a leader, but he is 'secondary' – not the immediate object of fascination and desire, but rather the symbolic representative of the idea that stands in place of the libidinal object in artificial groups.

The state, then, is not a group after all – certainly not a group in Freud's sense. Indeed, the state is the opposite of the *Masse*, its antidote and remedy; in Balibar's analysis, Kelsen's *Staat* installs not a form of identification but rather 'a limit of identification', 'neither "positive" nor "negative" but rather "empty"' (5). So much, it would seem, for psychoanalysis. Kelsen nonetheless does not fully abandon the affinity between group and state, finding in psychoanalysis the means of diagnosing a certain pathology that attends the state as an idea. 'The state', he writes, 'is the *guiding idea* [*die führende Idee*], which the individuals belonging to the variable groups have put in the place of their ego-ideal in order thereby to identify themselves with one another' (23; 123). As an 'idea' and not a person, the state indeed provides substantially more stability than can the atavistic, more viscerally libidinized image of a leader. Yet, insofar as the idea of the state is inserted in the mesmerizing place of the ego-ideal, it nonetheless retains the capacity to trigger magical thinking. The affective appeal of the state as a *führende Idee*, that is, can mobilize the same regressive tendencies associated with the mesmerizing personality of a *Führer*.

Writing in 1922 as a liberal constitutionalist, Kelsen has in mind not only the partisan dangers already arising on both the left and the right of the political spectrum, as well as the threat of nationalism in a multiethnic state, but also *any* projection of the state as a substance or body, as a 'trafficking active personality' other than and above its laws (35).[14] Throughout the essay, Kelsen adamantly seeks in order to destroy any substantialization of a 'crowd mentality' or 'group mind', habits of thought that he discovers in the authoritarian theories of Le Bon and Durkheim. He insists instead on individual psychology as the ground and limit to any responsible, post-metaphysical account of group psychology. Or, to echo Freud, *Massenpsychologie* must always proceed by way of *Ich-Analyse*. For Kelsen, there is no body politic, no substance accruing to and personalizing the state; what he finds himself working towards is a 'politics without a state', to match modern science's 'psychology without a soul' (36). The idea of the state is the totem of modern politics:

> In principle therefore it is the same thing when for primitive totemistic thought social unity, the combining of a multitude of individuals into a unity, can only be expressed in the visible and palpable substance of the sacrificial (totem) animal devoured in common, and when modern politics and law can only conceive of this abstract social code, this system of legal and compelling norms, the unity that is to say of the limiting social community (and the community consists solely in this code), as something of the nature of a substance, as a 'real', entirely anthropomorphically constructed 'person' ... If modern politics is primitive in this respect, the totemic system is just the politics of primitives. (35–36)

Politics, in other words, began as theology, and it *remains theological* insofar as thinkers continue to make the error of attributing a separate personified reality to the state.

A reality separate from what? A reality separate, jurist Kelsen answers, from *law*: 'Actually, there is only one object [the law], where theory has endeavoured to define and differentiate two and their relationships [the state and the law]. The state as a code of human conduct is precisely identical with the compelling code which is understood as the law or legal code' (35). For the variable 'group', Kelsen substitutes the artificial 'state'; but because 'state', as a galvanizing and ultimately libidinal idea, threatens to trigger its own totemic and regressive processes, he insists that 'state' is nothing more nor less than 'law'. 'Law' for Kelsen, unlike 'state', is not 'an idea'; it is a code of norms, of organizing obligations, rules 'in consequence of which the conduct of one individual is oriented according to the will and nature of another' (33). But is law exempt from the regressive movement into totemism that haunts the state as idea?

In a long critique of Durkheim, Kelsen argues that Durkheim and Freud, equally students of totemism, both 'hit upon the intimate relationships between man's social and religious experience' (32). Durkheim, concerned above all to justify the structure and necessity of authority, rested content with the maxim equating 'society' and 'God'. For Durkheim, the totem animal incorporates the social bond with the image of a deity; under secularization, society itself becomes the totem that embodies authority for a community bound together by shared forms of compliance. In Kelsen's account, however, Freud goes much further, into an analysis of causes: 'Freud has not contented himself with explaining that God and society are identical. He has exposed the psychic root to which both the religious and the social bond goes back, and this precisely by his endeavour to explain totemism on individual-psychological lines' (32). Kelsen then goes on to retell the story of the primal horde, culminating with the totem meal: 'The father who has been slain and devoured is in repentance elevated to a deity and that which in his lifetime he actually and practically prevented by force – the sexual intercourse of his sons with the females of the group – is transmuted in the way discovered by psycho-analysis (the so-called "postponed obedience") into the content of social and religious norms' (33). The story of the primal horde serves to explain the transformation of external coercion (exerted by the primal father on all the brothers) into internal obligation (enforced by each brother against himself, in alliance with his *semblables*). The totem meal also institutes membership in the group, formed by the consumption of common meat, and is thus an *Urszene* for citizenship. The *idea* of the state takes the place of the *idea* of the father (each understood as meaning-bodies, psychic images or *Vorstellungen* capable of libidinal investment). The system of norms, on the other hand, transmutes and displaces the externally imposed prohibitions of the father. Norms constitute a system and structure rather than an image and idea. Norms *take shape* on the other side of the totem meal, and norms *give shape* to the artificial groups that we call 'states', without themselves filling up the place of the ego-ideal. Norms provide the structuration that institutes limits to group identification and thus stabilize its disastrous mobility. It is the code, and not the image of leader or the idea of the state, that, according to Kelsen, establishes the *politeia* (a word I substitute here for 'state' in order to emphasize its constitutional nature and its distinction from mesmerizing ideas).

Yet norms, not unlike the state idea, continue to derive the power of their prescriptive *ought* from the recollected force of the father. In 'The Law as a Specific Social Technique', an essay from 1941, Kelsen continues his anthropological reflections, though no longer directly keying them to Freud. He argues that the sanctions that give laws their force in the modern state are elaborations of religious and communal forms of punishment that checked behavior in more primitive societies. Whether the threat of coercion comes from fear of 'dead ancestors', shame before one's neighbors, or the sanctions delivered by a judicial body, the *ought* of the legal norm gains its special energy from the possibility of retribution. 'Voluntary obedience', Kelsen writes, 'is itself a form of motivation, that is, of coercion, and hence is not freedom, but it is coercion in the psychological sense' – we might say, coercion in the *psychoanalytic* sense, since the totem meal names the moment that digests the despotic injunctions of the 'dead ancestor' into the systematic internal force of moral law.[15]

The moment in the legal code that marks this point of transformation is what Kelsen calls the *Grundnorm* or 'basic norm'. The *Grundnorm* is the founding point of a system of norms, the apex of validation where recourse to higher principles of authority cease, a not fully rational point kept in place by ascription, faith, or presumption. (Recall here the function of the primal signifier, S_1, or *point de capiton* in Lacanian psychoanalysis.) The recurrent example in Kelsen's writings is the giving of rules by a father to a son. A father tells his son, 'You must go to school', and the son asks him why. The correct answer according to Kelsen is not 'Because school is good for you' (a rational explanation based on the content of the precept) or 'Because I told you so' (an explanation that refers to the father's personal authority), but rather, '"Because you ought to obey the commands of your father"'. This answer refers to the principle of validity that gives prescriptive force to the words of a parent. But why should I obey the commands of my father? Because obeying parents is part of the Ten Commandments. But why should we follow the Ten Commandments? Here we approach the *Grundnorm*, the vanishing point of the legal system that derives its validity from God. According to Kelsen, the correct answer to this final question does not refer to a fact or act ('because God commanded it') but rather to the ultimate principle of validity, the norm of norms, for this system: 'because we ought to obey the commands of God'. And this founding, grounding norm, 'the normative foundation of God's authority to issue norms', is not a positive norm, but rather a 'norm presupposed in our thinking'; 'it is the basic norm of this moral-religious system'.[16]

The authority of the father's command is validated by the authority of God. Although this may sound on the face of it a bit like Durkheim and before him Filmer, Kelsen's emphasis is on the system of norms *as a system*, and not on the personal authority of any parent or legislator. This is, definitively, the rule of laws, not men, indeed, of laws and not God, even when God Himself appears as a legislator. Thus the *Grundnorm* still functions to refer the system to itself, reinforcing the circuit of validity without direct reference to an external fact. The choice of the Decalogue is not arbitrary; not only does it provide a memorable and familiar example of a moral code, but its delivery at Sinai, following the totem meal of the Pesach lamb, is also a membership-forming event, the moment in which Israel was created as a sacred nation in relation to the singularity of God and His commands as well as the events of Exodus. (The *Grundnorm* of the Decalogue is embedded in its opening statement, counted as the first commandment in the Jewish tradition: 'I am the Lord thy God, who took you out of Egypt'.[17]) Kelsen's choice of the Fifth Commandment, to honor thy father and mother, as his example of a positive norm echoes his early conversations with Freud about paternal authority. We are here very close to the totem meal, where coercion turns into obligation, the horde becomes a primitive *politeia*, and law gains its peculiar force, a force that depends not only on the possibility of judicial sanction ('law makes the use of force a monopoly of the community'[18]), but also on more distributed and dematerialized social and psychological sanctions involving shame, praise, conscience, and guilt. The Decalogue itself famously ends with the injunction against covetousness, arguably an internal failing rather than an outright transgression against the neighbor, and thus begins the work of turning the law inward.

As the end point of questioning, the *Grundnorm* presses beyond the norm into the exceptional ground or *Grund* of its own institution, while remaining a normative declaration that asserts a principle of validity, not truth. In Balibar's gloss, 'Kelsen believes that the law is an *a priori* synthesis of obligation and coercion, and that this synthesis supports itself, even if it must defend itself from the resurgence of the archaic and the theological' (7). The opacity and finality of the *Grundnorm*, its emergence at the end of the line of recursive questioning, points beyond the system it orients, disclosing the prepolitical moments of force and violence that the law is designed to overcome, yet it remains the definitive constitution of the fixed or artificial group, a protection against rather than an instance of regressive identification to the condition of the *Masse*. As such, the *Grundnorm* exists at the conceptual border between the group and the state, policing the division between theology and politics while remaining firmly on the side of the latter. Granting the state its special consistency, the *Grundnorm* keeps both the ghosts of dead fathers and contagious identifications with new ones at bay. Thus the *Grundnorm* marks for Kelsen the limit of constitutionalism, but never its abrogation or exception.

Carl Schmitt: from totem to taboo

Put otherwise, Kelsen is not Schmitt. The agnostic Kelsen was a Jew who converted to Catholicism in order (he hoped) to better pursue his work as a scholar and judge. As a liberal jurist and legal positivist, he strived to shield the purity of the law from atavistic returns to totemism, while insistently comparing secular codes to religious ones and diagnosing political thought as itself ruled by totemic impulses. His major opponent during the Weimar years was Carl Schmitt, architect of the state of exception, critic of constitutionalism, and an exegete of the theological impulses of the political. Rather than focusing on Schmitt's legal and philosophical works (especially his famous Weimar tracts, *Political Theology* and *The Concept of the Political*), I use Schmitt's foray into Shakespeare studies, *Hamlet or Hecuba: The Intrusion of the Time into the Play*, as a specifically literary scene on which to develop further the differences between Kelsen and Schmitt on questions concerning the tragic, conceived broadly as a political and philosophical as well as strictly dramatic category, and to consider the significance of both figures to a reading of *Hamlet* in the light of political theology.

Schmitt published his *Hamlet* essay in 1956 after leading a seminar on the topic at the Volkshochschule in Düsseldorf the year before. In *Hamlet or Hecuba*, Schmitt explains Hamlet's inhibitions as well as the play's wavering on the question of Gertrude's guilt by mapping the play's scenario onto the family romance of James I. James, like Hamlet, was the son of a father (Henry Stuart, Lord Darnley) who had been murdered by the man (the Earl of Bothwell) who would then marry James' mother Mary Stuart.[19] Schmitt's reading is by no means simply aberrant; Alvin Kernan, in *Shakespeare, the King's Playwright*, arrives at a similar reading in his analysis of *Hamlet* in the court of James. Schmitt's topical analysis, pursued with the shameless single-mindedness of an amateur, is simultaneously directed towards potentially more interesting terrain, namely 'the source of the tragic'.[20] Tragedy, argues Schmitt, finds its source at the juncture where 'the time', in the form of historical scandal and taboo, intrudes or breaks into 'the play', understood as the generically regulated spectacle of the theater as *Spiel* or aesthetic play.[21] Such intrusions of real time into the self-referential world of a drama create hermeneutic ambiguities and confusion (they form part of the 'problem' of problem plays); such encounters can also, however, push the work of art into a higher level of articulated complexity by bringing the drama into creative contact with troubling historical undercurrents that remain difficult to confront by other means.

Schmitt takes his title, of course, from Hamlet's First Soliloquy:

> O what a rogue and peasant slave am I!
> Is it not monstrous that this player here,
> But in a fiction, in a dream of passion,
> Could force his soul so to his own conceit
> That from her working all his visage wann'd,
> Tears in his eyes, distraction in his aspect,
> A broken voice, and his whole function suiting
> With forms to his conceit? And all for nothing!
> For Hecuba!
> What's Hecuba to him, or he to her,
> That he should weep for her?
>
> (II.ii.544–545)

For Schmitt, the Player's tears for Hecuba issue from the ducts of a continental theater shaped by both the tenets of classical representation and the state formations of modern absolutism, a world in which what

Hamlet calls 'the dejected havior of the visage' (I.ii.81) is both produced and distanced by the decorum of theatricality. Hamlet cannot weep for Hecuba because he does not brook the kinds of mediation that solicit an actor to recite the story of a Trojan queen as if it were his own, but in full knowledge that it is not, via an idiom that signals his difference from his subject matter. (Schmitt is responding here, in his own way, to the archaic stylistic signature with which Shakespeare renders the Player's Speech.) Elizabethan theater, argues Schmitt, uses methods more brutal and direct than those of Seneca or Racine; Shakespeare's plays take place on an open stage where contact among audience, players, roles, and contemporary politics remains raw and unpredictable. Hamlet, writes Schmitt, 'does not weep for Hecuba'; so, too, we spectators of the play are not 'meant to weep for Hamlet as the actor wept for the Trojan queen. We would, however, in point of fact weep for Hamlet as for Hecuba if we wished to divorce the reality of our present existence from the play on the stage' (41–42). For Schmitt, writing in response to Benjamin, *Hamlet* is no *Trauerspiel* – no metatheatrical play of mourning – but rather a genuine tragedy, a work that hits on the realities of contemporary existence, not through aesthetic reflection of what is normative but rather through the symptomatic avoidance of what is taboo.[22]

In the most powerful passage of his *Hamlet* essay, Schmitt writes:

> It is then all the more crucial to recognize that this drama, which never ceases to fascinate as a play, does not completely exhaust itself as play. It contains components that do not belong to the play and, in this sense, is imperfect as play. There is no closed unity of time, place, and action, no pure internal process sufficient unto itself. It has two major openings through which historical time breaks into the time of the play, and through which this unpredictable current of ever-new interpretive possibilities, of ever-new, yet ultimately unsolvable, riddles flows into the otherwise so genuine play. Both intrusions – the taboo surrounding the guilt of the queen and the distortion of the avenger which leads to the Hamletization of the hero – are shadows, two dark areas. They are in no sense mere historical-political connections, neither simple allusions nor true mirrorings, but rather two given circumstances that are received and respected by the play and around which the play timidly maneuvers. They disturb the unintentional character of pure play and, in this respect, are a *minus*. Nevertheless, they made it possible for the figure of Hamlet to become a true myth. In this respect they are a *plus*, because they succeeded in elevating *Trauerspiel* to tragedy. (44)

Distinguishing the entrance of historical scandal into the play from more conventional forms of allusion and allegory, Schmitt suggests that the mapping of James' unspeakable family romance onto Hamlet's situation *breaks into* and *breaks up* the literariness of the fiction. The essay's subtitle, *Der Einbruch der Zeit in das Spiel*, translated by Pan and Rust as 'The Intrusion of the Time into the Play', indicates the special role played by the real in Schmitt's account of the tragic: 'the time' is not a positive content reflected by the play in the manner of a representation, but rather a negative quotient that interrupts the play as precisely what cannot be directly depicted in it. Schmitt speaks of the real in this regard as 'the shadow of objective reality' (51) and as a 'surplus value' (45). In Schmitt's evaluation, the seriousness of tragedy and myth, unlike the playfulness of mere *Trauerspiel* and fiction, is founded on the livid collision of the real with representation, a real only provisionally kept at bay by the operation of generic conventions and their *Grundnormen*. Although Schmitt dismisses psychoanalysis as 'the death spasm of the purely psychological phase of *Hamlet* interpretation' (50), surely his reference to Freud demonstrates the operations of denegation, since there is something akin to psychoanalysis in his pursuit of taboo and its symptomology.

Such a description of literature may sound hopelessly naïve, based on an unexamined opposition between the real and representation. As Schmitt's most careful readers, who include Carlo Galli, Johannes Türk, and Adam Sitze, have argued, however, 'the time' that breaks into the play is always itself defined by conflict, and is thus structured as a negativity and not a positivity.[23] The real of *Hamlet* involves for Schmitt not only the scandal of James' family romance, but also the wars of religion that helped trigger that scandal; born of a Catholic mother, raised by Protestants, and forced to play both sides of the confessional game in order to secure his own power, James is in Schmitt's analysis 'a king who in his fate and character was the product of the strife of his age' (30). Hamlet, too, mirroring James, 'stands between Catholicism and Protestantism, Rome and Wittenberg. Even his doubt about the ghost of his father that appears to him is determined by the antithesis between Catholic and Protestant demonologies that arose from differing dogmas of Purgatory and Hell' (61). In Schmitt's judgment, Shakespeare was able to sublimate Hamlet's position 'in the middle of the schism that has determined the fate of Europe' into a modern myth, equivalent in intensity to Quixote and Faustus (52), precisely by overlaying the prince with the king. Anticipating contemporary interest in Shakespeare's relationship to Catholicism, Schmitt intuits the anxiety and formal disturbances of the play as arising in part from confessional conflict and its spectral symptoms.[24] 'Mary Stuart', writes Schmitt, 'is still for us something other and more than Hecuba' (52). By this he means that Mary cuts an historical figure whose existence as something other than literature in turn allows her to become a magnet of conspiracy, fantasy, and longing whose affective undertow reveals the constitutive fault lines in European religious and geopolitical identity.[25] The real is raw for Schmitt not in the manner of a carrot (natural, immediate, and authentic) but in the manner of a wound: exposed to view and to the elements by the fact of epochal conflict. (So Claudius to his counterpart in England: 'Since yet thy cicatrice looks raw and red / After the Danish sword' [IV.iv.63–64]).

If Schmitt's Hamlet is a priest in Protestant drag, he also likes to dress up as a pirate in his mother's closet. The Hecuba soliloquy begins with Hamlet's confession, 'O what a rogue and peasant slave am I!' A rogue is a vagrant and a vagabond, a villain and a masterless man. Rogues are *rash*, a word that haloes all of Hamlet's precipitous acts, from the slaughter of Polonius in the closet (III.iv.26) to his entrapment of Rosencrantz and Guildenstern. When 'a pirate of very warlike appointment' boards the ship that is taking Hamlet to England (IV.vii.14–15), the prince steals the occasion for his own purposes: 'Rashly – / And praised be rashness for it' (V.ii.6–7), he discovers the commission ordering his death and switches the names of his companions for his own. This romance episode unfolds on the watery stage that joins England's burgeoning maritime economy to Denmark's epic history of piracy. In 'Appendix Two: On the Barbaric Character of Shakespearean Drama: A Response to Walter Benjamin's *The Origin of the German Tragic Drama*', Schmitt distinguishes English drama from continental *Trauerspiel* by virtue of the island nation's very different geography and economy. English drama, he writes, 'belongs to the thoroughly peculiar historical evolution of the island of England, which had then begun its elemental appropriation of the sea' (59). Echoing the arguments he makes at greater length in his major postwar work, *The Nomos of the Earth*, Schmitt argues that England skipped over the 'constricted passage of continental statehood' by jumpstarting an industrial economy oriented by the sea, whose unregulated vastness disclosed a region free from law within the increasingly legal ordering of European territory. Such a geopolitics eschews the political *politesse* of continental courtiers in favor of the barbarism of pirates: 'Following the lead, first of seafarers and pirates, then of trading companies, England participated in the land appropriation of a New World and carried out the maritime appropriation of the world's oceans' (65). The barbaric character of Shakespeare's dramaturgy – mixed with respect to modes and genres, dismissive of classical unities, and refusing to weep for Hecuba – reflects in Schmitt's analysis England's at once exceptional and definitive contribution to the epochal drama of modern spatial ordering. Schmitt's Hamlet – rogue, peasant slave, and pirate too – is inhibited not by thinking too precisely on the event (IV.iv.41) but by his terrifying proximity to the very wellsprings of contemporary action.

Kenneth Reinhard argues that Schmitt's sovereign resembles the primal father in Freud: 'The primal father and the sovereign occupy the position of extreme dictators whose word both violates the rule of the total state and promises it *totality*, closure, drawing the line between the inside and the outside.'[26] Schmitt's conception of the real borders on the Lacanian in its affiliation with originary conflict, placing Schmitt opposite to Kelsen at the totem table set by Freud. Kelsen is more concerned with *totem* than *taboo*: for Kelsen, the *totem meal* represents the moment when the primal father is consumed and digested into law, while the *totem animal* represents the continued political-theological desire to incarnate the state in a personal figure, emblem, or idea, rather than accepting its status as pure law. What is at stake here in part is the tension drawn by Graham Hammill in the next essay between sovereignty (the primal father) and constituent power (the band of brothers). For Schmitt, taboo signals the proximity of historical scandal and hence the intrusion of time into the play; for Kelsen, taboo functions to install prohibition, and thus falls on the side of norm, not exception.[27]

Although Kelsen does not speak of *Hamlet*, he might have argued instead that the play forces itself to occupy the digestive side of the totem meal, despite and in response to the infernal regurgitations manifested by the Ghost.[28] The play lurches toward constitutionalism through its insistence on the elective character of monarchy and its move toward the lateral relations among 'brothers' (Hamlet, Fortinbras, Laertes, Horatio) at the expense of father-piety. Perhaps Kelsen would have suggested that the Ghost's commandment to 'remember me' becomes the *Grundnorm* of the play, equivalent to the relational statement, 'I am the Lord thy God' at the head of the Decalogue. Like Moses at Sinai, Hamlet inscribes the Ghost's words in his tablet:

> Remember thee!
> Ay, thou poor ghost, whiles memory holds a seat
> In this distracted globe. Remember thee?
> Yea, from the table of my memory

> I'll wipe away all trivial fond records,
> All saws of books, all forms, all pressures past,
> That youth and observation copied there,
> And thy commandment all alone shall live
> Within the book and volume of my brain,
> Unmix'd with baser matter.
>
> (I.v.95–104)

We extrapolate 'Remember me' as a commandment to revenge, itself a primitive instance of justice, yet its actual content in Hamlet's speech is more purely formal, instituting the father as the orienting referent in a system of new behaviors, but not himself dwelling within the world his command sets up. To be remembered is also *to be dead*, to have already been served up, and to be served up repeatedly, in the funeral baked meats of the totem meal and its ritual iterations.[29] And to remember is to erase, to remove former traces from the tables of memory in order to make room for a new inscriptive order. It may indeed be, as I have argued elsewhere, that Hamlet 'remembers' his father not by executing the command to revenge (when he does finally kill Claudius, it is part of a reactive sequence of behaviors in a script written by others rather than the execution of a deliberate sanction), but rather by moving, like Orestes, beyond revenge toward the more constitutional – and more mediated – forms of the political represented by election in the play.[30] Whereas Schmitt insists on the existential contact between art and life, between norms and exception, as the truth of *Hamlet*, Kelsen might instead have located the play in the system of validating norms that constitute a *politeia*, but with a sense of the play's special emphasis on the ambiguities and fascination – also bordering on the existential – of the *Grundnorm* as the remaining gnomic link between constituting and constituted power, between the living father and the dead one.[31] In such a scenario, Hamlet is more prince than pirate – not the New Prince of Machiavellian dictatorship, but the *princeps* as First Citizen, renewer of constitutionality as a potentiality of the political. There is something inherently 'secondary' in Hamlet's special form of princeliness, bordering on Kelsen and Freud's 'secondary leader', a mere representative, who, unlike the king or sovereign, is not himself a lightning rod of identifications, but rather a switch point in the mediating circuitry that distributes and buffers libido throughout the *politeia*.

In the spirit of Kelsen, we might also imagine the play as executing a critique of any substantialization of the state in the form of a sovereign body above and beyond the law, an exposure that culminates in Hamlet's declaration that 'The King is a thing of nothing'.[32] At the end of his Freud essay, we witnessed Kelsen linking the hypostatizing tendencies of political thought to the primitive politics of totemism. In what amounts to an anti-political-theological declaration, Kelsen declares, 'If modern politics is primitive in this respect, the totemic system is just the politics of primitives' (36). When Hamlet tells Claudius that the body is with the king, but the king is not with the body, he might be heard, with Kelsen's help, to be making a similar point. That is, Hamlet both evokes and rejects the theory of the king's two bodies associated with the writings of another political theologist, Ernst Kantorowicz. Just as Hamlet approached by way of Kelsen works to ward off the Schmittian sovereign by inscribing the *Grundnorm* of memorialization as such into the tablets of his brain, notebook, and state papers, so too, he insists on the 'nothingness' of the sovereign's mystic body, and hence questions the legitimacy not only of Claudius' rule, but also of the medieval iconography of sacred kingship that drifts around the ramparts of Elsinore like so much radioactive dust.

Of course, Hamlet's mordant flippancy, like Kelsen's stalwart positivism, may not deal such a lethal blow after all; Žižek argues, following Kantorowicz, that it is precisely the nothingness of the king's mystic body that makes him so very hard to kill.[33] Hamlet may have named the problem, but he has by no means come up with a solution. So too, Kelsen, like Hamlet, may not be able to keep the primal father off the political stage; his obscene demands and rotting charisma are with us still, and we need the thinking of Schmitt as well as Kantorowicz to understand the many forms of his returns. But this by no means renders the achievement of Kelsen superfluous or irrelevant. Acknowledging Kelsen's place at the totemic symposium allows us to raise crucial questions for and about political theology. Is there, for example, a political theology native to liberalism and to the broader tradition of constitutionalism to which liberalism belongs – an idea increasingly stressed by Miguel Vatter and other readers of biopolitics beyond sovereignty?[34] Does the citizenship tradition that stems from the *politeia* first thought by the Greeks simply encrypt bio-power by constitutively excluding *zoē* from the *polis*, or does the Western political tradition harbor antidotes and reserves in response to biopower? When and why – under what conditions – do norms become creative and emancipatory rather than simply disciplinary and coercive, developed locally and collectively rather than by appointment from above? And what is the continuing contribution of psychoanalysis to the project of defining and re-zoning the parameters of life, love, and law? In his Kelsen essay, Balibar follows Kelsen in criticizing Freud for always 'regressing to an arche' – to the myth of the primal father and his disastrous appeal – rather than trying to account for the 'synthesis or unity of assemblages' that constitutes the plural circles of modern citizenship.[35] *Politeia* means both *constitution* and *citizenship*, both an order of norms and a body of persons, capturing the objective and subjective dimensions of civic belonging. If we decide to continue thinking the possibilities of *politieia* in a liberal-democratic direction (as I certainly intend to do), we will need a figure of Kelsen's integrity as well as Schmitt's rashness in order to even begin carrying it off.

Notes

1. I would like to thank Etienne Balibar for sharing the typescript of his Kelsen essay, 'The Invention of the Superego: Freud and Kelsen 1922' (forthcoming) with me in the context of a seminar he taught at the University of California, Irvine on voluntary servitude (2005), and for his inspiration in all matters concerning the political. I would also like to thank members of the Political Theology Reading Group, the University of California, Irvine, Summer 2008, for our lively discussions of Kelsen, Schmitt, Strauss, and Kantorowicz. This essay was written in the context of my collaborations with David Pan and Jennifer Rust, who translated Carl Schmitt's *Hamlet or Hecuba: The Intrusion of the Time into the Play* (New York: Telos Press, 2009). In our introduction to that volume, Jennifer Rust and I sound the limits as well as the strengths of Schmitt's *Hamlet* reading for contemporary criticism. My comments on Schmitt in this piece are indebted to the formulations that Jen Rust and I arrived at together in the summer of 2008.
2. See the special issue of *Religion and Literature* (38:3 [2006]), ed. Graham Hammill and Julia Reinhard Lupton, on 'Sovereigns, Citizens, and Saints' for a sampling of positions and possibilities, as well as the 2000 review essay by Ken Jackson and Arthur Marotti, 'The Turn to Religion in Early Modern Studies', *Criticism* 46:1 (Winter 2004), 167–190. For a broader view of political theology beyond Renaissance studies, see Hent de Vries and Lawrence E. Sullivan, eds, *Political Theologies: Public Religions in a Post-Secular World* (New York: Fordham University Press, 2006).
3. Eric Santner's *Psychotheology of Everyday Life* (Chicago: University of Chicago Press, 2001) and *On Creaturely Life: Rilke, Benjamin, Sebald* (Chicago: University of Chicago Press, 2006) have done much to create links among political theology (with its conservative and often Christian genealogy), psychoanalysis, and Jewish philosophy around questions of life.
4. See especially Giorgio Agamben, *Homo Sacer: Sovereign Power and Bare Life* [1995] (Stanford, CA: Stanford University Press, 1998) but also his analysis of Messianism in *The Time That Remains: A Commentary on the Letter to the Romans* [2000], trans. Patricia Dailey (Stanford, CA: Stanford University Press, 2005), and the recent collection, Matthew Calarco and Steven DeCaroli, eds, *Giorgio Agamben: Sovereignty and Life* (Stanford, CA: Stanford University Press,

2007) which includes a short essay by Agamben on the life of the multitude from Aristotle to Dante.

5. On Schmitt and Benjamin, see Samuel Weber, 'Taking Exception to the Decision: Walter Benjamin and Carl Schmitt', *Diacritics* 22:3/4 (1993), 5–23. On Schmitt and Kelsen in Weimar, see David Dyzenhaus, 'Legal Theory in the Collapse of Weimar: Contemporary Lessons?', *American Political Science Review* 91:1 (1997), 121–135. For a series of conceptual and historical comparisons between Schmitt and Kelsen, largely conducted in favor of Kelsen, see Dan Diner and Michael Stolleis, eds, *Hans Kelsen and Carl Schmitt: A Juxtaposition* (Gerlingen: Bleicher, 1999). Schmitt's *The Nomos of the Earth in the International Law of the Jus Publicum Europeaum* was translated by G. L. Ulmen (New York: Telos Press, 2003).

6. Biographical background on Kelsen is taken from Nicolletta Bersier Ladavac, 'Hans Kelsen (1881–1973): Biographical Note and Bibliography', *European Journal of International Law* 9 (1998), 391–400; Rudolf Métall, *Hans Kelsen: Leben und Werk* (Vienna: Deutike, 1968), and Clemens Jabloner, 'Kelsen and His Circle: The Viennese Years', *European Journal of International Law* 9:2 (January 1998), 368–385.

7. Hans Kelsen, *Die Staatslehre des Dante Alighieri* (Vienna and Leipzig: F. Deuticke, 1905).

8. I have attempted to forge a more constitutional direction for political theology in *Citizen-Saints: Shakespeare and Political Theology* (Chicago: University of Chicago Press, 2005). In 'Citizenship Beyond the State: Thinking with Early Modern Citizenship in the Contemporary World' (*Citizenship Studies* 11:2 [2007], 117–133), Andrew Gordon and Trevor Stack have taken the mixed and emergent paradigm of early modern citizenship as a template for thinking about citizenship in the contemporary world. On Shakespeare and citizenship, see also Oliver Arnold, *The Third Citizen: Shakespeare's Theater and the Early Modern House of Commons* (Baltimore: Johns Hopkins University Press, 2007). Like Kelsen and Freud, Arnold is interested in how 'the right to consent *as* freedom per se joins liberty and subjection in a novel political-psychic calculus: because the subject consents to the law, he must obey them; the subject knows he is free because he obeys' (34).

9. The German and English versions of the text, published in 1922 and 1924 respectively, are not identical, and may reflect further conversations between Freud and Kelsen. I am using the English text as my main source in order to clarify or illuminate the use of words such as 'group'. Hans Kelsen, 'The Conception of the State and Social Psychology With Special Reference to Freud's Group Theory', *International Journal of Psycho-Analysis* 5 (1924), 1–38. First published in German as '*Der Begriff des Staates und die Sozialpsychologie. Mit besonderer Berucksichtigung von Freuds Theorie des Masse*', Imago 8:2 (1922), 97–141.

10. The phrase that appears as 'social groups' in the opening line of the English text (1) appears as '*soziale[n] Gebilde*' in the German text. In *Group Psychology and the Analysis of the Ego* [*Massenpsychologie und Ich-Analyse*], Freud usually uses the word *Masse*, which, as the translator explains, imperfectly combines the French word '*foule*' (crowd) from Le Bon and the English word 'group' from McDougall, texts that Freud adumbrates in his study. Sigmund Freud, *Group Psychology*, in *Standard Edition of the Psychological Works of Sigmund*

Freud, trans. James Strachey, 24 vols. (London: Hogarth, 1974) [*SE*] XVIII, 69n. Elsewhere, Kelsen poses the question in more directly Freudian terms, the punctuation indicating the distance between their approaches: 'ob der Staat als eine "psychologische Masse" der durch die Freudsche Psychanalyse aufgehellten Strucktur angesehen werden kann' (114; cf. 109, 118).

11. Freud, *Group Psychology*, 113 (hereafter cited parenthetically in the text).

12. 'What is called the state', Kelsen writes, 'is quite different from the phenomenon that was described by Le Bon as the "group" and explained psychologically by Freud' (20). 'Was man "Staat" nennt, ist etwas gänzlich anderes als jenes Phänomen, das als "Masse" von Le Bon geschildert und von Freud psychologisch erklärt wurde' (Kelsen 1922, 119).

13. Thus Kelsen writes: 'it is precisely these [regressive affective] ties that are *lacking* in the individual who is a member of the groups called "organized" or "artificial" by McDougall and Freud, since in them the characteristic regression, for whose sole explanation these affective ties, this libidinal structure, had to be adduced, is *wanting*' (22).

14. See Raphael Gross, *Carl Schmitt and the Jews: The 'Jewish Question', the Holocaust, and German Legal Theory* (Madison, WI: University of Wisconsin Press, 2007): 'The backdrop to Kelsen's efforts at a formalization of law and his fight against any efforts at substantializing constitutional law – say, through monarchs, or through the *Reichspräsident* as a substantial embodiment of the political unity of the *Volk* – would appear to be the historical situation of multi-ethnic Austro-Hungary. Its continued existence was best assured through a purely formalized type of rule, and not through the construction of a substantially-determined public interest' (103). Gross' piece compares the political theologies of both Schmitt and Kelsen in order to demonstrate the anti-Semitic and anti-Jewish elements not only of Schmitt's critique of Kelsen, but also of his rejection of law in general and his substitution of the more substantialized *Nomos*, an argument developed further in his book, *Carl Schmitt and the Jews*. Although Gross' emphasis on 'substance' in Schmitt should be supplemented by Carlo Galli's emphasis on conflict and *polemos*, the very polemic of his own stance is a bracing and important corrective for contemporary readers attracted to Schmitt. Carlo Galli, 'Presentazione dell'edizione italiano', *Amleto o Ecuba*, trans. Simone Forti and ed. Carlo Galli (Bologna: Il Mulino, 1985), 7–35.

15. Hans Kelsen, 'The Law as a Specific Social Technique', *The University of Chicago Law Review* 9:1 (December 1941), 80, 79.

16. This version of the scene is taken from Kelsen, 'On the Basic Norm', 108, an address delivered to the Boalt Hall Students Association in 1959 (*California Law Review* 47:1 [1959], 107–110). He uses the same example in 'The Function of a Constitution', *Essays on Kelsen*, ed. Richard Tur and William L. Twining (Oxford: Clarendon Press, 1986), 112. On the *Grundnorm* in relation to Kelsen's democratic theory, see Andreas Kalyvas, 'The Basic Norm and Democracy in Hans Kelsen's Legal and Political Theory', *Philosophy and Social Criticism* 32:5 (2006), 573–599.

17. On the Ten Commandments and psychoanalysis, see Kenneth Reinhard and Julia Lupton, 'The Subject of Religion: Lacan and the Ten Commandments', *Diacritics* 33:2 (2003), 71–97.

18. Kelsen, 'The Law as a Specific Social Technique', 81.

19. In her beautiful study of Britain's romance with Mary Queen of Scots from the reign of Elizabeth to the reign of Victoria, Jayne Lewis suggests that Freud's theory of totemism, reoriented around the guilty mother, can eluci-date 'the schism of desire' passed from the Elizabethans to their children and grandchildren to become, for many, a 'paradoxical condition of their shared fictions of integrated collective and personal identity' (*Mary Queen of Scots: Romance and Nation* [London: Routledge, 1998], 7).

20. Schmitt, *Hamlet or Hecuba*, 5. Subsequent references are given parentheti-cally in the text.

21. For expansions of this and other points in this summary of Schmitt, see Rust and Lupton, 'Schmitt and Shakespeare', in Schmitt, *Hamlet or Hecuba*, xv–li.

22. By *Trauerspiel* (literally 'play of mourning'), Schmitt has in mind not only the German vernacular word often translated as 'tragedy' and roughly syn-onymous with the Greek loan word, *Tragödie*, but also Walter Benjamin's polemical reclamation of the specificity of *Trauerspiel* as the form taken by tragedy in modernity. See Benjamin, *The Origin of the German Tragic Drama* [1928] (New York and London: Verso, 1998) and Schmitt's explicit responses to Benjamin in *Hamlet or Hecuba*.

23. Jennifer Rust and I summarize these positions, 'Schmitt and Shakespeare', xxiii–xxxv. Victoria Kahn criticizes Schmitt for opposing 'real action' to 'mere literary invention' ('Hamlet or Hecuba: Carl Schmitt's Decision', *Representations* 83 [2003], 67–96, 69). Johannes Türk argues that the aesthetic in Schmitt ultimately serves to capture and display the difference between politics and play ('The Intrusion: Carl Schmitt's Non-Mimetic Logic of Art', *Telos* 142 [2008], 73–89). Carlo Galli argues that Schmitt's Hamlet is not a myth of origins but a figure of fissure ('Amleto o Ecuba', 34).

24. On Shakespeare and Catholicism, see especially Richard Wilson, *Secret Shakespeare: Studies in Theatre, Religion and Resistance* (Manchester: Manchester University Press, 2004); Wilson's monograph is of interest here not least because it deploys a topical method not unlike Schmit's own. See also Richard Dutton, Alison Findlay, and Richard Wilson, eds, *Theatre and Religion: Lancastrian Shakespeare* (Manchester: Manchester University Press, 2003); and Stephen Greenblatt, *Will in the World: How Shakespeare Became Shakespeare* (New York: W. W. Norton, 2004) and *Hamlet in Purgatory* (Princeton: Princeton University Press, 2001).

25. Cf. Lewis, *Mary Queen of Scots*.

26. Kenneth Reinhard, 'Toward a Political Theology of the Neighbor', in Slavoj Žižek, Eric Santner, and Kenneth Reinhard, *The Neighbor: Three Inquiries in Political Theology* (Chicago: University of Chicago Press, 2005), 11–75 (56).

27. 'Primitive peoples' behavior conforming to the social order, especially the observance of the numerous prohibitions called "taboos", is determined principally by the fear that dominates the life of such peoples.' Hans Kelsen, *General Theory of Law and State* (New York: Russell and Russell, 1945, 1961), 18.

28. 'The sepulcher …Wherein we saw thee quietly inurn'd, / Hath oped his pon-derous and marble jaws, / To cast thee up again' (*Hamlet*, I.iv.48–51).

29. Kelsen gives revenge as an early example of the 'social technique' of law: 'Only later, at least within the narrower group itself, do there appear, side by side with transcendental sanctions, sanctions that are socially immanent,

that is to say, socially organized, to be fulfilled by the individuals according to the provisions of the social order. In relations among the groups, revenge appears very early as a reaction against an injury considered unjustified and due to a member of the group.' 'The Law as a Specific Social Technique', 77.

30. Julia Reinhard Lupton, 'Hamlet, Prince: Tragedy, Citizenship, and Political Theology', in Diana Henderson, ed., *Alternative Shakespeares 3* (London: Routledge, 2008), 181–203.

31. For a reading of Schmitt that emphasizes its normative – Kelsenian – dimensions, see Andreas Kalyvas, 'Hegemonic Sovereignty: Carl Schmitt, Antonio Gramsci and the Constituent Prince', *Journal of Political Ideologies* 5:3 (2000), 343–376.

32. On the doctrine of the king's two bodies in *Hamlet*, see Jerah Johnson, 'The Concept of the King's Two Bodies in *Hamlet*', *Shakespeare Quarterly* 18:4 (Autumn 1967), 420–434. For critiques of Kantorowicz, see Lorna Hutson, 'Not the King's Two Bodies: Reading the Body Politic in Shakespeare's *Henry IV*', in *Rhetoric and Law in Early Modern Europe* (New Haven: Yale University Press, 2001), 166–198; and David Norbrook, 'The Emperor's New Body? Richard II, Ernst Kantorowicz, and the Politics of Shakespeare Criticism', *Textual Practice* 10:2 (1996), 329–357. Hamlet's phrase has become a leitmotif in recent reappropriations of Kantorowicz for analyses of contemporary politics; see Slavoj Žižek, *For They Know Not What They Do: Enjoyment as a Political Factor* [1991] (New York and London: Verso, 2002), 229–277 and Claudia Breger's response to Žižek, 'The Leader's Two Bodies: Slavoj Žižek's Postmodern Political Theology', *Diacritics* 31:1 (Spring 2001), 73–90.

33. Žižek, *For They Know Not What They Do*, 253–277.

34. Miguel Vatter, 'Strauss and Schmitt as Readers of Hobbes and Spinoza: On the Relation between Political Theology and Liberalism', *New Centennial Review* 4:3 (2004), 161–214.

35. A further question I cannot even begin to broach here would involve the role that internationalism plays in the thinking of both Kelsen and Schmitt; a key link here for the purposes of early modern studies is the text of Dante's *De Monarchia*, the subject of Kelsen's earliest publication and a recurrent figure in Schmitt's *Nomos of the Earth*.

6

The Marlovian Sublime: Imagination and the Problem of Political Theology

Graham Hammill

Political theology might best be understood to designate a link between sovereignty and constituting power. The twentieth-century German jurist Carl Schmitt implies this understanding in his chapter, 'Political Theology', in his book with the same title. Schmitt adduces political theology in order to ague against the eighteenth-century French writer, Emmanuel Joseph Sieyès. Sieyès attempts to legitimate the role of the Third Estate in the French Revolution by drawing a fundamental distinction between constituting power, the power to make a new political order, and constituted power, the power to preserve the already established legal system. Associating constituting power with the nation, Sieyès goes on to propose that the nation inhabits a state of nature that precedes any social or legal bonds and, therefore, has the authority and legitimacy to remake already existing social and legal bonds.[1] In Schmitt's analysis, political theology disproves the location of constituting power in what he disparagingly calls the 'organic unity' of the people.[2] 'All significant concepts of the modern theory of the state are secularized theological concepts', he asserts, 'not only because of their historical development ... but also because of their systematic nature'.[3] Just as theology needs a God who can intervene in and suspend the natural order, so too does the state need a sovereign who can stand above the law, suspend it, and decide exceptional emergency situations – that is, 'decide where there is an extreme emergency as well as what must be done to eliminate it'.[4] The 'systematic nature' of this analogy between God's and the sovereign's prerogative enables Schmitt to argue that the people do not and cannot have the same agency, or what Schmitt calls the decisionist character, as the sovereign. Constituting power must be located in the personal authority of the sovereign because the people as an organic unity cannot make decisions outside the bounds of the law.

But Schmitt's argument is reversible. His strenuous effort to contain constituting power with the figure of the sovereign also suggests that the theologico-political sovereign is a kind of reaction formation that attempts to resolve a prior rupture between sovereignty and constituting power. Developing this implication through a critique of Schmitt, Antonio Negri argues that constituting power should be understood as a crisis that displaces the dominant terms of the constituting order. For this reason, constituting power cannot and should not be understood within the terms of that order. Neither sovereign (Schmitt's answer) nor nation (Sieyès' answer) will suffice in explaining constituting power as 'absolute process'.[5] Rather, constituting power is nourished by the very absence of terms within the constituted order by which it might find determination or finality. In *Insurgencies*, a survey of radical political thought that runs from Machiavelli to Lenin, Negri argues that this definition of constituting power as crisis emerges in the Renaissance. Once Machiavelli argues for the legitimate constitution of the state *ex novo*, he poses the problem of constituting power even before Sieyès names it. As Machiavelli proposes, political making cannot be judged by the moral protocols of the established order – Romulus cannot be considered immoral for murdering Remus even though the murder of a brother was an affront to Roman law – because the moral protocols of an established order are there to ensure obedience. Political making, by contrast, effectively breaks with older systems to create a new order.[6] Negri's primary objective is to show how constituting power assumes a utopian horizon, a yet-to-be realized telos that exceeds the grasp of individual political thinkers but also takes shape in the history of radical political thought that he develops. But for Renaissance studies, Negri's more significant insight is that constituting power gets associated with imagination. Especially in his reading of the seventeenth-century English political writer James Harrington in *Insurgencies* and in his longer analysis of Baruch Spinoza in *The Savage Anomaly*, Negri shows how imagination turns against sovereignty, exposing the link between sovereignty and constituting power that defines political theology to be something more like an aporia.[7]

In this essay, I will explore the role that imagination plays in late sixteenth-century Catholic political writing, which uses imaginative means to outmaneuver and recuperate relations between the sovereign decision and the public enemy, and then in some of Christopher Marlowe's works, particularly but not solely his erotic poem *Hero and Leander*. Marlowe formulates sovereignty and the political subject through Catholic critiques of Elizabethan prerogative. This does not mean that Marlowe was particularly sympathetic with the Catholic position. Whereas the Church aimed to place moral limits on prerogative, thereby claiming its proper role in relation to the monarchy, Marlowe uses critical accounts of the sovereign's capacity to decide to explore constituting power as both a political and an aesthetic form of making that emerges through crises in moral thought. My argument will proceed in three stages. First, I will show how Catholic propaganda from the early 1580s duplicates Elizabethan structures of sovereignty, unwittingly disclosing the place of imagination and fiction-making within those structures. Then, reading *Hero and Leander* in the context of Catholic critiques of unlimited prerogative, I will argue that Marlowe focuses on bad decision-making in order to posit a political subject driven by passion and imagination. Finally, through a more formal reading of the poem and other of Marlowe's works, I will suggest that Marlowe repeatedly turns imagination against sovereignty in ways that go beyond Elizabethan and Catholic theories of sovereignty and show aesthetics itself to be a competing figure for constituting power.

Imagination and the public enemy

In Elizabethan controversial literature, prerogative was intimately bound up with the figure of the public enemy. As a legal term, prerogative names the monarch's right to make political, ecclesiastical, and judicial decisions outside the bounds of law and parliamentary advice. At issue in the concept is legitimacy. Marlowe's word for prerogative in *The Jew of Malta* is 'arbitrament', which suggests that what prevents arbitration from seeming arbitrary is the self grounding authority of the arbitrator, the one who decides.[8] But for English Catholics like Cardinal William Allen, and for Marlowe as well, it would be more accurate to say that what prevents prerogative from seeming arbitrary is the naming and condemning of the public enemy. In his *Defense of English Catholics*, published in 1584, Allen argues that Elizabeth wrongfully consolidates her position as head of church and state by condemning English Catholics as traitors. Elizabeth's principal adviser, Burghley, justifies this treatment of Catholics in *The Execution of Justice* (1583/4) on the grounds that Elizabeth's decision is both prudent and legal. Not only are Catholics really 'seditious' sowers of 'rebellion', but also the decision to treat them as such is based on Elizabeth's prerogative which is granted by 'God's goodness'. In response, Allen offers a subtle account of the way in which unchecked prerogative is authorized by the very fiction of the public enemy which prerogative also authors, arguing that

Catholic treason is nothing less than a 'fiction' manufactured by 'the politiques of our country' to secure 'their new state'.[9]

Allen understands very well the use of fiction to produce political truth because in his own *True Report*, published the year before his exchange with Burghley, he uses the fiction of the public enemy in a similar way. The real enemy, Allen argues, the most dangerous sower of 'fiction' and 'forgerie', is the intelligencer who, like Judas, betrays the Church to the state for money. What follows in the *True Report* is a series of written confessions by John Nichols, Laurence Caddey, Edward Osberne, and Richard Baines, printed and made public in order to confirm what Allen calls 'the singular benefits of [God's] truth'.[10] These confessees admit their efforts to subvert English Catholicism and proclaim their conversion or return to the faith. A glance at Allen's private letters shows that at least some of these confessions were produced by torture, but the *True Report* presents decorative letters as the site where the body is made to produce political truth.[11] Both the decorative 'A' that begins Baines' confession and the decorative 'I' that begins Osberne's hold confessors' bodies in place as their hands follow the shape of the letter, suggesting confession as enforced conformity to the truth that is being uttered (Figures 5 and 6).

I don't mean to suggest that these are actual portraits of Baines or Osberne. The decorative 'A' is taken from the Rheims *New Testament*, published a year before the *True Report*. It is tempting to speculate that the decorative 'I' was made specifically for the *True Report* as an image of a confessee held in place by the first-person pronoun, but it's likely that the letter was cut from an earlier book as well. More pointedly, I mean to argue that in the *True Report*, these decorative letters duplicate Allen's later response to Burghley that the public enemy is, first of all, a political fiction which is subsequently made to be true. In the context of the *True Report*, the letters stand as chastening fictions, serving the local function of correcting linguistic duplicity, ornamental remedies to intelligencers' efforts to seduce 'some of the yonger sort ... into discontentment & to mislike of rule and discipline' by using 'novelties of wordes joyned with pretty proverbs, termes, and mocking taunts'.[12] More broadly, as instruments of Catholic propaganda, these letters counteract the use of portraits of Tudor monarchs to decorate initial characters in a variety of legal and other documents, showing the majesty of the sovereign body as the force authorizing the text that follows. For example, in the 1563 edition of *Acts and Monuments*, John Foxe begins his discussion of Constantine with a decorated 'C' which encloses Elizabeth, whose throne rests on the torso of the Pope.

THE
CONFESSION OF RICHARD
BAINES PRIEST AND LATE STV-
dent of the Colledge of Rhemes, made after
he vvas remoued out of the common
gaile to his chamber.

S my miferie & vvickednes vvas greate
which I vvill now fet downe to the
publifhing\of my ingratitude to God,
the Church, and my fuperiors, fo vvas
Gods iuftice, mercy and prouidence
meruelous towards me to faluation as
I vereiv hope. Of al vvhich to the glory of Chrift, and
fatisfaction of the holy Church and all her children
whom I haue offended or fcandalized,& to mine owne
vvorthy confufion temporall, I intend to make this my
publike confeffion, that al that ftand, may by my exaple
beware of a fall,and fuch as be fallen may thereby make
haft to aryfe againe.

The very ground of my fall and of al the wickednes
ether committed or inteded, was my pride which droue
me to a lothfomenes to liue in order and obedience, to
conceipts of mine owne vvorthines and manifold dif-
contentement of the fchollarlike condition vvherein I
liued, to an immoderat defire of more cafe, welth, and
(which I fpecially alfo refpected) of more delicacie of
diet and carnal delits then this place of banifhment was
like to yeld vnto me, though (vvo vnto me that could
not fee fo fare before)the ftudets ftate in the Seminarie,
vvhere

Figure 5   William Allen, *A True Report of the Late Apprehension and Imprisonment of John Nichols* (Rheims, Printed by John Fogny, 1583), 24v. © The British Library Board. 4903.aa.27

As Richard McCoy describes it, this 'C' becomes an emblem of the presence chamber, which authorizes and is authorized by Elizabeth's sacred role as supreme governor of the English Church.[13] Responding to Foxe and others, the initial characters in the *True Report* suggest

THE
SATISFACTION OF EDWARD
OSBERNE PRIEST, TOVCHING
his frailties, and fall from the Catholike
Church, at his being in
England.

 T is almightie Gods great wisedom &
mercy, that he hath besides his other iu-
ste iudgements either téporall or eternall vpon sinners, ioyned also for most
parte some great affliction and tor-
ment of minde for a continuall inward
checke and chasticement of their offences euen in them
selues, and in this life: which caused a great Clearke to
saie, *Thou haste commaunded good Lord, and so it is, that eue-*
ry vnordinate appetite should be a punishment to it self.

But no sinne breadeth this internall vexatió so much,
as that which is committed against a mans owne skill
and conscience, specially the voluntarie forsaking of
that faith, truth, and religion vvhich God by his spirit
in the *Holy Church,* and sacred word, hath made him par-
taker of. I speake not of his case that impugneth of ma-
lice the knowne truth, as many arch-heretikes haue
done, & wickedly do, for that is a sinne against the holy
Ghost, & such often carieth about with them such hel-
like torments of conscience and desperation, that they
may be thought to beginne their dánation euen in this
life : but I meane of others only, vvho by frailtie of the
flesh, feare of worldly distresses, doubte of temporall
torments, and disgraces, or somme other humane infir-
mitie, be often driuen to yeld in somme points to the
threates,

Figure 6   William Allen, *A True Report of the Late Apprehension and Imprisonment
of John Nichols* (Rheims, Printed by John Fogny, 1583), 27v. © The British Library
Board. 4903.aa.27

that it is not the majesty of the sovereign but the fictional body of
the enemy that justifies the text's claims to truth, an enemy that – for
Allen – the Church and not the state has the ability and authority to
identify.[14]

It's not particularly surprising that Allen relies on the very structure he will soon charge Elizabeth with unjustly producing, since in the early 1580s he and other English Jesuits were arguing for shared or divided sovereignty. By duplicating the ability of the monarch to identify the enemy while *correctly* identifying the *true* enemy, Allen is able to argue that the Church should share sovereignty with the state and offer moral counsel when the monarch makes bad or inappropriate decisions. What is surprising is that, in duplicating the ability of the monarch to identify the enemy, Allen suggests an imaginative component to the sovereign's decision. Although Allen would be loath to admit it, the decorative letters in the *True Report* imply that what substantiates the enemy as truly a public enemy is the routing or channeling of imagination. The *True Report* opens with the image of an intelligencer, sneaking behind the letter 'G' of the opening phrase, 'Good Christian reader', the relation between image and text showing the intelligencer to be a counterfeiter who dissimulates by creeping around or behind language, using the naïvety and bad reading practices of good Christians as his cover (Figure 7).

The function of this decorative letter is to raise the general possibility in the imagination of the book's readers that the good and Christian world which they inhabit is threatened by some unseen and untrustworthy enemy. This is the lesson of the *True Report*'s opening sentence: 'Good Christian reader, the children, and specially the Priests of Gods Church, have ben manifoldly assailed by their adversaries in our countrey these later yeres.' As the argument proceeds, that imagined threat becomes increasingly real as Allen offers up specific examples of the 'conscienceles men' who threaten English Christendom.[15] The pleasure of reading the *True Report* (a slim pleasure, to be sure) is the pleasure of discovering this 'true' enemy, the pleasure of moving from imagination to reality, a movement which substantiates the authority of the Catholic Church as a partner in government while simultaneously exposing the limitations – and implicitly, the hypocrisy – of the Elizabethan state for trusting such unseemly characters.

Marlowe would have been aware of these accounts from his work as an intelligencer at Allen's seminary at Rheims in the early 1580s, when he was spying on Catholics for the government, and in the later 1580s and early 1590s, through his association with some of the intelligencers whom Allen exposes, Baines in particular. Allen and other Catholic controversialists were among Marlowe's main sources for thinking about prerogative and the state more generally, but Marlowe also established a unique and innovative relation to that literature. Rather than outing the enemy in the service of sovereignty, church or

THE PREFACE.

ood Chriſtian reader, the children, and ſpecially the Prieſts of Gods Church, haue ben manifoldly aſſailed by their aduerſaries in our cou_ trey theſe later yeres : firſt by the vvriting and preaching of the Sea_ maiſters, vvhich made no great im_ preſſion : Secondly by authoritie of the Ciuil Magiſtrate , vvhich vvas more forceible, but yet preuailed no further then to the loſſe of ſome rich_ mens tranſitory goods,&a fevv poore mens temporal liues, neither the one nor the other periſhed to the ovvners but both laid vp vvith Chriſt and be_ ſtovved vpon him to the hundreth fould aduantage in the next, and to

A ij

Figure 7   William Allen, *A True Report of the Late Apprehension and Imprisonment of John Nichols* (Rheims, Printed by John Fogny, 1583), sig. Aii r. © The British Library Board. 4903.aa.27

state, Marlowe explores prerogative from the position of what Allen calls 'Falsebrethren', double agents who falsify brotherly love on both sides of the divide, 'traiterously' slandering English Catholics while abusing the honor of the Privy Council by producing false evidence.[16]

By this, I don't just mean that Marlowe was a spy, which indeed he was; I also mean that Marlowe approaches the conceptualization of the emerging state as a kind of literary double agent, writing neither in support of the Elizabethan state nor in support of Catholic political critique, but mobilizing Catholic critique to reconceptualize Elizabethan prerogative by provoking and staging imagination. We might take the decorative letter 'G' as emblematic of Marlowe's strategy. Marlowe assumes a fundamental division between Catholic and Protestant claims; rather than taking sides, however, he stages that division in order to play one side against the other, sneaking in and around moral language, writing both more and less than 'good Christian readers' would care to read.

Hero's decision

One reason Marlowe's strategy has not been particularly obvious in *Hero and Leander* has to do with textual editing. All editions of the poem agree on the narrative sequence up to the very end, when Hero and Leander 'obtaine' a 'truce' through 'gentle parlie' (762). Afterward, in the earliest edition of the poem, published by Edward Blount in 1597/8, Hero decides to have sex with Leander *before* Leander attempts to persuade her. In his 1910 edition of Marlowe's *Works*, Tucker Brook reverses the order, having Leander speak before Hero makes her choice, so that, instead of choosing to have sex with Leander, Hero is forced into it. Since Tucker Brooke's 1910 edition of Marlowe's *Works*, almost all editors of *Hero and Leander* have rearranged the final sex scene between the two lovers, making a simple scene of coercion out of what in Blount is a complex scene of decision-making and consent.

The textual problem has to do with two lines in Blount's edition which I have italicized that make up only half of a conceit:

> Loue is not ful of pittie (as men say)
> But deaffe and cruell, where he means to pray.
> *Euen as a bird, which in our hands we wring,*
> *Foorth plungeth, and oft flutters with her wing.*
> And now she wisht this night were neuer done,
> And sigh'd to think upon th'approching sunne.

In his 1821 edition of *Hero and Leander*, Samuel Weller Singer calls these lines an 'awkward excrescence' and moves them so that they come twenty lines earlier, at the beginning of the sex scene instead of at the end, serving now as a conceit for Hero's imagination:[17]

> *Even as a bird, which in our hands we wring,*
> *Foorthe plungeth, and oft flutters with her wing,*
> She trembling strove, this strife of hers (like that
> Which made the world) another world begat,
> Of unknown joy.

(772–777)

Tucker Brooke accepts Singer's emendation (as have all editors after Singer) and, in an act of editorial bravado, goes on to speculate that the lines were misplaced in Blount's edition because Blount got the pages of Marlowe's manuscript mixed up.[18] Based on this theory, Tucker Brooke moves ten lines from the end of the sex scene to the beginning. With the exception of Richard Sylvester's anthology of *English Sixteenth-Century Verse*, every modern edition of *Hero and Leander* that I know follows Tucker Brooke's re-ordering – including L. C. Martin's 1933 *Complete Poems*, Stephen Orgel's 1971 *Complete Poems and Translations*, Fredson Bowers' 1973 *Complete Works*, Roma Gill's volume in the 1987 *Complete Works*, and the widely circulating *Norton Anthology of English Literature*, which follows both Martin and Bowers.[19] This, even though the very few editors who actually address the problem agree that the conjectures upon which Tucker Brooke's decision rests are not particularly compelling. Gill's judgment can stand in as a statement of editorial consensus. 'Whatever the reasons for the error, there is no doubt about the need for correction, and the appropriateness of Tucker Brooke's action.'[20] The obvious problem here is that the only reason to accept the 'appropriateness' of Tucker Brooke's action is to assume that Marlowe really wanted to write a rape scene.[21] Gill's affirmation is a product of critical interpretations of the poem, themselves based on Tucker Brooke's emendation, which argue that Marlowe's poem variously authorizes, critiques, or subverts predatory male desire by showing its deleterious effects on women.[22]

Focusing on Blount doesn't reverse these readings so much as shift the question from male desire to imprudent decision-making. In Blount, Hero decides to have sex with Leander, driven by imagination to commit herself to a decision which turns out to be against her own self-interest:

> She trembling strove, this strife of hers (like that
> Which made the world) another world begat,

Of unkowne joy. Treason was in her thought,
And cunningly to yeeld her selfe she sought.
Seeming not woon, yet woon she was at length,
In such warres women use but halfe their strength.
Leander now like Theban *Hercules*,
Entred the orchard of *Th'esperides*,
Whose fruit none rightly can describe, but hee
That puls or shakes it from the golden tree:
Wherein *Leander* on her quivering brest,
Breathlesse spoke some thing, and sigh'd out the rest;
Which so prevail'd, as he with small ado,
Inclos'd her in his armes and kist her to.
And everie kisse to her was as a charme,
And to *Leander* as a fresh alarme.
So that the truce was broken, and she alas,
(Poore sillie maiden) at his mercy was.
Love is not ful of pittie (as men say)
But deaffe and cruell, where he meanes to pray.
Even as a bird, which in our hands we wring,
Foorth plungeth, and oft flutters with her wing.
[   ]
And now she wisht this night were neuer done,
And sigh'd to thinke upon th'approching sunne.

(Blount 763–786)

What makes Blount's version of this passage so fascinating is that Hero's defeat is more a matter of her own decision than it is of Leander will. At the very moment that Hero asserts the capacity to decide whether or not to have sex with Leander, she also becomes her own worst enemy, internalizing the enmity bound up with prerogative as 'treason ... in her thought'. Hero's decision is informed by the complex role of imagination that early modern political writers foreground in discussions of prudence. Because prudence is a form of reasoning based on experience and not pure rationality, so the argument goes, it often demands the capacity to imagine situations which have not been experienced even as it is also susceptible to being overruled by the imagination of those experiences. That is, while imagination enables prudence, it also misdirects prudential reasoning toward false conclusions. For this reason, political writers encourage reading as a way to limit imagination. Realizing that 'Proper Prudence ... can hardly be tied to precepts', in his *Six Bookes of Politickes* Lipsius goes on to cite sententia gleaned from a wide array of literary and historical works in order to offer, if not precepts, then statements that might strike and rein in the imagination by tying it to wisdom derived from previous, practical experiences.[23] In focusing on Hero's bad decision, Marlowe raises ethical considerations that go beyond the strictly prudential. What are the conditions and scenarios in which one might act contrary to one's own self-interest? And what combination of imagination and passion might drive one to betray oneself? At the same time, by casting Hero's decision in terms of treason and war, Marlowe introduces political considerations as well. Under what conditions might a sovereign break her oaths and truces? Under what conditions might a sovereign be charged with treason? And what are the political models that might explain this paradoxical situation?

We can find an answer to the last two questions in the *Admonition to the Nobility and People of England and Ireland* (1588), plausibly written by Cardinal Allen. The *Admonition* casts Elizabeth as a tyrant by portraying her rule as a series of imprudent decisions driven by imagination and passion, and then goes on to argue that Elizabeth's exercise of unlimited prerogative legitimates and, in fact, demands Spanish intervention. Most significant for my own purposes is that the *Admonition* justifies these claims by focusing on a scene of sexual pleasure. After insinuating that Elizabeth 'abused her bodie' with a long list of 'divers' persons, Allen explains why Elizabeth refused to marry. 'Her affection is so passinge unnatural', he writes, 'that she hath bene heard to wishe, that the day after her death she might stand in sum high place betwene heaven and earthe, to behold the scramblinge that she conceyved wold be for the croune; sportinge herself in the conceyte and foresight of our future miseries, by her onlie unhappines procured: not unlike to Nero, who intending for his recreation to set Rome on fier, devised an eminent pillar wheron he might stand to behold it.'[24] Elizabeth is a tyrant like Nero, Allen proposes, because she courts a fantasy version of England's demise, an image of future destruction which she takes so much enjoyment in imaging that it explains and drives her present decision to sleep with diverse persons.

Pedro de Ribadeneyra offers a broadly theoretical justification of Allen's argument in his *Religion and the Virtues of the Christian Prince Against Machiavelli,* published in Madrid in 1595 and subsequently translated into Latin, Italian, French, and English. A companion to the Spanish ambassador to England during the death of Mary and coronation of Elizabeth, Ribadeneyra maintained a strong interest in

Elizabeth's claims to sovereignty and the effects of those claims on the Catholic Church in England. Like Allen and other Jesuit controversialists, Ribadeneyra represents Elizabeth as an absolutist prince, seeing her claims to be supreme governor of church and state to be evidence of her absolutism. Instead of leveling a sustained attack on Elizabeth, however, Ribadeneyra attacks the proto-absolutist French writer, Jean Bodin, and his theory of indivisible sovereignty. Because Bodin defines sovereignty as the right to command others absolutely, he also argues that it cannot be shared. In response, Ribadeneyra argues that reason of state – his phrase for the sovereign decision – is divided in itself and necessitates a choice that cannot be grounded in political thinking. While 'all princes ought to hold it ever before their eyes, if they want to succeed in governing and conserving their states', nevertheless 'reason of state [razón de Estados] is not one alone, but two': 'one, false and unreal, the other, sound and true; one, deceitful and diabolical, the other, certain and divine; one, that has Religion from the State, the other that has the State from Religion'.[25] By arguing for a decision that precedes political decision-making, Ribadeneyra effectively turns Bodin's understanding of sovereignty upside down. Instead of locating the indivisibility of sovereignty in the purity of the sovereign's unbound will, Ribadeneyra subsumes the sovereign within an opposition between true and false paradigms of decision-making in which the decision to accept Bodin's theory of sovereign power is already a bad decision, a decision against religion as the justifiable limit on royal prerogative. The cunning of Ribadeneyra's account is that it produces divided sovereignty as a necessary situation, the unavoidable if also unexpected outcome of political absolutism.[26] For a ruler to choose what for Ribadeneyra is the right reason of state means accepting the councillary role of the Church in sharing sovereignty by placing limits on royal prerogative, while to exercise unlimited prerogative means confronting the limits of one's capacity to decide through just war with monarchs loyal to the Church.

Like Ribadeneyra, in *Hero and Leander* Marlowe explores sovereignty through a critical reassessment of proto-absolutist political thought. At the festival of Adonis, Marlowe describes Hero's beauty by associating it with the kind of extra-legal will associated with absolutist political thought and exemplified by the well-known phrase from the Justinian Code, what pleases the prince has the force of law.[27] Writing within this tradition, Bodin notes that the tag occurs at the end of all French edicts and laws, to show that all law depends 'upon nothing but [the sovereign's] meere and franke good will'.[28] In a particularly jarring simile, Marlowe compares Hero's beauty to an overwhelming enemy force and her onlookers to soldiers who '[a]wait the sentence of her scornefull eies' (123). At first, Marlowe seems to suggest that Hero's beauty inspires a 'fear of death' that immobilizes and disarms (121). But this simile ends up being a form of misdirection, one that translates an effect into a cause. It is, as subsequent lines make clear, because of the soldiers' 'faithfull love' and not Hero's that her beauty has a chastening and civilizing effect, inspiring a fear of Hero's capacity to decide which transforms soldiers and princes into obedient subjects doomed to experience the bond of love through imaginary death: some, 'seeing great princes were denied, / Pyn'd as they went, and thinking on her died' (128–130). That is, Marlowe suggests that Hero's decision is largely a fantasy, driven by the pleasure that her onlookers take in imagining themselves to be the object of her 'scornefull eies'. By underscoring an affective, passion-driven complicity between subject and sovereign, Marlowe discloses the role of imagination in a way that legal and juridical reason cannot. Whereas Bodin argues that legal relations are supported by a sovereign will, Marlowe prioritizes the aesthetic bond between subject and sovereign – a bond rooted in the effects of beauty on the act of judgment – in order to indicate that the force of the sovereign decision is in no small measure derived from the paradoxical pleasure that the political subject takes in imagining the effects that might result from that decision.

When Marlowe returns to Hero's capacity to decide in the bedroom scene at the end of the poem, he supplements this imaginary version of prerogative with one in which Hero's decision divides her against herself. While Marlowe only minimally represents Hero's decision in the context of religious division, nevertheless the structure of her decision is informed by Catholic critiques of Elizabeth's unlimited prerogative. Singer is most likely right that the couplet about the bird 'which in our hand we wring' implies a conceit whose second half is missing. Whether this means that the lines are out of place or that the manuscript is incomplete, in either case the couplet recalls the turtle-dove that Hero was sacrificing when Leander first saw her, indicating that the ending of the poem unfolds the implications of that initial religious act. That is, Hero's decision to commit 'treason' is an effect of the religious difference first introduced by Leander when he calls Hero's sacrifice a 'sinne' and 'sacrilege' (306, 307). Through Leander's metaphor Marlowe translates religious conflict into sexual difference in order to change the terms within which to represent decision-making. For Ribadeneyra, prerogative implies its own limitations and, therefore, necessitates shared sovereignty between church and state. Rather than fully exploring religious conflict as he does in *The Massacre at Paris*, here Marlowe expresses the limitations of prerogative through a crisis between morality and imagination, emphasizing the temporality of Hero's decision as a limit to the ideal, 'unknown' pleasure that she pursues. Following 'unknowne joy' in spite of the demands of chastity, Hero's decision involves commitment to 'another world' that remains imaginary because foreclosed by social norms. At the same time, Hero is also subject to another, darker version of the future. Committing herself to 'another world' of 'unknowne joy' prevents her from seeing the probable future to which she also commits herself. Hero's striving after 'unknown joy' doubles and is doubled by an unknown future which Marlowe portrays as an ominous, cruel and deadly judgment. 'Love is not ful of pittie (as men say) / But deaffe and cruell, where he meanes to pray'. Wittingly or not, Hero's decision commits her to a possible future that she imagines and a probable future that she cannot or will not see.

Aesthetics as political thought

Catholic political thought can deepen our sense of Hero's dilemma by offering paradigms that help us to see why Marlowe would focus on her capacity to decide, but it doesn't fully explain Marlowe's understanding of sovereignty. For that, we need to turn in a more focused way to aesthetics. From Plato to Spenser (and, indeed, beyond) writers have often associated beauty with the common good. When an end in itself, beauty is merely representation and is therefore fleeting and corrupting, so the argument goes, but when associated with virtue, beauty leads to moral goodness. Francis Bacon expresses this understanding quite well. 'Beauty is as summer frults', he writes, 'which are easy to corrupt, and cannot last; and for the most part it makes a dissolute youth.' But, he continues, if beauty 'light well, it maketh virtues shine, and vices blush'.[29] What distinguishes corrupting beauty from the beauty that makes virtue shine is beauty's relation to holiness or divinity, 'the soveraine light', as Spenser puts it at the end of his *Hymne of Heavenly Beautie*, 'From whose pure beams al perfect beauty springs'.[30] Rather than reflecting sovereign power, for Marlowe beauty only offers an illusory retreat from it. Dr. Faustus turns to an imaginary Helen of Troy in order to defend against the conundrum of sovereign power in which he finds himself. Either he accepts God's 'tribunall seate', in which case Mephostophilis will 'arrest' his soul for disobeying his 'soveraigne Lord', or he accepts Mephostophilis' injunction to 'Revolt', in which case he exiles himself from the kingdom of heaven (DF 12.103, 57–58, 59). Helen's beauty covers the scandal of Dr. Faustus' position by allowing him momentarily to imagine himself entering another world. Marlowe suggests a similar dynamic when Tamburlaine reflects on beauty, much to his surprise, after commanding the murder of the Damascan virgins and displaying their corpses on the city walls. Like Dr. Faustus, Tamburlaine's reflection on beauty offers him some distance from sovereign power, here his own efforts to ground his authority in massacre and slaughter.

Instead of aligning sovereignty with beauty, Marlowe thinks sovereignty through self-ravishment, or what I would like to call the Marlovian sublime, the experience of self-desecration in which characters attempt to protect themselves against the effects of absolute power by imagining what Cynthia Marshall has called self-shattering, a desire for the undoing or betrayal of the self.[31] In Marlowe, suicide is an extreme and literalizing form of this phenomenon, when, for instance, Bajazeth, Zabina, and Agydas kill themselves to prevent further punishment by Tamburlaine, or when Olympia sets up her own death to prevent her forced marriage to Theridamas. But Marlowe is more interested in the role of imagination than he is in the body *per se*. 'O it strikes, it strikes', Dr. Faustus exclaims when the hour of judgment arrives, 'now body, turne to ayre / Or *Lucifer* wil beare thee quicke to hel' (DF 13.110–111). Dr. Faustus' fantasy of corporeal disintegration oddly defends against imminent death and punishment by staging a more severe death and punishment in imagination. That is, Dr. Faustus claims punishment *before* judgment in order to pre-empt and therefore oddly prevent judgment and, perhaps more significantly, the presence of the sovereign who judges. Fully aware that Lightborne has come to kill him, Edward responds by reimagining his death as a form of religious ecstasy:

> I see my tragedie written in thy browes,
> Yet stay a while, forbeare thy bloodie hande,
> And let me see the stroke before it comes,
> That even then when I shall lose my life,
> My minde may be more stedfast on my God.
>
> (Ed 2, 22.73–78)

In this his final speech, Edward moves from recognition of imminent catastrophe, 'I see my tragedy', to a fantasy doubling of that catastrophe that imitates the punishment he faces as a way to pre-empt it: 'let me see the stroke before it comes'. In *The Jew of Malta* Barabas considers indulging in this kind of ravishment. 'Thinke me so mad as I will hang my selfe', he briefly soliloquizes, 'That I may vanish ore the earth in ayre, / And leaue no memory that e're I was' (JM 1.2.261–263). But he suspends this fantasy in pronouncing that he will 'rouse' and 'awake' himself in order to claim the very authority by which his wealth was taken away in the first place (267).

As the first translator of Book One of *The Parsalia* into English, Marlowe may well have derived his interest in the sublime from Lucan's poem which, as David Norbrook describes it, 'displays a manic delight in its own iconoclastic creativity', and which, Norbrook shows, was a touchstone in seventeenth-century republican political and literary culture.[32] But this does not mean, as Patrick Cheney has recently argued, that Marlowe espoused republicanism either as a positive political program or even as a political ideal at all.[33] When Marlowe focuses on the experience of the sublime, the complex relation between sovereignty and imagination that makes up this experience shows that it gives form to the central aporia of early modern political theology between sovereignty and constituting power. As political thinkers from Hannah Arendt to Negri have argued, if the central insight of early modern political thought is that 'power becomes an immanent dimension of history', to quote Negri, then the central problem is coming up with a model of sovereignty equal to that insight.[34] For Arendt, the theory of the sacred monarch was a partial answer, one that acknowledged the state as a made apparatus while obscuring that very insight. If the monarch has the right – like God – to make and remake the state outside the bounds of law, then this is as much a denial of constituting power as it is an acknowledgment of it. 'Theoretically speaking', Arendt writes, 'it is as though absolutism were attempting to solve this problem of authority without having recourse to the revolutionary means of a new foundation; it solved the problem, in other words, within the given frame of reference in which the legitimacy of rule in general, and the authority of secular law and power in particular, had always been justified by relating them to an absolute source which itself was not of this world.'[35] The fact that the sovereign is authorized through a transcendental framework distresses that framework (in the sense of distressed furniture) and, in the process, looks forward to a new order in which the immanent force of constituting power can be actualized as such. Negri reframes the problem as a general principle. Once a political order is made, constituting power is 'conceptually reconstructed not as the system's cause but as its result. The foundation is inverted, and sovereignty as [supreme power] is reconstructed as the foundation itself.'[36] For Marlowe, the experience of the sublime acknowledges this aporia and struggles to express it by reclaiming the role of making as imagination. Instead of legitimating imagination on the grounds that it is analogous to sovereignty – as, for example, Sidney does in the *Defence* – in the pre-emptive, temporal dimension of self-shattering Marlowe suggests an antagonist relation between the two. For Marlowe, ravishment becomes a compromise position in which imagination reclaims the force of making against sovereignty by staging the very effects of sovereign power in and as imagination.

In aesthetic terms, the movement from the festival of Adonis at the beginning of *Hero and Leander* to the bedroom scene at the end of the poem is a movement from beauty to the sublime. In the scene at the festival of Adonis, imagination solidifies the bond between subject and sovereign as Hero's lovers submit in anticipation of her judgment. Hero's beauty implies but also displaces a more ominous picture of sovereignty, one in which Hero stands as conquering general to her lovers who 'Await the sentence of her scornefull eies'. At the end of the poem when Hero chooses ravishment, Marlowe opposes sovereignty and imagination while acknowledging that the two cannot be separated. Hero asserts her paradoxical status as sovereign subject by giving herself over to the 'treason' in her 'thought', claiming prerogative through the internalization of enmity and by choosing an imaginary future. This opposition between sovereignty and imagination culminates in Hero's blush, when she stands naked and ashamed, between her decision to follow 'unknowne joy' and the social judgment which that decision risks.

> Thus neere the bed she blushing stood upright,
> And from her countenance behold ye might,
> A kind of twilight breake, which through the heare,
> As from an orient cloud, glympse here and there.
> And round about the chamber this false morne,
> Brought foorth the day before the day was borne.

> (801–806)

In part, Hero's blush looks forward to the kind of moral judgment that awaits her. At the same time, as a 'false morne' her blush paradoxically suspends that moment, bringing the day 'before the day was borne' as a way of holding onto 'another world' of imagination. Like Bajazeth or Zabina at the end of *Tamburlaine*, Hero's blush shows her to be abandoned to her fate while imaginatively suspending that fate, paradoxically 'Preserving life', as Bajazeth puts it, 'by hasting cruell death'

(T1 4.4.102). Bajazeth is describing a phenomenon whereby, in order to survive, his empty stomach draws humors from the rest of his body, which enfeebles his body and draws death nearer.

The relation between life and death that concerns Marlowe at the end of *Hero and Leander* is more aestheticized and ornamental. The life being preserved is the life of imagination, which metaphorically feeds on Hero's doom. At least, this is one way to explain why Marlowe ends *Hero and Leander* with such a highly artificial gesture. Confusing Hero's 'false morn' with the real morning, Apollo sounds his harps, which causes Night to pass away:

> Which watchfull *Hesperus* no sooner heard,
> But he the day bright-bearing Car prepar'd.
> And ran before, as Harbenger of light,
> And with his flaring beames mockt oughly night,
> Till she o'recome with anguish, shame, and rage,
> Dang'd downe to hell her loathsome carriage.

> (813–818)

The point is not just that Hero cannot control time. More precisely, Hero's investment in imagination jumpstarts and brings about the moment that her shame attempts to suspend. It is as if through the resources of imagination and ornamentation Marlowe tries to break the sovereign's hold on time in order to imagine the possibility of a more open, if also undecided and rather anxious image of the future.

***

It is worth noting that the other essays in this volume that take up the problem of political theology focus on Shakespeare. In her reading of Schmitt, Hans Kelsen, and *Hamlet*, Julia Reinhard Lupton shows how Shakespeare imagines new norms of citizenship while still accounting for the rash supplement of the sovereign decision. And both Ken Jackson and Gary Kuchar emphasize the Pauline and Christian aspects of Shakespeare's political and theological commitments. Shakespeare can create new experiences of time (Jackson) and new forms of subjectivity (Kuchar) because he is a deeply Christian playwright. Marlowe tells a different story. He is more interested than Shakespeare in probing and exploiting the differences, divisions, and aporias that emerge when state and church become one. Rather than worrying about religion *per se*, Marlowe shifts the central problems that define political theology from Christianity to the body so that, as I have argued, the aporia between sovereignty and constituting power gets played out in and as sexual feeling and erotic imagination.

We have become accustomed to thinking of the monarch's body as the central figure of political theology, largely due to the powerful influence of Ernst Kantorowicz's groundbreaking study *The King's Two Bodies*.[37] As the historical process by which the 'prince stepped into the shoes of Pope and Bishop', political theology involves the mystification and sanctification of the king's body.[38] Moreover, if we follow Schmitt, then the other significant figure that emerges is the enemy. Schmitt recognized the significance of the enemy in a fairly limited way, from the perspective of the sovereign. Whereas in *Political Theology* he defines the sovereign as 'he who decides on the exception', Schmitt's discussion of the enemy in *Concept of the Political* suggests that among the most important decisions that the sovereign makes is the decision between friend and foe.[39] For Schmitt, the state and the state alone decides who is the enemy with reference neither to 'a previously determined general norm nor by the judgment of a disinterested and therefore neutral third party' but to the political interests of the state.[40]

Marlowe's work offers a powerful answer to Kantorowicz and Schmitt because he engaged with political theology not out of a commitment to religion or to the public good but to explore the constituting forces of imagination and enmity through the affective body. Hero's decision to follow her fantasy places a troubling aesthetic pleasure in conflict and treason at the core of political sovereignty. One consequence is that fantasy, desire, and imagination take precedence over virtue and moral goodness. It is difficult not to read *The Rape of Lucrece* as a corrective to *Hero and Leander*. Shakespeare follows a long tradition of humanist and classical political thought when he locates the origin of republican Rome in Lucrece's decision to commit suicide rather than endure the stain of an impropriety which she did not choose. Lucrece's decision defers to a republican order of self-regulation that Hero finds herself unable to follow. It would be a mistake, however, to infer from Shakespeare's response that Marlowe was somehow in favor of absolutism. Instead of suggesting the boundless absolute authority of the state, Marlowe repeatedly stages the sense of enmity that underwrites and delimits this assertion, opening the possibility for rethinking political sovereignty through a version of the sublime that is grounded in the expansion of prerogative against absolutist visions of sovereignty, an acute sense of political making, the boundlessness of imagination, and what Machiavelli calls the capacity to be 'al tutto cattivi', altogether bad.[41]

Notes

This essay was originally written for a plenary session at the Shakespeare Association of America in 2007. I'd like to thank the members of the audience for responses I received there. And I'd especially like to thank Heather Hirschfeld and Susanna Monta for reading earlier drafts of this essay and offering incisive and insightful responses.

1. 'La constitution n'est pas l'ouvrage du pouvoir constitué, mais du pouvoir constituant', Sieyès explains. And subsequently, 'on doit concevoir les nations sur la terre comme des individus hors du lien social ou, comme l'on dit, dans l'état de nature'. Emmanuel Joseph Sieyès, *Qu'est-ce que le Teirs état?* (Geneva: Librairie Droz, 1970), 180–181, 183.
2. Carl Schmitt, *Political Theology: Four Chapters on the Concept of Sovereignty*, trans. George Schwab (Cambridge, MA: MIT Press, 1985), 49.
3. Ibid., 36.
4. Ibid., 7.
5. Antonio Negri, *Insurgencies: Constituent Power and the Modern State*, trans. Maurizia Boscagli (Minneapolis: University of Minnesota Press, 1999), 13.
6. Niccolò Machiavelli, *Discorsi*, in *Tutte le Opere di Niccolò Machiavelli*, ed. Guido Mazzoni and Mario Casella (Florence: G. Barbèra Editore, 1929), 1.9.72–74.
7. *Insurgencies*, 120–1, and especially *The Savage Anomaly: The Power of Spinoza's Metaphysics and Politics*, trans. Michael Hardt (Minneapolis: University of Minnesota Press, 1991; rpr. 2000), 86–98.
8. *The Jew of Malta*, 1.2.81. Unless otherwise noted, references to all other works by Marlowe are cited from *The Complete Works of Christopher Marlowe*, ed. Edward J. Esche, David Fuller, Roma Gill, Richard Rowland et al., 5 vols (Oxford: Clarendon Press, 1987). Subsequent reference will be cited parenthetically in the text.
9. William Cecil, *The Execution of Justice in England*, with William Allen, *A True, Sincere, and Modest Defense of English Catholics*, ed. Robert M. Kingdom, Folger Documents of Tudor and Stuart Civilization (Ithaca: Cornell University Press, 1965), 7, 80, 79.
10. William Allen, *A True Report of the Late Apprehension and Imprisonment of John Nichols* (Rheims: Printed by John Fogny, 1583), sig. Aiiii r.
11. See David Riggs, *The World of Christopher Marlowe* (New York: Henry Holt, 2004), 135–137. It was through Riggs' excellent study of Marlowe's life that I first came across the decorative letters in Allen's *True Report*.
12. Allen, *True Report*, 25r.
13. Richard McCoy, *Alterations of State: Sacred Kingship in the English Reformation* (New York: Columbia University Press, 2002), 18.
14. Erna Auerbach gives a comprehensive study of Tudor uses of decorative letters in the Plea Rolls in *Tudor Artists: A Study of Painters in the Royal Service and of Portraiture on Illuminated Documents from the Ascension of Henry VIII to Elizabeth I* (London: University of London, Athlone Press, 1954). Alexandra Walsham discusses Catholic propaganda and print in '"Domme Preachers"? Post-Reformation English Catholicism and the Culture of Print', *Past and Present* 168 (August 2000), 72–123.
15. Allen, *True Report*, sig. Aii r–v.

16. William Allen, *An Apologie and True Declaration of the Institution and Endevours of the Two English Colleges* (Rheims: Jean de Foigny), 14v.

17. *Select English Poets*, ed. S. W. Singer, no. 8 (London: Chiswick Press, 1821), 39.

18. *The Works of Christopher Marlowe*, ed. C. F. Tucker Brooke (Oxford: Clarendon Press, 1910), 511.

19. *English Sixteenth-Century Verse: An Anthology*, ed. Richard S. Sylvester (New York: Norton, 1984); *Marlowe's Complete Poems*, ed. L. C. Martin (London: Methuen, 1933); *Complete Poems and Translations*, ed. Stephen Orgel (Harmondsworth: Penguin, 1971; rpr. 1986); *Christopher Marlowe: The Complete Works*, ed. Fredson Bowers (Cambridge: Cambridge University Press, 1973); *The Complete Works of Christopher Marlowe*, ed. Roma Gill; *The Norton Anthology of English Literature*, gen. ed. M. H. Abrams, 2 vols (New York: Norton, 2000).

20. Gill, *Complete Works*, 1:186–187.

21. Judith Haber, '"True-loves blood": Narrative and Desire in *Hero and Leander'*, *ELR* 28:3 (Autumn 1998), 384. John Leonard also makes a case for Blount, giving a close reading of images, motifs, and metaphors of war in Blount's version of the poem's ending. See 'Marlowe's Doric Music: Lust and Aggression in *Hero and Leander'*, *ELR* 30:1 (Winter 2000), 67–73.

22. J. B. Steane reads the final scene of *Hero and Leander* as displaying 'a sort of sexual brutality'; Werner Von Koppenfels argues that Leander attacks Hero at the end of the poem, revealing a violent, predatory undercurrent to male desire; David Lee Miller argues that Hero is forced to bear the burden of male heterosexual desire and 'to constitute her sexuality by internalizing the discontinuity between her position as subject and as object of desire'; Gregory Bredbeck argues that Marlowe uses the blazon in the poem to fetishize and query the naturalness of male heterosexual desire; and Claude Summers, working from Miller's and Bredbeck's readings, agrees that *Hero and Leander* confirms male virility as a part of an 'Elizabethan sex-gender system' and adds that the poem also subverts this system by representing a variety of non-normative desires. J. B. Steane, *Marlowe: A Critical Study* (Cambridge: Cambridge University Press, 1964), 331; Werner Von Koppenfels, 'Dis-covering the Female Body: Erotic Exploration in Elizabethan Poetry', *Shakespeare Survey* 47 (1994), 129; David Lee Miller, 'The Death of the Modern: Gender and Desire in Marlowe's *Hero and Leander'*, *SAQ* 88 (Fall 1989), 773; Gregory Bredbeck, *Sodomy and Interpretation: Marlowe to Milton* (Ithaca: Cornell University Press, 1991), 113–114; Claude J. Summer, '*Hero and Leander*: The Arbitrariness of Desire', in J. A. Downie and J. T. Parnell, eds, *Constructing Christopher Marlowe* (Cambridge: Cambridge University Press, 2000), 147.

 Only by ignoring this segment of the poem can Georgia Brown argue that in the poem Marlowe 'explores a feminised form of authorship', acknowledging 'an analogy between the aesthetic and the feminine'. Playing Ovidian poetics against epic seriousness, Brown argues that Marlowe responds to cultural imperatives which demanded that literature be moral and didactic by using Hero to celebrate forms of literature and literary production which are trivial, prodigal, and wanton. Georgia Brown, 'Gender and Voice in *Hero and Leander'*, in Downie and Parnell, eds, *Constructing Christopher Marlowe*, 149. Brown substantially develops her reading of Elizabethan Ovidianism in

Redefining English Literature (Cambridge: Cambridge University Press, 2004), 102–177. In his study of the epyllion, Jim Ellis argues that in *Hero and Leander* Marlowe explores the intersections of rhetoric, sexuality, and male citizenship. See *Sexuality and Citizenship: Metamorphoses in Elizabethan Erotic Verse* (Toronto: University of Toronto Press, 2003), 94–108.

23. Justus Lipsius, *Sixes Bookes of Politickes or Civil Doctrine* (London: Ponsby, 1594), 4.1.59.

24. Allen, *An Admonition to the Nobility and People of England and Ireland Concerning the Present Warres made for the execution of his Holines Sentence, by the highe and mightie Kinge Catholike of Spaine* (Antwerp: A. Conincx, 1588), B2r, B3v.

25. Pedro Ribadeneyra, *Tratado de la religion y Virtudes que debe tener El Principe Cristiano para Goberar y Conservar sus Estados Contra lo que Nicolas Maquiavelo y los Politicos de este Teimpo Enseñan* (Buenos Aires: Editorial Sopena Argentina, 1942), 11; *Religion and the Virtues of the Christian Prince Against Machiavelli*, trans. and ed. George Albert Moore (Washington, DC: The Country Dollar Press, 1949), 253, translation slightly altered. For a discussion of Ribadeneyra, see Robert Bireley, *The Counter-Reformation Prince: Anti-Machiavellianism or Catholic Statecraft in Early Modern England* (Chapel Hill: University of North Carolina Press, 1990), 111–135.

26. For a discussion of divided sovereignty in the context of Spain and English Catholicism, see Jacques Lezra, '*Phares*, or Divisible Sovereignty', *Religion and Literature* 38 (Autumn 2006), 13–39.

27. The entire phrase includes an explanation that justifies this sovereign power. 'What pleases the sovereign has the force of law, since by royal law concerning his rule, the people confer on him and lodge in him all their rule and power [Quod principi placuit, legis habet vigorem: utpote cum lege regia, quae de imperio eius lata est, populus ei et in eum omne suum imperium et potestatem conferat].' It was standard practice among medieval jurists to refer to Justinian when addressing the problem of 'princeps legibus solutus', or the prince unbound or loosened from the law. Paul Vinogradoff, *Roman Law in Medieval Europe* (Oxford: Oxford University Press, 1961), 65–68; Thomas Gilby, *The Political Thought of Thomas Aquinas* (Chicago: University of Chicago Press, 1958), 51–54.

28. Jean Bodin, *The Six Bookes of a Common-weal*, trans. Richard Knolles (London: [Printed by Adam Islip] impensis G. Bishop, 1606), 1.8.92.

29. Francis Bacon, 'Of Beauty', in *The Essays*, ed. John Pitcher (London: Penguin, 1985), 190.

30. 'An Hymne of Heavenly Beautie', 295–296, in *Edmund Spenser: The Shorter Poems*, ed. Richard A. McCabe (New York: Penguin, 1999).

31. Cynthia Marshall, *The Shattering of the Self: Violence, Subjectivity, and Early Modern Texts* (Baltimore: Johns Hopkins University Press, 2002), 53. Both Michael Goldman and Heather James have explored this phenomenon in suggestive ways. For Marlowe, Goldman argues, pleasure often takes the form of abandonment in which characters forsake the world of social norms in order to experience a kind of ravishment in being overtaken. And as James shows, it is through the licentiousness of this pleasure that Marlowe speaks to and about the state. See Michael Goldman, 'Marlowe and the Histrionics of Ravishment', in Alvin Kernan, ed., *Two Renaissance Mythmakers: Christopher*

Marlowe and Ben Jonson (Baltimore: Johns Hopkins University Press, 1977), 22–40; Heather James, 'The Poet's Toys: Christopher Marlowe and the Liberties of Erotic Elegy', *MLQ* 67:1 (March 2006), 103–127.

32. David Norbrook, *Writing the English Republic: Poetry, Rhetoric, and Politics, 1627–1660* (Cambridge: Cambridge University Press, 1999), 33. For a brief discussion of Marlowe and Lucan, see my essay, 'Time for Marlowe', *ELH* 75:2 (Summer 2008), 301–303, 306–307.

33. Patrick Cheney, *Marlowe's Republican Authorship: Lucan, Liberty, and the Sublime* (Basingstoke: Palgrave Macmillan, 2009), 68–77.

34. Negri, *Insurgencies*, 11.

35. Hannah Arendt, *On Revolution* (London and New York: Penguin, 1990), 160.

36. Negri, *Insurgencies*, 13.

37. Ernst H. Kantorowicz, *The King's Two Bodies: A Study in Mediaeval Political Theology*, with a preface by William Chester Jordan (Princeton: Princeton University Press, 1957; rpr. 1997).

38. Ernst H. Kantorowicz, 'Mysteries of State: An Absolutist Concept and Its Late Medieval Origins', *Harvard Theological Review* 48 (January 1955), 67.

39. Schmitt, *Political Theology*, 5.

40. Carl Schmitt, *The Concept of the Political*, trans. and introduced by George Schwab (Chicago: University of Chicago Press, 1996), 27.

41. Machiavelli, *Discorsi*, 1.27.94.

7
Humanism and the Resistance to Theology

William N. West

> Nur noch ein Gott kann uns retten.
> – Martin Heidegger

> ... cuius deinde avulsa passim membra, sicuti fabulae ferunt, Aesculapius ille collegit, reposuit, vitae reddidit; qui tamen deinde fulmine ictus ob invidiam deorum narratur.
>
> – Angelo Poliziano

'Theory' has taken long aim at theology, from Friedrich Nietzsche's 'God is dead ... And we have killed him!' to Roland Barthes' 'We know that a text does not consist of a line of words, releasing a single "theological" meaning (the "message" of the Author-God).'[1] This thin description hardly establishes a theology that fifteenth- or sixteenth-century thinkers would have recognized, although they might well have been interested in some of the questions that poststructuralism raised. Robert Markley, for instance, claims that in the seventeenth century 'Theology ... marks the contested territory of what literary and cultural critics and some historians call theory', differently represented but similarly directed.[2] In the discourse of 'theory', all too often citing theology has served as little more than a signal of disparagement, but the overt rejection – resistance is too moderate a term – of the theological from 'theory' is recurrently tempered by reappropriations of its projects, or perhaps attempts to recover conceptual territory from it. For Nietzsche proclaims not God's disappearance, but the mode in which God's anxious absence continues to work; Barthes announces the death of the author-God with a line from 'Sarrasine', but his *S/Z* on the same story relies on a version of Christian allegoresis; in his later works

167

Derrida turns toward both thinking religiously and thinking within the religious traditions of Judaism; and in his final interview, the prophet against metaphysical thinking Heidegger laments that 'Only a God can save us now.'[3]

The zone of contact between the discourses of theory and theology continues to be vigorously contested. But lately things have changed: lines are being redrawn and the stone the builders refused looks more and more likely to be the capstone. 'What we are hearing today', remarks the theologian Graham Ward, 'is the theological voice that modernity repressed.'[4] Stanley Fish's prediction that following the death of Derrida religion would be 'where the action is' has been realized – and retrospectively, the passionate denials of theology and conflicted conversions toward it from both poststructural writing and Marxist-inflected historicist criticism suggest that the action has been there for a while.[5] The self-consciously theoretical work that gropingly followed poststructuralism and New Historicism has increasingly worked backwards against this critical version of the 'secularization thesis' about the disenchantment of the world, first in the increasing attention among later New Historicists to religious modes of thought and affect.[6] But Fish presciently saw religion displacing not only 'high theory' but 'the triumvirate of race, gender, and class', laying out in neat shorthand the defining critical struggle of the last decades of the twentieth century between historicizing and textualizing camps, and recognizing that a return of religious thinking would recast the map of cultural criticism.

The religion toward which some of these theories turn sometimes seems scarcely narrower than the 'theology' of Barthes. For one of the religious turn's most acute exponents, Julia Reinhard Lupton, religion is a field (not necessarily the only one) of human interaction

> that resists localization and identification with a specific place, time, nation, or language ... the religious turn in Renaissance studies represents the chance for a return to theory, to concepts, concerns, and modes of reading that found worlds and cross contexts, born out of specific historical traumas and debates, but not reducible to them.[7]

Lupton, like Markley and now like many others, reclaims religious thinking as a frame for the broad questions that have at times eluded recent early modern studies. Such critics turn again to writers like Max Weber, Schmitt, Benjamin, and Kantorowicz; to Derrida, Lacan, and Levinas; or to more recent work by Žižek, Agamben, and Badiou not only for their analyses of religion, but as examples of religious think-ing.[8] In this transformed critical landscape, Edward Said's still powerful distinction between secular and religious criticism, despite the terms he chose, does not seem to address adequately the place of religious think-ing in literary and cultural studies now.[9] For Said, secular criticism was the product of care – both diligence and engagement – local, detailed, open; in all these qualities 'religious criticism' was its opposite, serving 'as an agent of closure',[10] reducing the rich particularity of life to bare schemata, and, worst of all, doing it by accident. The contemporary religious turn is none of those things, least of all unaware. Nor is it clear what modes, fields, or practices might serve as its other in this develop-ing critical landscape.

Barthes' denunciation of the 'Author-God' marks one likely point of origin for contemporary breadth of what it might mean for a text to be 'theological'; Heidegger's rejection of the ontotheology of metaphysical thinking might be another. In the late return to religion in literary and cultural studies, it has not always been easy to prise theology from the other fields in which it has seemed to move in tandem: of modernity, still conceived of as a disenchantment or secularization of the world (Gadamer; Blumenberg; Habermas; Charles Taylor); of ethics (Levinas; Derrida; Eagleton); of poetics (Schwartz; Hawkes); of politics (Žižek; Agamben). Such multivalences reflect the richness of its traditions, and are one reason that returns have been possible from so many directions. As thought by Lupton, Badiou, or Derrida, religion can describe the phe-nomena of disruption and disjuncture so important to the discourse of poststructuralist theory, and reintroduce possibilities of *communitas* and sublimity, trauma and absolute risk, radical difference and revolution-ary change, to the unremarkable running of contemporary life. It also offers a corrective to an academy that has shown recurrent tendencies to overlook the irrational, the contingent, and the fortuitous in its dis-ciplinary practices, reducing its teaching to the mastery of a *technē* or an archive. Maybe contradictorily, the return to religion offers critique to thought and practice that continue in patterns structured by premises which they do not acknowledge.

The scope of the meaning of *religion* can be compared to the expan-siveness covered in the homonyms Renaissance *humanitas* and modern 'humanism'. I write *homonym* because there are few semantic grounds for assuming that these terms mean the same thing, and it is not even clear that they share a genealogy.[11] Of course, for many poststructural-ists, most notably Foucault, humanism and the 'sciences of man' were only specific examples of theological thinking.[12] But the most basic comparison of Renaissance and modern humanisms shows that the terms are almost incommensurable, and neither are their relations to such categories as the religious or the secular at all clear. As with religion and theology, transformations of the critical fields within which these terms initially developed have redistributed their positions with respect to other elements in those fields and thereby changed their functional meanings. This essay is not the place to ask why problems of such lasting moment that seem to reach beyond their specific cultural origins should be referred to a discourse of theology or religion rather than to another or others called, for instance, imagination, humanity, science, or, as before, 'theory'. But this does need to be asked; the name is not without consequence, certainly, recalling as it does, perhaps tactically, resonant but unspecified histories, institutions, and powers.

Somewhat surprisingly, then, central to the question of the relation of the religious and the secular on the contemporary scene is that of the relation of humanism and philology, which we might now distinguish from humanism – an instinct that I will later argue is suggestive – as these were practiced in the fifteenth century and in their subsequent histories. These terms seem to have no connection outside of historical accident of their initial coincidence, but they were used almost synonymously by their early modern practitioners, and it took centuries before they drew apart enough to be distinguished.[13] This essay tries to test whether there is anything in that Renaissance humanist project that might speak – almost as if by accident, certainly not by filiation – to our current interrogations of what humanism might be (if anything), and whether these capacities fall outside of a religious thinking of numinous rupture. I want to argue that humanist philology represents a viable secular mode of criticism in a time when religious thinking can be said to be emergent, in that it acknowledges but declines transcendence – in fact resists it, as earlier poststructuralist thinking did not. The force of this is to return agency to the human and to the world, not agency in the 'theological' sense rightly (I think) critiqued by Barthes and others, but in the sense of acknowledging the scandalous friction and purchase of things on things proposed by Latour and similar thinkers. This does not answer, but merely returns to the problem of, Habermas' sense that the modern world requires peculiar forms of 'secular legitimation', as well as Blumenberg's contention that to ask legitimation from modernity frames the question wrongly.[14] But it can pose the challenge of what a secular theory worth the name would look like in this developing system of positions.

In the 1990s a challenge like this would have made almost no sense, and even now one response might be to adopt theoretical models that do not substantially engage with the transcendent thinking of religion: those of Foucault, perhaps, or of Deleuze and Guattari. Certainly other positions exist that do not originate in a thinking that is avowedly religious. But in the particular context of the turn to religious thought in and as theory, to begin there would feel to me like an evasion of the question's force, overlooking and repressing the division of the sacred and the human in the willed error that Giorgio Agamben calls secularization. So I will instead pose my response on the terrain as the return to religion has reformed it, mainly through some writings of Agamben and the *Miscellanea* of Angelo Poliziano, to consider more fully some theoretical or speculative uses of religious thinking. For while religious thinking can ground both radical change and eternal rest, and take into account both grounds of sameness or indifference and absolute otherness, secularity may also include the possibility of the irrational, but without promising its recuperation, and that, I will argue, is the path taken by Renaissance humanism, and philology. Finally, religion's other is the subject of Renaissance, humanist philology – that propositional discipline which, of all formulations ending in -*logy*, is the only one that takes *logos* as its object rather than as its subject and form, as human rather than quasi-divine: love of the *logos* rather than the *logos* of another field.

An initial challenge is to determine what, in the current situation of 'theory', can even be said to constitute religious or theological discourse – or, put another way, what discourses might escape that sprawling field. It no longer seems adequate to limit religious thinking to creeds and practices that have been traditionally identified as religious or theological.[15] Looking to the work of scholars who have tried to think religiously, I propose this: Religious thinking parts the world; in parting it, it also departs from it; in other words, it is the thinking of transcendence.[16] But this gesture of overreaching, I think, suggests a way in which terms like religion or theology can be translated to cover the fields similarly sketched but differently valued by Barthes and Lupton. As Victoria Kahn notes, what is secular is not to be conceived of as by that token anti-religious. The secular world is itself a product of religious thought, the result of a distinction made first by Augustine, later again by Luther, of the two cities or kingdoms – the city of God and the city of man.[17] For Augustine, these coincide in the world we walk through, which is simultaneously the 'theologically neutral', unmarked and unremarkable world of recurring *saecula* and the powerfully teleological world of Christian salvation history. For Luther, the two kingdoms of the Christian and the worldly man are yet more

deeply involved, so that civil, secular bondage is paired with an experience of the freedom that is internal to Christian consciousness, which
can be treasured but never uttered. But transcendence is not limited to
fields of thinking associated with what are traditionally called religions.
In terms of looking for an absolute position outside a system to resolve
conflicts within it, the rationalist Jürgen Habermas and the Talmudist
and philosopher Emmanuel Levinas are equally religious thinkers.[18]

The instantiating religious act is a division of the world's demesne,
so that part can be set aside for human activity and part for the divine.
Thus for Giorgio Agamben, 'every separation ... contains or preserves
within itself a religious core'.[19] Graham Hammill and Lupton quote
Régis Debray's still more capacious claim that 'The work of organization
is by its very nature "religious"', where organization assumes partition
and division as a necessary antecedent.[20] This claim of division has a
history of its own, inscribed in the multivalence of the word *logos* as it
appears in the first sentence of the Gospel of John. John's translation
of the Genesis myth into the frame of Greek philosophy presented a
stumbling block for later translators into Latin, which seemed incapable
of grasping the Greek word *logos* in a single term. The usual alternatives
were *verbum* (word), *ratio* (reason, as in English *rational*), and *sermo*
(discourse), sometimes presented as a meditation on the ungraspable
semantic wealth – ungraspable in Latin, at least – of the word *logos*.[21]
No doubt this practical crux of translation suggested a comparable ineffability in the object it sought to grasp as well. The division between
Logos and *logos*, the one divine *ratio* or reason, the other human *sermo*,
conversation, or discourse, might be said to grasp in a single term the
gap between religion and whatever lies outside it. For Debray, organization in turn requires a system that is closed and finite, and such a
system can only be closed by action external to it.

Debray thus follows thinkers like, in their different ways, Luther and
Kant, in ascribing freedom only to the realm of the spirit, the world's
other, or outer.[22] In 'Humanism and the Resistance to Theory', Victoria
Kahn observes a similar disjunction between an ideal of knowledge as
Aristotelian *theōria*, which dealt with those fields and objects of study
within which there could be certain knowledge, like physics, metaphysics, or mathematics, and what had come by the 1980s to be called
'theory', a rhetorical field constituted by its object as capable of only
probabilistic, contestatory knowledge. In this she lent a history to Paul
de Man's 'The Resistance to Theory', which described in the resistance
to the theoretical project a symptom of its formal undecidability; resistance to theory, he pointedly observed, *was* theory.[23] Both were logically

prior to their contexts. For Kahn, such resistance could be observed historically in the approach taken by Renaissance humanists who distinguished fields of which absolute knowledge was possible from those in which knowledge was not, like ethics, politics, or poetics – what would later be called the human sciences or humanities – all of which were historically produced and limited, contingent, and mobile. While both de Man and Kahn observe a cleavage within theory between the rhetorical and the stative, de Man, one might say, takes the position of the stative, while Kahn shows the rhetoricity at work in such a statement. What Kahn described as humanism's resistance to theory, then, was not a constitutive or ontological hole that necessarily opened between two different functions of language and knowledge, but a particular response to the problems of activating knowledge toward particular but changeable ends – much as she understands the roots of the secular to appear in particular religiously motivated partitions of the world.

The interest of de Man and Kahn in the resistance of 'theory' to *theōria* can suggest a way toward a critique of that religious thinking which has taken up the ground that used to be covered by the theoretical project, and the difference between 'theory' and *theōria*, or within *logos*. In 'In Praise of Profanation', Giorgio Agamben proposes a similarly rich way of looking at the relations between the religious, the secular, and the profane. Like de Man, Agamben takes a structure of difference as obligatory; like Kahn, he sees it as historically instituted. Perhaps it is closest to simply being already operant. To sacralize something, according to Agamben, is to remove it from human use by or in sacrifice. But a sacred or religious object can be profaned: taken back to the common share and use of humankind through contact with the unsanctified. For Agamben, then, religion does not bind – as if it derived from the false but widespread etymology *re-ligare* – but picks out, distinguishes, cherishes, holds apart – all possible translations for another of its traditional etymologies, *re-legere*, a word that perhaps most remarkably in this context means 'to read' or 're-read'.

With the world divided and distributed between the City of God and the city of man, the realm of religion is manifested in transcendence, the myriad points of passage that irrupt from beyond into the ordinary or everyday. These irruptions are what religious thinking reads, or picks out, or cherishes. Agamben's description of the appearance of sacred and profane seems to take place in experienced or even historical time: what has once been set aside by an act of sacralization, an act of profanation then takes back for common use. But perhaps it makes more sense to look at it as a schema of relations that does not grant either side temporal or logical priority. As Agamben acknowledges, sacred and profane are mutually constitutive poles of the system that properly distributes the world among gods and men, and each sphere retains traces of the other (77–81). Ultimately, it is unclear which realm has precedence, or even priority. For Kahn, too, the first religious cut seems to produce the separate spheres of the sacred and the secular; one might then say that division is not really religious, but in fact secularizing.[24] The same division establishes both realms in one act. Profanation is the restoring of what has been set aside for the gods to the common, but it is not an undoing of the division of religion so much as support and continuation of it.

In many of the texts that trace the contemporary return to religion, the transcendence that characterizes the religious sphere appears as a calling from beyond the world.[25] Like the allotment of the world's demesne, such a call is religious in structure regardless of its content: Lupton and Hammill, for instance, give the moment of interpellation as described by Althusser as an example of a theological form, however against his expressed intent. Such a call must etymologically be extraordinary (or out of order), metaphysical (beyond nature), *après-coup*, evental, ecstatic, all of which define their realms in terms of their distance from the given: *ordo, physis, coup, e-venire, ek-stasis*; order, nature, cut, a coming-forth or standing-out. Religion is not merely opposite or other to the cycle of the secular, but stands in a determinate relation without/beyond/after/from it. Perhaps this religious hailing is the calling that enacts the category that bootstraps a hierarchy from the induction of specifics (*katagoreuō*, call out in public, or denounce), and thus the conceptualization that stands at the gate of logical *ratio*. This calling to transcendence can be taken up or refused, but this seeming human choice tends to conceal that it brooks no real option. It cannot be overlooked, nor hastened, nor deferred, nor swayed, for the divine is indifferent or impassible to any counter-call from the world. That is what transcendence means.[26] As Heidegger allows with resignation, 'There remains to us only the possibility of readying our readiness for God's appearance, in thinking and in poetry, or for God's absence, in decline … We cannot think him nearer, we may at best only waken the readiness of awaiting.'[27] But Heidegger's real anxiety is not his thought's powerlessness, but its finitude: 'I know nothing about how this Thinking takes effect.'[28]

In contrast to Heidegger's anguished and faithful waiting, Agamben describes another possible response to the calling. A space of secularity is instantiated along with the transcendent realm of religion by the binding cut of sacrifice. But the sacrificial act of division is another act of binding, though, as through it gods and humans are given their allotments and kept separate in their realms – the world and the worldly for us, what is beyond the world for the gods. Agamben calls profanation the *negligence* of this division. We need to read closely enough here to notice that Agamben's negligence is not an overlooking, but a careful not-reading, *nec-legere*: not distinguishing or cherishing, but also not ignoring. As well as being returned to the common share of humankind, what is profaned is also transformed in crossing from the sacred. As Agamben further observes, this category of common use is also readily sacralized in other terms and under a variety of ideologies. The possibility of transcendence tends to convert what lies outside it into its likeness, as if by contagion, which Agamben associates primarily with profanation but which applies at least as well to transcendence. While acts of profanation are always finite and particular – this touch, that word profanes that sacred object – the possibility of reading transcendence transforms in its entirety every field that it traverses. In contact with the category of transcendence, the array of lived experience readily responds by becoming attuned to it, more unified and organized and attendant, 'the everyday', *Erlebnis* rather than a group of experiences, even an immanence, all of which promise more than meets the eye, or the mind's eye, or the mind's motion. At a touch, the originating division cleaves across everything at once; all *sermo* becomes subject to the possibility of subordination to *ratio*, just as conversely what is unsanctified may profane the sacred.

But it is also possible to remain scrupulously within the newly opened field without referring outside that field.[29] Agamben here introduces another state that he calls *play*, an exploratory attention which diverts objects as they pass from sacred to worldly realms, suspending their reorganization into new sacralities and causing them to tarry, as 'pure means' the ends of which have been 'joyously forgotten' (86).[30] For Agamben, to secularize is to try too hard to forget the fact of the division of the sacred and profane, or to dwell in the human world by repressing the sacred. To profane something, on the other hand, recalls and supports the realm of religion even as it draws back from and suspends it. To play is to prorogue giving over to use and commonness what has become profane, to hold open the possibility of transforming it in activity.

Play mediates differently between the religious and profane worlds than the sacrifice that initiates and divides them. For one thing, play is extensive; as pure means and negligence, it resists the closures of reading (*legere*) and binding by dividing (*religere*). Its infinite extensibility across positions which are finite makes play ineffable, although not as the ungraspable totality of the sacred is ineffable.[31] It is non-hierarchical, in that it assumes no interior position of privilege. Everything within it operates by its own rules.[32] Play as a suspense of the religious is thus *horizontal*, in two senses. In early modern terms, the secular formation of play is represented as the *theatrum mundi*. This frequent metaphor is imagined in two distinct ways: as a stage fully visible only from outside it, or as a wholly encompassing space of display that entirely lacks, or ignores, an outside; as *ratio* when the *theatrum mundi* is the spectacle of the world presented to God or to those who know to behold it from outside, as *sermo* when it is the profane playing space within which humanity makes its life for itself.[33] Play is thus, finally, *negligent* in Agamben's sense: that is, as an intensive not-reading, a refusal to privilege by selection one class or category of cases, it is resistance to the re-ligious re-reading that reaffirms division so as to imagine its transcendence. Play emerges in response to the world divided between sacred and secular, as the possibility of its transformation.[34]

I want now to turn to an example of this sort of profane, playful reading in Angelo Poliziano's *Miscellanea*. From one perspective, the humanists of the Renaissance can be fitted fairly comfortably into the framework of the secularization thesis. Products of a period that emphatically did not worry that God was dead, they concentrated on the recovery and establishment of secular, ancient texts, most of them pagan. They avoided ideas that could easily be transferred from one field to another. Instead, they gave their attention to the specificities of their world, in particular to history as the product of human action, largely through a shared but incompletely methodized collection of practices and tactics that they most often called *philology*.[35] Philology was a field of study in the ancient world, too, but a relatively undistinguished one compared to the master discourse of philosophy. In his epistles, Seneca regretted the decline of the search for truth into flirtations with grammatical curiosities, so that 'what had been philosophy has become philology'.[36] Nietzsche slyly transvalued it in his inaugural lecture at the University of Basel: 'what has become philosophy had been philology'.[37] Even when what was at stake was the proper use of words, ancient writers tended to view philology as merely instrumental; in the antiquarian writings of Aulus Gellius, for instance, philological insight derived from and signified true philosophical knowledge.[38] But the humanists of the early Renaissance took up this rebarbative work for itself, working at hard knots of words that resisted translation, to which they learned to shape themselves rather than shifting them easily from context to context.

Renaissance humanists and their philology were far from secular, if that is taken to mean overlooking the field of religion; they were in constant converse with it. Lorenzo Valla's exposure on linguistic grounds of the Donation of Constantine as a late forgery, or Erasmus' edition of the Greek New Testament to replace the aging Latin Vulgate, are both one kind of example of the secularizing labor of the humanist philologist on the religious text. But Valla and Erasmus substitute a different external authority for a Christian one. Erasmus seeks to uncover the historical origins of Holy Scripture, and Valla looks to the changing usage of Latin to locate its variants within a particular space of time and show that the Donation was written long after Constantine. They recreate the structure of transcendence of religious thinking, although they do not call on a god for aid. They too distribute their materials between two different realms, privilege one of them, establish principles of order, and set a limit on the expansiveness of their field; it was the philology of the humanists that turned Latin from a lively tongue to a dead and static one. They do not, in Agamben's sense, profane the religious, so much as they extend its principles to the secular, or substitute secular terms for its.

Angelo Poliziano, scholar and poet in the brilliant circle surrounding Lorenzo de' Medici, offers another kind of example. Unlike Giovanni Pico della Mirandola or Marsilio Ficino, also members of Lorenzo's court, Poliziano avoided theology and philosophy, and wrote instead on philology, law, and literature.[39] Modern scholarship makes the appearance of Poliziano's *Miscellanea*, the first part in 1489, the second left unfinished at his death in 1494, a watershed in the history of philological scholarship.[40] Written in explicit imitation of Aulus Gellius' *Noctes Atticae* and Claudius Aelianus' *Poikilês Historias* (*Various Studies*), Poliziano's *Miscellaneorum Centuria Prima* discussed one hundred philological problems in essays ranging in length from a dozen or so lines to several pages. His *Centuria Secunda* was never completed, and in fact disappeared until it was rediscovered in 1961. In its loudly professed but carefully modulated carelessness of form the first century of *Miscellanea* somewhat resembles other humanist exercises in Latinity like Valla's *Elegantiae* (1441) or Filippo Beroaldo's *Annotationes centum* (1488): 'unordered and assorted, like a thicket (*silva*) or a stew (*farrago*), I have written it not continuously or continently, but by bounds, and piecemeal'.[41] Like these works, too, it poses its horizontal sprawl against both the comprehensive order of the scholastic *Summa* and the exhaustiveness of the humanist edition, two exercises of *ratio* and religious (in the broad sense) thought; indeed, Poliziano names Aquinas in his first chapter[42] and sets a series of readings of Catullan lines throughout the *Centuria Prima* against Antonio Partenio's recent edition of the poet.[43] Closure, order, hierarchy, a distinction between inside and outside, the qualities of the system of the secular and the religious, are all replaced in *Miscellanea* by a leveling horizontality – extensive, undifferentiated, flat, *negligent* in that it refuses to pick out or privilege particular sources, but nevertheless careful, scrupulous, and deeply invested.

The *Miscellanea* demonstrate their secularity in content as well as form. They cover not the Latin of the Church, but the Latin and Greek of pagan writers. They are self-consciously trivial, in every sense; they celebrate the minutiae of classical texts, but also the power of the *grammaticus*, as Poliziano proudly styled himself, the teacher of the academic *trivium* of the elementary language arts of grammar, rhetoric, and logic.[44] Aulus Gellius, one of Poliziano's models for a collection of loosely related anecdotes about usage, meaning, and history of language, is – like Seneca – relentlessly dismissive of the mere *grammaticus*, whose narrowness of knowledge is regularly shown up in conversation with Gellius and his friends, whose grammatical chops are outward signs of less visible philosophical depths. The *Miscellanea*, in contrast, refuse to subordinate grammar to philosophy, or even to distinguish them fully except in the terms set out by the *grammaticus*. As the knowledge of language that is prior to any work that makes use of it, grammar is the discipline most capable of adjudicating the discourses of all other branches of learning. In the contemporaneous *Lamia*, an introduction to a set of lectures of Aristotle's *Prior Analytics* that opened with *Panepistemon*, Poliziano acknowledged with false modesty: 'Whatever that animal may be that people call a philosopher, I am sure that you can figure out easily that I am not one.'[45] As a *grammaticus*, though, he alone is able to say who is meant by the otherwise inscrutable term *philosophus*:[46] while the *grammaticus* does not engage in philosophy himself, he serves as the arbiter of what sorts of practices fall rightly under the name.[47]

Poliziano advises that the necessary training for the *grammaticus* must be literally encyclopedic, encompassing 'the circle of learning'.[48] Similar claims had been made for the orator by Quintilian and the architect by Vitruvius, those favorite pedagogues of the Renaissance. But neither of these writers wants his ideal practitioner of what he chooses as the master art to be consumed by the trivial matters that engaged the *grammaticus*: Quintilian, for instance, does not want his orator to labor on 'ceratinous or crocodilian ambiguities' (1.10). But Quintilian's offhand reference also demonstrates that he does expect the orator to recognize such references, if only enough to know to overlook them. Poliziano, however, in fact dedicates two careful, negligent (in Agamben's sense) chapters of *Miscellanea* to defining precisely what ceratinous or crocodilian ambiguities are (1.54, 55). His willingness to take them up trumps Quintilian's studied superiority to them; not only can Poliziano explain Quintilian's references, he can also show that his detailed grammatical knowledge enables him to explain what baffles Quintilian's supposedly higher learning.

This negligence of hierarchy, even when he is carefully sorting lexemes or splitting dilemmas, is one of the qualities that distinguishes the profane humanism of Poliziano's text from similar ones by his contemporaries. A commentary that moves unpredictably through its text, picking out famous names, Greek words, or what it can handle easily, 'now at a meaning, now at a word, and then at either' – the *legere* in *religere* – whether correct or not, always distorts what it takes as it subject because it begins by splitting word and meaning.[49] It is not that word and sense are linked by any necessity – humanist thinkers almost universally accepted the conventionality of language.[50] But word and meaning exist on the same interpretive level and in the domain of the *grammaticus*, whose trivial knowledge turns out to be the legitimating ground for the philosopher's pretended seriousness. *Miscellaneorum Centura Prima* exposes how surfaces are misread as signs of depths: 'Never lacking are those to whom (as Marcus Tullius says) any alabaster box seems be full of strong perfume.'[51] Poliziano's metaphor neatly diagnoses the lure of exteriority in several dimensions: rather than being sensed directly and without distinction, involving sight and scent together as one extensive phenomenon, the alabaster box is, literally, analyzed by a division into exclusive parts. This partition in turn suggests imaginary relations of implication that cross from one part to another. The expensive outside of the box is mistaken to imply a correspondingly rich inside, but in a wholly different mode; the absence of connection and of scent alike are overlooked in the fidelity to what is mistaken as a sign. Poliziano's warning shows how an initial division into exterior and interior projects assumptions from the accessible into the hidden, which return in the form of a message from beyond appearances (or in this case, the within). Poliziano makes this point to check the automatism of such an analysis. Instead of imagining his text as a container for sense, Poliziano suggests that his readers think of it as a colored throw rug. The word he uses is στρώματα, a generic term for things that are spread out – in other words, for horizontal surfaces.[52]

The extensiveness, rather than comprehensiveness, that character-izes the *Miscellanea* makes them impossible to delimit entirely. This deterritorializing impulse lets Poliziano and his readers resist both division and totalization – seen as related, because both work through delimitations and hierarchizations – of the text into a form that transcends the array of its elements.

Probably the most famous of Poliziano's notes is as good an example of his approach as the first century provides. It is a gloss on Catullus' second poem, 'Passer, deliciae meae puellae' ('Sparrow, my girl's delight'), in which the poet describes his love playing with her pet bird, cuddling, stroking, and kissing it.[53] Poliziano cuts to the chase: 'That sparrow of Catullus, as I judge, allegorically figures a certain obscene understanding, which I don't know how to express while preserving my modesty.'[54] But the will to knowledge prevails, and Poliziano adduces a few comparable lines of Martial:

> Give me kisses, but Catullan ones,
> And if they end up being as many as he said,
> I will give you Catullus' sparrow.[55]

As Poliziano observes, though, only an inept poet would 'say that he is going to give the boy Catullus' sparrow after the kisses, and not, I suspect, something else more substantial'.[56] Bringing Martial's apparent understanding of Catullus' sparrow back to Catullus' poem, Poliziano advises, 'What this is, for the sake of shame I leave to everyone's own conjecture about the sparrow's native wantonness.'[57]

Despite describing the poem's meaning as 'allegorical', Poliziano nevertheless does not ascribe to it levels of meaning or any of the usual metaphors for the relation of sense to expression in allegory. Instead, a word in one context is used to interpret another in a different one. This becomes more striking still when we consider two analogues that Poliziano passes over, one a more explicit poem from about forty years earlier by Jacopo Pontano and based on Martial's epigram, the other a second-century CE gloss preserved in a medieval codex that assimilates *strutheum* (from Greek *strouthos*, 'sparrow') with the 'obscene male member'.[58] Both of these texts were known to Poliziano, and both make his so-called allegorical reading easier. But he mentions neither of them. For Poliziano to cite either a modern poet or an earlier grammarian on the sparrow risked making a distinction between Catullus' Latin and a metalanguage needed to decipher it. By bringing one ancient work into contact with another, Poliziano avoids the need to establish a metalanguage of individual authority or ancient grammatical theory. He remains within a shared textual Latinity. Meaning accumulates through the linking of equally weighted examples rather than deriving from a source outside the operations of Poliziano's ancient texts. His philology does not look for any answers within, beyond, or beneath its texts, but only beside them; the only relations it conceives are horizontal ones.

This is not a typical example from the text of the *Miscellanea*, but it is not unusual either. In fact Poliziano's text has no really typical chapters, no apparent patterns of organization, and nothing completely different from or particularly representative of the rest. Some chapters compare readings in various manuscripts; some, like this, compare words from different texts; some correct physical errors made in copying or lineation; some tell what seem to be jokes (which are not very good). Variety is part of the alogicality of the collection, and a set of demonstrations of the resistance of *sermo* to organization in the terms of any *ratio*. Nevertheless, Catullus' sparrow is a good example of the kind of work that comprises the *Miscellanea*: singular, unconnected, multiply related to other works. Poliziano offers no model of the world divided from itself or contained within itself, but the world itself, coterminous with itself and conceived as a discursive field. It is a lifeworld, not an entity; a condition of active, verbal be-ing rather than a state, or standstill, of being. Language is not only the tool through which the world is investigated; it determines – actually – the world. The world of which *Miscellanea* are a part, in fact, is something like the linguistic representation of what might be, inexhaustible rather than ineffable.[59]

The practices that I have been setting out as examples of Poliziano's profane and humanist criticism are given by Poliziano the ironically suggestive name of *divinatio*, a term that passes into the vocabulary of modern textual criticism through the foundational work of Karl Lachmann in the nineteenth century.[60] In *Panepistemon*, Poliziano's description of the different branches of science written at roughly the same time as the *Miscellanea*, *divinatio* is a 'mixed' form of knowledge, combining revealed, divine knowledge and invented, human knowledges; it refers to any of the methods of prophecy through the reading of signs of the gods' intent. But the second century of the *Miscellanea* begins with a very different account of *divinatio* as the mode of Poliziano's own philological work. The first chapter 'De Divinatione' begins with the example of the text of Cicero's *De natura deorum*, to which Poliziano applies a tale he takes from that text, the story of Hippolytus. Unjustly cursed by his father, Hippolytus is driving furiously away along the road in his chariot when a monster rises from the ocean and terrifies his horses; in their panic he is thrown from the chariot and dragged to death along the road. Obviously much could be said about this account of the traumatic encounter of the human and the divine, but Poliziano neglects the story of Hippolytus for that of his healer,

> that Aesculapius who, as the stories say, when those limbs which had been torn away in every direction, gathered, reconfigured, and restored them to life; and who nevertheless then is said to have been struck down by a thunderbolt because of the jealousy of the gods.[61]

This version extends past the death of Hippolytus, torn between the competing demands of several gods, and makes the gods' final victim the healer Aesculapius, who restores Hippolytus' torn body to life and thus sets himself against all gods at once. Aesculapius is not a transgressor like Phaethon, another mortal who takes the gods' power for his own. He extends the power of the human to reach to that of the gods. Grammatically, he restores Hippolytus' body 'to life' rather than restoring life to his body; that is, he does not draw his capability from another realm, but broadens the capacity of his own world. He is also the figure of the secular with whom Poliziano associates his philological work.[62] The chapter that follows restores the mutilated text of Cicero's *De natura deorum*, 'deprived of head and hands' like its writer when he was murdered by his political enemy Marcus Antonius after the death of Caesar.[63] Poliziano proceeds not by excavating a meaning that Barthes might call theological which could govern reconstruction, but by attending to the recurrence of a few phrases in the text. Gathering, reconfiguring, and restoring these to sense, Poliziano deduces a singularly insignificant cause: the physical dislocation of a sheaf of pages from a copy at some point during its transmission, when they were somehow misfiled into the rest of the manuscript. Reconstructing this history of accident and contingency is what Poliziano calls *divinatio*.

Divinatio is something like the philologist's last resort, a best guess at a reading of a text in the absence of decisive evidence. It is also, despite its name and because it proceeds 'not by the authority of antiquity but by conjecture', something like the mixed art of philology at its very purest.[64] For Poliziano at least, the name is ironic: where certain knowledge cannot proceed, the philologist makes a choice based on whatever signs he can. But unlike the revelatory calling that marks the advent of the religious, these signs originate within the horizons of the text. Philological *divinatio* relies on human intuition rather than on divine intervention, and leads to a qualified reading that is passionately affirmed, but that also resists legitimation through external means and is subject to possible revision in the light of new argument.[65] Nothing at all stands beyond it, only beside it and in addition to it. What is named a kind of revelation is not, and the gods are shown to be destructive, vindictive, and incommunicative, as if Aesculapius' death were the revenge of religion on the possibility of the secular. In this account, secular work comes to the aid of the sacred to restore its text and 'the parent of the Roman tongue and of Roman philosophy … deprived of head and hands'.[66] Every element that we might consider religious – paternity, philosophy, divinity, a traumatic interruption from beyond – is present in Poliziano's account of Aesculapius, but they are neglected in order to carry out another kind of work.

Religion is the category of the transcendent, beyond the given or apparent. But there can be no corresponding category of the secular, as the practice of *divinatio* shows. The contingent, incomplete horizontality of the realm of *sermo* is resistant to the smoothing appropriation of *ratio*, which seeks to supplement it with a hidden structure on a different level of organization. In an essay called 'Judgment Day', Agamben considers the effect of the ordinary through looking at some photographs that he says 'I love'. This is meant to be a category of sheer contingence, clearly, and it shows more yet: for instance, Agamben loves best those photographers who take pictures 'walking without any goal and photographing everything that happens' (23). This would seem to be the very type of the secular, if there were one. His longest look is at a daguerreotype of the Boulevard du Temple, said to be the first photographic image that reveals a human being, a small figure who accidentally has remained still long enough while getting his shoes shined to be recorded by the slow shutter of the camera. In the lyrical description that follows, though, the image is elevated to transcendence as a vision of the Day of Wrath.

In Agamben's choice of this image, something ordinary is recast – not without Agamben's awareness, to be sure – as something no longer of the world. It is generalized into a symbol for the ordinary, indeed 'the most banal and ordinary gesture', which Agamben rightly notes invests it with 'the weight of an entire life', a numinousness alien to both that life and that gesture, to the extent that the image itself calls from beyond: 'the subject shown in the photo demands something from us' (24–25). Here Agamben's act of selection, *legere*, is religious. It ontologizes its singularity, transcending each image, each moment, and passing into religious reverie on what is sometimes misleadingly called the everyday – misleadingly because this word tends to reach past the singularity of moments that closely resemble each other but never fully replicate each other, shaping them into a category that already, and necessarily, transcends the manifold of their small differences. The photograph is an instrument that subverts secular things.

In contrast, Poliziano's effort is recording and comparing particulars, the unglamorous, inhuman letter devoid of spirit, failing to be distinct enough even for abjection, 'imitating the counters of actuaries'.[67] His philology is not subversive, but elusive. Refusing to draw lines between here and a beyond, it singularizes, dissolving the world into uncountable instances. This philology does not look for, or wait for, revelation. It follows signifiers in the uncertain but commonly available ground of language, *sermo*, other signifiers – in the text as a signifier of the present rather than in an author or meaning projected past it. What is secular neither stages nor answers the spectacular call from beyond. It does not call out. It does not wait to be found. It is what is uncalled for, what must be traced by a specifically human and worldly attention. But it equally resists the recovery of its profaneness into another religiosity. When de Man noted in 'The Resistance to Theory' that 'Technically correct rhetorical readings may be boring, monotonous, predictable and unpleasant, but they are irrefutable',[68] the penitential rigor of deconstructive readings serves rhetorically to underscore their seriousness.[69] Poliziano's readings are, in contrast, readily refutable, certainly arguable; they rely on no forms of external legitimation and cannot set bounds for contestation. Perhaps a philology of fragments that neither promise wholes nor gesture toward a deconstructive abyss, but simply preoccupy an undefined field of possibility, can only appear exciting within the horizon of a society that is not surprised by its own deep religiousness, in which the absolute and absolution are already everywhere. Such extensive, ungrounded, decisive but inconclusive work does not ask for the sublime glamor that the predication of religion bestows on an ordinary thing, and preserves the things of this world in their secularity.

Poliziano's humanism does not set itself the task of assimilating the unassimilable, but simply of exposing what is not assimilated – not, though, the unassimilable, as if that could be a category *per se*, like the secular, and not with the idea of recuperating it or making it speak, but simply to mark it, to note it where one might equally have passed over it without noticing it, which is to say, to make it stand out, dissimilate it, profane it, to score difference where it was invisible or unseen, again not with the idea of naming or identifying it (from *idem-facere*, to make the same), but to proliferate it.[70] Poliziano's philological praxis resists framing itself against an imaginary mirrored *alter ego*, or an act of communion with another agent; it extends itself in language both inhuman and utterly ordinary.

What Agamben calls the religious act of *re-legere* has an early parallel, at least in structure, in a word that puzzles Poliziano's model Gellius, *rescire*. The suffix *re-*, Gellius notes, almost always means 'again'. What, though, is its force in *rescire*, a rare word combining *scire*, 'to know' and *re-*? He decides that '*rescire* is properly said of one who learns of something that is hidden, or unlooked for and unexpected'.[71] In formation *religere* and *rescire* should be synonyms, but in practice they promise something different, and additional; both are modes of *negligence* as Agamben saw it. A philologist like Poliziano finds this out by bringing the words together, so that the uncalled-for word reveals something else unlooked for. This inhuman practice of philology grounds a secular criticism, the subject of which is literally made of humanity but not in its image: 'something that is hidden, or unlooked for and unexpected'.

Notes

I am grateful to Victoria Kahn, Julia Reinhard Lupton, Regina Schwartz, Alicia Sands, and Sam Weber for the conversations, and criticisms, they have shared with me on this topic.

1. 'Gott ist todt! Gott bleibt todt! Und wir haben ihn getödtet!' Aphorism 125, *Die Fröhliche Wissenschaft*; cf. Aphorism 108: 'God is dead: ... And we – we must still overcome his shadow!' ('Gott ist todt: ... Und wir – wir müssen auch noch seinen Schatten besiegen!'; in *Werke: Kritische Gesamtausgabe*, gen. eds. Giorgio Colli and Mazzino Montinari, vol. 5.1 of 9 vols. [Berlin: Walter de Gruyter, 1967–], 159, 145); see also Lacan's comment that God is not dead, but Unconscious (Jacques Lacan, *Four Fundamental Concepts of Psychoanalysis*, trans. Alan Sheridan [New York: W. W. Norton and Company], 27, 59). Roland Barthes, 'The Death of the Author', in *Image/Music/Text*, trans. Stephen Heath (New York: Hill and Wang, 1977), 142–148 (146). Examples of shakily defined 'theology' being set up by critical discourses in order to be demolished could be multiplied almost at will: Heidegger's rejection of ontotheology; early Derrida; Lyotard's suspicion of *grands récits*; Foucault's anti-humanism. Except as noted, all translations are my own.
2. Robert Markley, *Fallen Languages: Crises of Representation in Newtonian England, 1660–1740* (Ithaca: Cornell University Press, 1993), 7.
3. On Barthes' *S/Z*, see Bruce Holsinger, *The Premodern Condition: Medievalism and the Making of Theory* (Chicago: University of Chicago Press, 2005); Martin Heidegger, '"Nur noch ein Gott kann uns retten": Der Philosoph Martin Heidegger über sich und sein Denken', *Der Spiegel 50, Sonderausgabe 1947–1997* (1997, p. 285), interview date 1966, first publication 1976.

4. Graham Ward, *Theology and Contemporary Critical Theory*, 2nd edition (Basingstoke: Macmillan, 2000), viii; see also his 'Introduction', ix–xx.

5. Bruce Holsinger's edited collection 'Literary History and the Religious Turn' (Special Issue, *English Language Notes* 44 [2006]) directed me to Stanley Fish, 'One University, Under God?' *Chronicle of Higher Education*, 1 July 2005. Derrida in his late works explicitly turned toward questions of transcendence, but Fish accurately describes the symbolic weight of the name 'Derrida' as a sign of a particular kind, and time, of theory.

6. Ken Jackson and Arthur F. Marotti, 'The Turn to Religion in Early Modern Studies', *Criticism* 46 (2004), 167–190. The secularization thesis is the claim made by Max Weber and elaborated by many others that modernity is marked by increasing secularization; see Jürgen Habermas, 'Notes on a post-secular society', 18 June 2008 (http://www.signandsight.com/features/1714.html, accessed 18 February 2010) and Jane O. Newman, 'Enchantment in Times of War: Aby Warburg, Walter Benjamin, and the Secularization Thesis', *Representations* 195 (2009), 133–167.

7. Julia Reinhard Lupton, 'The Religious Turn (to Theory) in Shakespeare Studies', *English Language Notes* 44 (2006), 145–149 (146).

8. See the recent collections: Holsinger, ed., 'Literary History and the Religious Turn'; Graham Hammill and Julia Reinhard Lupton, eds, 'Sovereign, Citizens, and Saints: Political Theology and Renaissance Literature', Special Issue, *Religion and Literature* 38 (2006); Victoria Kahn, ed., 'Early Modern Secularism', Special Issue, *Representations* 105 (2009).

9. Edward Said, *The World, the Text, and the Critic* (Cambridge, MA: Harvard University Press, 1983). See Gourgouris for what remains useful in Said's early work after the redistribution of the critical terrain: Stathis Gourgouris, 'Transformation, not Transcendence', *boundary 2* 31 (2004), 55–79.

10. Said, *The World, the Text, and the Critic*, 290.

11. Christopher Celenza, 'Humanism and the Classical Tradition', *Annali d'Italianistica* 26 (2008), 25–49; Paul Oskar Kristeller, *Renaissance Thought: The Classic, Scholastic, and Humanistic Strains* (New York: Harper and Brothers, 1961), 8–13; Anthony Grafton and Lisa Jardine, *From Humanism to the Humanities: Education and the Liberal Arts in Fifteenth- and Sixteenth-Century Europe* (Cambridge, MA: Harvard University Press, 1986).

12. Michel Foucault, *The Order of Things* ([1966] London: Tavistock); but long before, Heidegger in his 'Letter on Humanism'.

13. See Julia Haig Gaisser, 'Some Thoughts on Philology', *Transactions of the American Philological Association* 137 (2007), 477–481; Helmut Müller-Sievers, 'Reading Evidence: Textual Criticism as Science in the Nineteenth Century', *The Germanic Review* 76 (2001), 162–171; Grafton and Jardine, *From Humanism to the Humanities*.

14. Habermas, 'Notes on a post-secular society'; Hans Blumenberg, *The Legitimacy of the Modern Age*, trans. Robert M. Wallace (Cambridge, MA: MIT Press, 1985); and most recently, Stanley Fish, 'Are There Secular Reasons?' *New York Times*, Opinionator blog, 22 February 2010 (http://opinionator.blogs.nytimes.com/2010/02/22/are-there-secular-reasons/?scp=1&sq=fish%20secular&st=cse, accessed 23 February 2010).

15. Theology is not concerned with whether or not God exists, which it takes as a premise; it is concerned with the effects of divine existence. For a statement

of traditional theology see Ward, *Theology and Contemporary Critical Theory*:
'We believe first and theology is the ongoing reflection upon that belief …'
(3); Mark Taylor presents a theology founded not on belief but on seeking (*Erring: A Postmodern A/Theology* [Chicago: University of Chicago Press,
1987]). In neither case, though, is the existence of God established through
theology. Eagleton gracefully addresses the issue of belief in critical religious thinking by analogizing it to belief in psychoanalysis: Terry Eagleton,
Trouble with Strangers: A Study of Ethics (Chichester: Wiley-Blackwell, 2009),
322–324.

16. I base this reduction particularly on Taylor, *Erring*, 7–10, Markley, *Fallen
    Languages*, 7, and Ward, *Theology and Contemporary Critical Theory*, 1, but
    more generally in conformity with the works cited here as examples of the
    return. See Niklas Luhmann's posthumous work on religion, discussed in
    Rudi Laermans and Gert Verschraegen. '"The Late Niklas Luhmann" on
    Religion: an Overview', *Social Compass* 48 (2001), 7–20.
17. On Augustine, see Kahn, 'Early Modern Secularism': 'Introduction'. On
    Luther, see Newman, 'Enchantment in Times of War', 135–136. Kahn rightly
    observes that not all secular thinking relies on a religiously made distinction.
    But to be recognized as a field of action, the secular must be separated from
    something, and it seems fair to generalize that the distinction that yields
    the secular also yields the religious. There are, of course, other distinctions –
    between appearance and being, to give only one example.
18. A 1994 *LA Times Magazine* headline called Habermas the 'Theologian of Talk'
    (Mitchell Stephens, 'The Theologian of Talk', *Los Angeles Times Magazine*, 23
    October 1994). By Habermas' religious thinking, though, I refer to his sense
    of a realm of shared human reasonability. Recently Habermas has guard-
    edly expressed that secularism may need to learn from religions; see Joseph
    Ratzinger (later Pope Benedict XVI) and Jürgen Habermas, *Dialektik der
    Säkularisierung: über Vernunft und Religion* (Freiberg: Herder, 2005). Another
    name for the desire for transcendence, whether traditionally religious or not,
    might be 'the displacement of politics', the desire to find a way around the
    conflicts of human interaction; see Bonnie Honig, *Political Theory and the
    Displacement of Politics* (Ithaca: Cornell University Press, 1993).
19. Giorgio Agamben, *Profanations*, trans. Jeff Fort (New York: Zone Books,
    2007), 74. Subsequent references are cited parenthetically in the text.
20. Régis Debray, *Critique of Political Reason*, trans. David Macey (London:
    New Left Books, 1983), quoted in Hammill and Lupton, 'Sovereign, Citizens,
    and Saints', 3.
21. In *Adversus Praxean*, Tertullian discusses *ratio* and *sermo*, together with *ver-
    bum*, as possible Latin translations of *logos*; and translated it as *Sermo atque
    Ratio* in *Apologeticum* 21.10 (T. F. Pollard, *Johannine Christianity and the Early
    Church* [Cambridge: Cambridge University Press, 1970], 60–65). In his trans-
    lation, Erasmus notoriously began the book of John *in principio erat sermo*
    rather than following the Vulgate's use of *verbum* (C. A. L. Jarrott, 'Erasmus'
    In Principio Erat Sermo: A Controversial Translation', *Studies in Philology* 61
    [1964], 35–40).
22. Debray's premises are challenged in Luhmann, among other sources (Niklas
    Luhmann, *Social Systems*, trans. John Bednarz, Jr., with Dirk Baecker; fore-
    word by Eva M. Knodt [Stanford: Stanford University Press, 1995]). But

they remain valuable in describing what the sphere of religion is claimed to present and the problems it offers to solve.

23. Paul de Man, 'The Resistance to Theory', in *The Resistance to Theory* (Minneapolis: University of Minnesota Press, 1986); Victoria Kahn, 'Humanism and the Resistance to Theory', in Patricia Parker and David Quint, eds, *Literary Theory/Renaissance Texts* (Baltimore: Johns Hopkins University Press, 1986).

24. Kahn, 'Early Modern Secularism'.

25. For example: Kenneth S. Jackson, '"More Other than you desire" in *The Merchant of Venice*', *English Language Notes* 44 (2006), 151–156; Philip Lorenz, 'Notes on the "Religious Turn": Mystery, Metaphor, Medium', *English Language Notes* 44 (2006), 163–172; Lupton, 'The Religious Turn (to Theory)'. See also Ward, who similarly insists that this call distinguishes theology from other disciplines: 'If theology is not to dissolve simply into psychology or, more generally, anthropology, it must have its origin in revelation' (*Theology and Contemporary Critical Theory*, 1).

26. Perhaps this explains the recurrent interest in the figure of Abraham, from Kierkegaard through Auerbach and Levinas, as one whose religion is passing over into transcendence rather than already calling from beyond.

27. Heidegger ('"Nur noch ein Gott kann uns retten"', 285): 'Uns bleibt die einzige Möglichkeit, in Denken und im Dichten eitne Bereitschaft vorzubereieten für die Erscheinung des Gottes oder für die Abwesenheit des Gottes im Untergang ... Wir können ihn nicht herbeidenken, wir vermögen höchstens die Bereitschaft der Erwartung zu wecken.'

28. Ibid.

29. To this one might compare Bourdieu's concept of the *habitus*, although Agamben's play seems to require consciousness of the *habitus* it draws on. Bourdieu's *habitus* is only visible to the outside observer, and so reinstates the hierarchy of the religious over the secular and the detached observer over the involved actor.

30. Gourgouris ('Transformation, not Transcendence') describes a similar pattern regarding Edward Said's 'secular criticism', where the critic essays to draw out the manifold positions that coexist between moments when complexes freeze into distinctly differentiated identities.

31. Roland Barthes, 'From Work to Text', in *Image/Music/Text*, 155–164 (158); Emmanuel Levinas, *Totality and Infinity: An Essay on Exteriority*, trans. Alphonso Lingis (Pittsburg: Duquesne University Press, 1969) – although Levinas' infinity requires a reference to the sacred.

32. It thus could be said to oppose a system of the kind theorized by Luhmann to the kind proposed by Debray. See also Laermans and Verschraegen, '"The Late Niklas Luhmann" on Religion'.

33. See William N. West, 'Knowledge and Performance in the Early Modern *Theatrum Mundi*', in Flemming Schock, Oswald Bauer, and Ariane Koller, eds, *Dimensionen der Theatrum-Metapher in der Frühen Neuzeit: Ordnung und Repräsentation von Wissen/ Dimensions of the Early Modern Theatrum-Metaphor: Order and Representation of Knowledge* (Hanover: Wehrhahn, 2008), 1–20.

34. See Gourgouris, 'Transformation, not Transcendence'.

35. Victoria Kahn, *Rhetoric, Prudence, and Skepticism in the Renaissance* (Ithaca: Cornell University Press, 1985); Victoria Kahn, 'Habermas, Machiavelli and

the Humanist Critique of Ideology', *PMLA* 105 (1990), 464–476; Kristeller, *Renaissance Thought*, 17–18.

36. Epistle 108.23, 'quae philosophia fuit, facta philologia est'.
37. Nietzsche, 'Homer und die klassische Philologie', *Werke*, gen. eds. Colli and Montinari, vol. 2. 1, ed. Fritz Bornmann and Mario Carpitella, 268, '...ich einen Satz des Seneca also umkehre "philosophia facta est quae philologia fuit"'.
38. Aulus Gellius, *The Attic Nights*, trans. John C. Rolfe, Loeb Classical Library, 3 vols (London: William Heinemann, 1927); examples abound. See also Erik Gunderson, *Nox Philologica: Aulus Gellius and the Fantasy of the Roman Library* (Madison: University of Wisconsin Press, 2009), 60–80, on the social position of the *grammaticus* during the Second Sophistic, and Gellius' treatment of him; Aldo Scaglione, 'The Humanist as Scholar and Politian's Concept of the *Grammaticus*', *Studies in the Renaissance* 8 (1961), 49–70 (61–64).
39. Vittore Branca, *Poliziano e l'umanesimo della parola* (Turin: Einaudi, 1983); Scaglione, 'The Humanist as Scholar', 51–52; Kristeller, *Renaissance Thought*, on the complex of fields that became *humanitas*.
40. For example: Anthony Grafton, 'On the Scholarship of Politian and its Context', *Journal of the Warburg and Courtauld Institutes* 40 (1977), 150–188; Mario Martelli, 'La Semantica di Poliziano e la *Centuria Secunda* dei *Miscellanea*', *Rinascimento* 13 (1973), 21–84; Scaglione, 'The Humanist as Scholar'; Vittore Branca's introduction to Angelo Poliziano, *Miscellaneorum Centuria Secunda*, ed. Vittore Branca and Manlio Pastore, 4 vols (Florence: Alinari, 1972), vol. I; Vincenzo Fera, 'Il Dibattito Umanistico sui "Miscellanea"', in *Agnolo Poliziano: Poeta Scittore Filologo: Atti del Convegno Internazionali di Studi Montepulciano 3–6 novembre 1994* (Florence: Le Lettere, 1998), 333–364. Peter Godman offers a contrary view: *From Poliziano to Machiavelli: Florentine Humanism in the High Renaissance* (Princeton: Princeton University Press, 1998).
41. 'At inordinatam istam, & confusaneam, quasi silvam, aut farraginem perhiberi, qui non tractim, & continenter, sed saltuatim scribimus, & vellicatim...', Angelo Poliziano, *Opera Omnia Angeli Politiani, et alia quaedam lectio digna ...* (Venice: Aldus Manutius, 1498), 157r.
42. Ibid., 167r.
43. Julia Haig Gaisser, *Catullus and his Renaissance Readers* (Oxford: Clarendon Press, 1993), 70–78; Fera ('Il Dibattito Umanistico sui "Miscellanea"') on other arguments in it. *Castigatio* or correction and 'chastening' of texts, which I would call a religious ordering, is exemplified by Ermolao Barbaro's *Castigationes Plinianae* (1492) and analyzed by Stephanie Jed, *Chaste Thinking: The Rape of Lucretia and the Birth of Humanism* (Bloomington: Indiana University Press, 1989); see also Grafton, 'On the Scholarship of Politian and its Context', 152–157.
44. See Scaglione, 'The Humanist as Scholar'.
45. 'Videamus ergo primum, quod nam hoc sit animal, quod homines philosophum vocant, tum spero facile intellegetis, non esse me philosophum', Poliziano, *Opera Omnia Angeli Politiani*, 327v.
46. Ibid., 333v.
47. Kant will repeat this move in his *Streit der Fakultäten*, albeit in a world that, unlike Poliziano's, assumes a religious divide between public and private reason and will.

48. Poliziano, *Opera Omnia Angeli Politiani*; see also Godman, *From Poliziano to Machiavelli*, 81.

49. Poliziano, *Opera Omnia Angeli Politiani*, 157v, 'Idque nunc ad sensum, nunc ad verbum, quod hic utrunque...'

50. See Lodi Nauta, *In Defense of Common Sense: Lorenzo Valla's Humanist Critique of Scholastic Philosophy* (Cambridge, MA: Harvard University Press, 2009), and Richard Waswo, *Language and Meaning in the Renaissance* (Princeton: Princeton University Press, 1987).

51. Poliziano, *Opera Omnia Angeli Politiani*, 157v, 'Nec enim desunt, quibus etiam (ut ait Marcus Tullius) alabastrus unguenti plena putere videatur'.

52. Ibid., 157v, 'στρωμάτεις, quasi stragula picta'. Roughly contemporaneously with Gellius, Clement of Alexandria had used the title Στρώματα for a work of theology.

53. Gaisser, *Catullus and his Renaissance Readers*, 74–78. Poliziano's interpretation remains current and in discussion today.

54. Poliziano, *Opera Omnia Angeli Politiani*, 160v, Cap. 6: 'Passer ille Catullianus, allegoricos ut arbitror osceniorem quempiam caelet intellectum, quem salva verecundia, nequimus enuntiare'. The verb *caelet* is literally to cut or engrave, preserving the sense of a single level of interpretation.

55. Martial, *Epigrams* 11.6.14–16: 'da nunc basia, sed Catulliana: / quae si tot fuerint quot ille dixit, / donabo tibi passerem Catulli'. The 'Catullan kisses' refer to another famous poem in which the lover counts the kisses he plans to give to his beloved.

56. Poliziano 1498, 160v 'si Catulli passerem denique, ac non aliud quippiam, quod suspicor, magis donaturum se puero post oscula'.

57. Poliziano, *Opera Omnia Angeli Politiani*, 160v: 'Hoc quid sit, equidem pro stili pudore, suae cuiusque coniecturae, de passeris nativa salacitate relinquo'.

58. Gaisser, *Catullus and his Renaissance Readers*, 76.

59. Cf. Salvatore Camporeale's understanding of Valla on *res*, *verba*, and the *sensus communis*, Nauta, *In Defense of Common Sense*.

60. Lachmann, for instance, picks up Poliziano's term, Paul Maas, *Textual Criticism*, trans. Barbara Flower (Oxford: Clarendon Press, 1958). See also Müller-Sievers, 'Reading Evidence'.

61. Poliziano, *Miscellaneorum Centuria Secunda*, 4.1r.

62. Branca, introduction to Poliziano, *Miscellaneorum Centuria Secunda*; Thomas M. Greene, *The Light in Troy: Imitation and Discovery in Renaissance Poetry* (New Haven: Yale University Press, 1982), 169. Giamatti misses this instance but looks at Hippolytus as a model of humanist recovery (A. Bartlett Giamatti, 'Hippolytus Among the Exiles: The Romance of Early Humanism', in *Exile and Change in Renaissance Literature* [New Haven: Yale University Press, 1984]).

63. 'romanae vel linguae vel philosophiae parentem ... truncatum capite et manibus', Poliziano, *Miscellaneorum Centuria Secunda*, 4.1r.

64. 'Hoc autem loco non vetustatis auctoritate sed coniectura nitimur duntaxat', Poliziano *Miscellaneorum Centuria Secunda*, 4.1r.

65. On openness to revision as a form of irony, see Richard Rorty, *Contingency, Irony, and Solidarity* (Cambridge: Cambridge University Press, 1989); Jonathan Sheehan, 'Sacrifice Before the Sacred', *Representations* 105 (2009), 12–36,

shows the undecidability of figures that 'unmask' ideology, for instance Hobbes (18–20).

66. 'romanae vel linguae vel philosophiae parentem... truncatum capite et manibus', Poliziano *Miscellaneorum Centuria Secunda*, 4.1r.

67. 'Imitabor igitur sectiones illas medicorum, quas Anatomas vocant. Imitabor & tabulariorum calculas', Poliziano, *Opera Omnia Angeli Politiani*, 334v.

68. Paul de Man, 'The Resistance to Theory', 19.

69. Müller-Sievers ('Reading Evidence') on the lure of the text that does not require reading.

70. Jonathan Goldberg and Madhavi Menon, 'Queering History', *PMLA* 120 (2005), 1608–1617, show that even the concept of identity is scored with difference.

71. Gellius, *The Attic Nights*, I. 2.19.

8
'Grace to boot': St. Paul, Messianic Time, and Shakespeare's *The Winter's Tale*

Ken Jackson

The Winter's Tale provides an extraordinary dramatic rendering of Paul's conception of the way time contracts or begins to come to an end following the messianic moment of 'grace' (*charis*) that is Christ's resurrection. In making this claim I rely on a great deal of late twentieth-century scholarship on Paul but I am responding, in particular, to Giorgio Agamben's *The Time that Remains*.[1] Simply put, Agamben's discussion of Paul and time helps begin to make clear how deeply Shakespeare read the apostle – although Agamben's view certainly cannot account for the full range of Shakespeare's response to Paul. I also would suggest, then, that the play is 'religious'.

But I would ask the reader to hold that term in abeyance and consider the argument before deciding what that term might mean in the context of current debates surrounding literary critical theory. As David Hawkes suggests in his provocative contribution to this collection, we are all (too) enamored at the moment by 'materialist' criticism and thus, I would suggest, are all too quick to find 'transcendences' to quash en route to this materialist heaven.[2] This is a bad and tedious intellectual habit, I think, and one that is particularly troubling to those interested in religion because part and parcel of this habit is a tendency to equate the religious with the transcendent. Things are a bit more complicated than that as I hope to show here. And it is in fact unwise, as current critical theory seems intent on doing, to hastily reinstitute old 'Enlightenment' divides between faith and reason, secular and religious, immanent and transcendent – especially since critical theory has just spent several decades critiquing the Enlightenment. The primary assumption of my argument here is that Shakespeare read Paul closely. In that, for example, I am in concert – rather than at odds – with Gary Kuchar's essay here, '"Love's Best Habit": Eros, Agape, and the Psychotheology of *Shakespeare's Sonnets*', even though his reading produces a 'non-religious' Shakespeare. Indeed, like Kuchar, I think it is critical to consider genre here. What I have to say about Shakespeare's religion ultimately depends on the participatory nature of drama, something quite distinct from the author/reader dyad of the sonnet form. In general, critical theory is much better off, I think, looking for continuities in approaches rather than old divides in order to develop new terminology and paradigms to deal with a 'post-post-Enlightenment' world.

Agamben and Paul

Agamben emphasizes the fact that Paul is not a prophet who predicts the coming of the messianic event (again, a moment of grace) but an apostle who speaks *after* the messianic event has happened. This basic distinction is often overlooked because subsequent developments in Christianity turned the religion's attention to the second coming (*parousia*) and thus to the future. To be more precise: primarily because of subsequent *non*-developments in Christianity – the fact that the messianic event did not seem to change all that much in a Christian's lived experience and the fact that the second coming has been deferred longer than Paul expected – the religion's attention rather naturally turned to the second coming (*parousia*) and thus to the future.

At the beginning of his book, Agamben cites Jacob Bernays' ironic remark that 'to have the Messiah behind you does not make for a very comfortable position'.[3] Indeed, from the earliest moments of Christianity to the present day priests and preachers have struggled to remind their followers that the messianic event already *has happened*, that one is living in Christ. Early Christian apologists generally had a tough time of it in many respects but, on this point, they probably had it a bit easier than those preaching today. As time went on and the expected second coming never came it became all the more difficult for church leaders of one sort or another to argue the point. Simply put, living in Christ is as hard as any other kind of living so 'Christians' have come to focus even more on what is yet to come and how 'to come' might more drastically change things. In brief, then, it is relatively easy to understand the tendency to 'forget' that Christ in fact has come and look for some other – better and more magnificent – event yet to come as the focus of the faith.

All this involves, as suggested, no complex theological apparatus to understand. A cursory scan of everyday experience in the 'Christianized' West is sufficient. The fact that Easter morning comes every year diminishes the holiday's larger significance whereas Christmas seems to expand infinitely subsuming even the 'secular'. Ask any self-identified Christian to articulate the distinction between the messianic event and the *parousia* and you will find, more often than not, a tendency to see the two as duplicates, part one and part two, a coming and a second coming – even though for Paul the two 'events' are quite, quite different. As the passage reads in *The Book of the Common Prayer*, 'Christ died. Christ is risen. Christ will come again.' That this is the 'mystery' of faith tends to be overlooked. It is not just the Anglicans or Episcopalians who focus on the coming again part as if it is a second, and more important, act.

Even counter-examples reveal the same tension: If a contemporary cultural critic is looking to explain the energy and power of Evangelical Christianity in America they could do a lot worse than to point simply to the fact that various Evangelical movements have been extraordinarily successful in reminding those predisposed to Christian belief that the Christ event already has happened. The simple Evangelical argument that one need not wait but one can live in Christ now can be remarkably effective – all the more so because it does, whether one likes it or not, genuinely point back to the earliest Christian moments and the significance of the Christ event rather than the *parousia*. Depressingly, one has to admit there is something fundamental about Christian fundamentalism.

Both Christians and those seeking to understand Christianity from a critical distance are admonished, then, to remind themselves that Paul spends most of his energy not on what is to come but on convincing listeners and readers that they live in the time that remains, the 'messianic' time that is left to us after the messiah has come but before the second coming and the end of time (*eschaton*). He really does not have all that much to say about the second coming. For Paul, that is on the immediate horizon and presumably it will take care of itself. Living in messianic time, in contrast, requires some explication.

For the sake of clarity, then, let me start again with Agamben, and the apostle Paul and time. Time begins to come to an end with the Christ event. Time begins to contract at that point. When Paul says things like 'time is short' (1 Corinthians 7:29) and 'the facion of this world goeth away' (1 Corinthians 7:31) he is not issuing simple repetitive announcements about the end of time or some apocalyptic warnings about the (non-)future – although, importantly, he does believe the end time is very near – but struggling to articulate the practically imperceptible shift in which time already has collapsed into itself, has contracted into what he calls the time of the now (*ho nyn kairos*).[4] Messianic time is thus not a new time or a 'supplementary time added on from outside' regular, chronological time but is, instead, 'like a time within time ... not ulterior but interior'.[5] The time of the now is then, for better or for worse, very, very much like the time we routinely experience. In fact, Paul makes clear that messianic time cannot be completely disentangled from chronological time. In some sense, the two need each other.

Given, again, that the 'vast majority of today's literary theorists claim allegiance to materialism' (Hawkes) it is perhaps useful to cast this 'religious' ontology in the terms of the philosopher of immanence *par excellence*: Gilles Deleuze. One could say that the Pauline Christ event brings about a Deleuzian awareness, at least for Paul, that our lasting 'ground of experience is not the present' or the particular 'but the non-present, i.e., the whole continuum of time itself' (or what Henri Bergson calls the 'pure past'), an awareness that imagined at its limit evacuates any individual's sense of himself or herself as distinctive.[6] As Paul might put it, we are all 'one' in Christ. In Deleuzian and Bergsonian terms, the reality we experience as real or 'actual' is really ephemeral while the unrepresentable or unthinkable of the 'whole continuum of time itself' – the 'virtual' – is substantial, material, immanent, what is. The Deleuzian virtual, then, is not unrelated to Pauline messianic time. The virtual is what truly 'is', but we can experience it only through the actual just as, following the Christ event, we are, in fact, in messianic time but we only come to 'know' that messianic time through chronological time.

The Deleuzian virtual is very much like a 'field of energies or potentialities' that only comes to be in the actual. As Peter Hallward puts it, 'A virtual creating is the reality that lives in any actual creature', but we cannot avoid or jettison the actual to get to the virtual.[7] The virtual and actual, the creating and the creatural, exist in a constant tension. Thus, for Deleuze, the virtual certainly does not *transcend* the actual in some higher plane of being. As critical theorists so often repeat these days, there is a univocal ontology posited here.[8]

Importantly, and perhaps surprisingly for some who persist in simply equating all religious thought with a push for transcendence, this is more or less the case for Paul, too. In Paul's messianic time, what we could call the 'new' virtual experienced as messianic time and the 'old' actual chronological time are given as one, the former only revealing itself in the latter. There certainly may be, again, a transcendence to come implied in Paul's *parousia* or second coming, a transcendence that would obliterate the actual for something 'other' – but that is not an issue for understanding messianic time or Shakespeare's 'religion'.

But, to the point: This feature of messianic time – the fact that it does not seem to differ from the chronological time we routinely experience – caused Paul certain rhetorical headaches just as it has caused modern Christianity no end of confusion in terms of its concentration on the *parousia* rather than the messianic Christ event. In brief, people were expecting something new with the messianic event and Paul found himself explaining that the call (*klesis*) of the resurrection has no new, positive content. Nothing new is called into Being, no new time is generated. Nor is there a new world created. In fact, Paul becomes deeply concerned with what happened in Corinth when 'enthusiasts' did try to embrace or create something 'new', mainly by celebrating and acting like the law literally no longer applied, like they were somehow completely disconnected from the world of chronological time. He had no interest in encouraging such enthusiasts to seize the spirit in this world.

On the contrary, Paul argued the messianic vocation or calling is the revocation of every current or existing vocation. The old is not so much supplanted by something new, Agamben reminds us, as it is 'nullified'.[9] This messianic nullification takes the form of the famous Pauline 'as if' or 'as not' (*hos me*). When, for example, the new Church in Corinth asks how one is to live after the messianic event Paul produces some of his most famous statements on living in messianic time, statements that rely on the formula of the 'as if' or 'as not' (*hos me*). Circumcision is the critical figure because this marks one's compliance with Jewish law. Jewish Christians insisted that those gentiles who wanted to be saved by Christ needed to honor this law while Paul, the apostle to the gentiles, argued they did not. But in arguing against the need for circumcision he was not abandoning the law, or suggesting anyone break the law, or suggesting anyone create a new law. He was only making his case for the 'nullification' of the law.

> But as God hathe distribute to everie man, as the Lord hathe called everie one, so let him walke: and so ordeine I, in all Churches.
>
> Is Anie man called being circumcised? Let him not gather his uncircumcision: is anie called uncircumcised? Let him not be circumcised.
>
> Circumcision is nothing, & uncircumcision is nothing, but the keeping of the commandements of God.
>
> Let everie man abide in the same vocation wherein he was called.
> (1 Corinthians 7:12)

The distinction between 'law' and 'faith' is nullified.

Agamben argues, in part, etymologically. For Paul, faith or *pistis* retains something of its ancient juridico-political meaning in that it refers to a pact or understanding between individuals that exists, in fact, prior to the law. *Pistis* is bound up in an archaic 'pre-juridical sphere in which magic, religion, and law are absolutely indiscernible from one another' and suggests something more like 'trust' between individuals who would 'give themselves over' to one another than it does *belief* in some transcendence that would supplant this trust.[10] Faith or *pistis*, Agamben says, suggests for Paul the 'unconditional self abandon to the power of another, which obliges the receiver as well' and the messianic event, in some sense, returns us or brings us into contact with that pre-juridical state of being.[11]

This is a fascinating argument, I think, and, as I will try to show later, illuminates Paulina's oft-cited call to 'awake your faith' at the end of *The Winter's Tale*. But, for the moment, let me point to a weakness or a gap in Agamben's account – a gap or weakness which is really a gap in Paul's thought. All that I have been describing does not tell us much about what it means to live under such 'nullification' in messianic time. The best one can say, perhaps, is that living in messianic time – a state of exception brought about by grace – involves living as not (*hos me*), or as if things had not changed much at all when, in fact, everything has changed.

Many of Paul's interlocutors through the years, including major current, critical theorists such as Slavoj Žižek recently, have suggested that the Pauline 'as if' is a subjective stance that one might assume, a certain critical distance toward the world one could somehow voluntarily adopt.[12] But it is not at all clear what or if Paul thought about that possibility either. His main concern in developing the 'as if' seems to have been in preserving an attention to the law, Jewish law, presumably to keep some sort of communal order in the case of Corinth and to keep relative political peace with Jewish Christians who already resented his mission of including gentiles as God's chosen people. Paul is adamant on preserving the law: 'What shall we say then? Is the Law sinne? God forbid' (Romans 7:7). Whether or not he thought a particular 'as if' subject position could be developed toward the world is left a matter for speculation or imagination.

While Paul exhorts his listeners and readers to remember that 'time is short', he just does not provide any psychological roadmap for responding to contracted messianic time. That has been left to the imaginations of those, like Shakespeare, who have had to live in messianic time much longer than Paul imagined. While current theoretical engagement with Paul is considered quite sophisticated and very much in line with what is known by historians about the apostle, that theoretical engagement tends not to consider Paul's apocalypticism which 'provided him with a teleology much different' than our own.[13] The significant gap between current readers of Paul and Paul himself on understanding messianic time seems to be that, for the apostle, the end time was near. For that reason, perhaps, he did not concern himself with the everyday particulars of living in messianic time. But those of us that live in 'extended' messianic time have to do so if they are to follow Paul with any accuracy. One has to imagine, in short, what life in messianic time involves and, in turn, one has to 'imagine' others – like Shakespeare – imagining what Paul meant.

This is not to suggest that Paul left his interlocutors utterly without guideposts to follow in understanding a psychological response to messianic time. Before turning to Shakespeare's religious imagination, I want to consider one of these guideposts in Paul, one that is particularly interesting when trying to understand the always confusing opening of *The Winter's Tale*, an opening that, as James Knapp has argued, is much more like the famous 'resurrection' scene that concludes the play than is generally acknowledged.[14]

Briefly, while he does not elaborate much on the psychological experience of messianic time for any given individual, Paul clearly assumed some would experience messianic time more comfortably than others. He suggests married men, in particular, will struggle with experiencing messianic time because they are still overly concerned about this world, the actual world we experience as 'real', and chronological time as figured in their relationship to their wives. Here, at greater length, is the critical language from Corinthians:

> And this I say, brethre, because the time is short, here afer that both they which have wives, be as togh they had none:
> And they that wepe, as togh they wept not: and they that rejoice, as though thei rejoyeced not; and they that bie, as togh the possessed not: And they that use this worlde, as though they used it not: for the facion of this worlde goeth away.
> An I would have you without care. The unmarried careth for the things of the Lord, how he may please the Lord.
> But he that is married, careth for the things of worlde, how he may please his wife. (1 Corinthians 7:29–33)

As suggested, Paul is urging listeners here to live as they were, as if nothing had changed with the messianic event. If you were married, stay married. If you were not married, do not seek to marry. But know that married men, more so than unmarried men, struggle to recognize the way in which the divisions of chronological time – past, present, and future – have been compressed into the messianic time of the now. Married men struggle to understand that the time of the now is 'what is' and that the chronological time has contracted into that messianic time. Having to attend to one's wife in the time that remains apparently intensifies the nullified distinctions that there is a past, present, and future for particular individuals and, correspondingly, specific relationships that somehow precede and distort the new being one in one time, in Christ.

Shakespeare and Paul

I want to suggest that Shakespeare seizes on this implied and complicated experience of messianic time in creating the notorious unmotivated jealousy of Leontes that inaugurates the action of the *The Winter's Tale*. The reaction of Leontes is, I will argue, set in this 'imagined' Pauline context, neither unmotivated nor, strictly speaking, jealousy. Instead, what unnerves and unhinges Leontes is the experience of messianic contracted time implied, although not fully described, in Paul's letters. And this experience, in turn, explains much of the rest of the play.

Shakespeare is not particularly subtle in drawing our attention to the link between the *tremor cordis* and the Grace (the Christ event) that has contracted time and began to join all as one. In this early scene Shakespeare works conspicuously through auditory imagery to draw our attention to 'Grace' and, specifically, its relationship to compressed time. As critics have long noted without fully grasping its import to the play as a whole, the word appears three times, beginning with Hermione's playful and casual prayer ('Grace to boot!' or, to modernize, heaven help me) upon hearing that Polixenes in part blames wives for interrupting an idyllic boyhood friendship. Hermione is praying here to be included in, rather than blamed for, the post-Edenic world Polixenes and Leontes occupy after meeting their wives (1.2.80). Having left the timeless Eden of boyhood, the three now struggle with a 'communal' expression of love. A few seconds later, still in a playful and casual tone, Hermione hopes that her 'first' well spoken words were 'Grace' (1.2.97). The echo of Grace culminates in her identifying her betrothal to Leontes as 'grace indeed'.[15]

When Hermione, Leontes' wife, takes the hand of their friend Polixenes at the same instant she is holding her husband's hand (1.2.108) – having just recalled their betrothal some years earlier – Leontes experiences what he calls a '*tremor cordis*', (1.2.110). This experience, Shakespeare makes clear, is both physical and mental and not necessarily connected to a particular emotion or feeling such as jealousy. Leontes eventually will interpret his reaction as jealousy, of course, and so will many critics and scholars, but Shakespeare clearly wants us to focus on its short-lived and sudden pre-interpretive visceral reception – and its strange relationship to 'Grace'. Shakespeare distinctly marks this initial reception of what most have been calling 'jealousy' as ambiguous. Leontes could just as well be experiencing joy as jealousy in this instant; in fact, he has to work to deny as much saying 'not for joy, not joy' (1.2.110).

The point here is not to suggest that Leontes misrecognizes an opportunity to embrace messianic joy and an enhanced reality brought about by some transcendent or trans-descendent interruption. Such voluntarism is not, as suggested, an option in the Pauline scheme of things. If time has begun to contract in Shakespeare's Pauline playworld and Leontes, Hermione, and Polixenes are thus all living in Pauline messianic time, then the chronologically determined distinctions of love (marital love vs. love in friendship, love in friendship vs. love for God, etc.) would have contracted to the point where they exist *only* in some ephemeral form. The moment when Leontes took Hermione's hand in marriage, the moment the husband and wife had just been lovingly and flirtatiously just recalling, is in this framework no different than – or simultaneous with – this immediate moment, *now*, in the play, when Hermione takes Polixenes in hand in a rather simple gesture of friendship or politeness. All three are linked together hand in hand as they will be at the end of the play – another, in some sense, simultaneous moment.

But because ephemeral forms of individuation such as marriage are still active, particularly for the married man, Leontes struggles painfully with this experience of time contraction, the fact that all are now all one in Christ. In struggling with messianic time the married Leontes only can process the *tremor cordis* as something known in chronological time: jealousy. In the Pauline framework, again, it is not as if one could choose messianic time over chronological time. The two 'times' coincide and one experiences them simultaneously. One must try to make sense of this coincidental experience. That Shakespeare chooses to render this experience as so excruciatingly painful for Leontes initially might seem odd. The messianic event is generally perceived by both Christians and observers of Christianity as a 'good thing', something to be desired. But it might help to recall that most, if not all, of Paul's writings on living in messianic time are addressed to people in psychological and spiritual distress, people wrestling with the split between the 'old' and the 'new', and his formulations of the 'as if' emerge in part to explicate and ameliorate that distress. In brief, if one is tracking Paul as closely as I am suggesting Shakespeare is tracking Paul then an experience of messianic time would be, at best, confusing to any given individual.

In the confusion of messianic time, it is quite natural that any given individual, and particularly a married man, would turn to something known in chronological time to explain this distress. What else could Leontes be experiencing with the *three* of them joined in love, a 'love' that – as he experiences it for an instant – subsumes the moment of romantic love and betrothal he had been so fondly recalling? One way to understand this is to turn again to 'materialist' thought, this time in an early modern context. One could say, perhaps, that following the *tremor cordis* Leontes submits (as all of us surely must) to what Jonathan Gil Harris calls a 'national sovereignty model of temporality', the notion or sense that we are at any given time citizens 'solely of one moment-state'.[16] From the vantage point of the present, the past is a threat, a foreign state, incongruous with the present. We cannot understand ourselves, Harris suggests, as 'polytemporal' beings. Nonetheless, from a Pauline perspective, the perspective I think of Shakespeare, we are in fact 'polytemporal' beings – with married men having a particularly distorted perspective of that polytemporality. The *tremor cordis* involves some distorted – and painful – access to that polytemporality.

Shakespeare, again, draws considerable attention to the fact that Leontes does not understand his own reaction – a reaction that is not apparently caused by anything – at this moment. My take is simply this: Leontes experiences messianic time in a brief instant but cannot reconcile this experience of messianic time with his still pressing everyday, ordinary, chronological experience of time in the present and the corresponding individual human relationships that occur in actual time. He thus translates this experience as something 'beyond commission' or unlawful: infidelity.

> Affection, thy intention stabs the center.
> Thou dost make possible things not so held,
> Communicat'st with dreams – how can this be? –
> With what's unreal thou coactive art,
> And fellow'st nothing. Then 'tis very credent
> Thou mayst cojoin with something; and thou dost,
> And that beyond commission, and I find it,
> And that to the infection of my brains
> And hardening of my brows.
>
> (1.2.137–145)

Leontes must provide, employing considerable linguistic labor, the emotional interpretation of this strange experience in messianic time. He must translate this experience of messianic time into something he understands (jealousy); he must 'cojoin' this strange experience of messianic time to a more familiar experience of chronological time. In the course of doing so, however, he utterly subordinates the experience of messianic time to chronological time. For Leontes in Act One, the two cannot coincide as Paul suggested they must.

Rather remarkably, Shakespeare later puts an audience in a position comparable to that of Leontes by startling them with the movement of the 'statue'. At the end of the play Shakespeare's 'art' creates an unsettling, unnerving experience not just for a character on stage but for those watching and listening in the theater.[17] That the experience is comparable is testified to by the fact that Leontes calls for the same 'madness' of the *tremor cordis*. At the appearance of the statue, he calls again for the experience knowing that it will throw him back to that horrifying moment sixteen years ago:

> O sweet Paulina,
> Make me to think so twenty years together!
> No settled senses of the world can match
> The pleasure of that madness.
>
> (5.3.70–73)

For all the trauma his *tremor cordis* caused him in Act One it seems that he calls for something very much like that *tremor cordis* when Paulina first reveals the statue and then threatens to draw the curtains because Leontes seems moved enough to believe the statue lives.

This time, however, an audience is forced, as was Leontes in Act One, to either immediately accept and embrace the experience elicited by the statue's movement without a clear interpretation or violently reject it as a sudden and unexpected act of deception by someone they have put their 'faith' in – not a spouse, but the playwright. When Paulina says, 'Either forbear, / Quit presently the chapel, or resolve you / For more amazement' (5.285–286) and 'those that think it is unlawful business / I am about, let them depart' (5.2.96–97) she is speaking to the actors on stage and the audience. The actors on stage are bound by the sovereign command issued by Leontes: 'Proceed. / No foot shall stir'. But the command sharply reminds us that an audience, in contrast, is not bound in the same way. An audience ultimately could reject the statue as a trick, a deception, something unlawful (in theatrical terms), as Shakespeare is acutely aware. In short, an audience could storm off, literally or figuratively, much as

Leontes did in Act One, denying the strange experience of messianic time provided for them by 'cojoining' this experience to something plausible, something that makes sense. Just as Leontes cojoined his experience of messianic time to jealousy in Act One an audience in Act Five could simply cojoin or translate their experience of Hermione's resurrection to something (very) familiar and plausible in chronological time: a deception, the playwright's trick, or a clumsy and contrived theatrical device. This sort of 'cojoining', then, would allow an audience – like Leontes in Act One – to choose either the plausible or the implausible, the sensible or the impossible, etc. To put this in Deleuzian terms, an audience could choose the actual or the virtual.

Somewhat surprisingly, however, even supposedly secular twentieth-century criticism and twenty-first-century audiences 'cojoin' the experience of the resurrected statute in an entirely different sense. Audiences accept the strange and startling experience of Hermione's resurrection *simultaneous with* the plausible explanation provided. For an audience, the odd but plausible world of chronological time suggested by the 'Paulina visits house everyday to keep Hermione' scenario coincides almost seamlessly with the utterly strange world of messianic time figured by the resurrection; and, again, rather than respond by denying this whole and complete truth as Leontes did in Act One we accept it. We immediately accept the incomprehensible and the comprehensible, the mysterious and the mundane, as one. We 'cojoin' the two, messianic time and chronological time, and experience both at once.

This all happens, as Paul might put it, in the 'twinkling of eye'. Or, to use the Shakespearean formulation, with 'every wink of an eye' that takes place upon experiencing Hermione's (non-) resurrection some new experience of 'grace' and messianic time would be born (5.2.110). An audience accepts Hermione's theatrical resurrection as such alongside its prosaic explanation (Paulina caring for Hermione in the 'outhouse') in much the way Paul told the Corinthians to accept the fact of actual death as it coincides with the messianic event. The Corinthians came to Paul and complained that death was still happening after the resurrection. How could this be? Paul responded:

> Now if Christ be preached that he rose from the dead, how say some among you that there is no resurrection of the dead? (I Corinthians 15:12)
>   ...
> Behold, I shew you a mystery; We shall not all sleep, but we shall all be changed.

> In a moment, in the twinkling of an eye, at the last trump: for the
> trumpet shall sound, and the dead shall be raised incorruptible, and
> we shall be changed. (I Corinthians 15:51–52)

A brief consideration of another critical moment in the play, the scene
involving the response of the oracle at 'sacred Delphos' (2.1.184), might
illuminate the 'religious' circumstances of the audience at the end of
the play. When Leontes sends Cleomenes and Dion to find the 'truth'
about Hermione they return with anything but the traditional, riddling
oracular profession an audience might expect. Instead, as Knapp says,
'the oracle is the epitome of clarity':[18]

> Hermione is chaste, Polixenes blameless, Camillo a true subject,
> Leontes a jealous tyrant, his innocent babe truly begotten, and the
> King shall live without an heir if that which is lost be not found.
> (3.2.132–135)

It should be clear now why Leontes utterly rejects this 'cojoining' of
the sacred and the everyday. His rejection of the clear oracular truth
corresponds exactly to his rejection of the coincidence of messianic and
chronological time in Act One. Shakespeare's Pauline married man cannot tolerate the juxtaposition of the two because, in his mind, the two
are separate. Conversely, an audience's easy and immediate acceptance
of this Delphic, divine truth (*not*) clouded in the sacred corresponds to
its acceptance of the 'resurrection' of Hermione and its plausible, real
world alternative: the staged art of Hermione literally acting in collusion with Paulina and perhaps Romano over a very long time. In the
twinkling of an eye – before a supposedly secular audience has time to
think it through – we accept unquestioningly as absolute truth the word
of (a) god.

Generally speaking, the dramaturgy has been so carefully orchestrated
that we do not even for a moment, like Leontes in Act One, contemplate
the possibility we have been deceived or cheated. An audience might
feel something like a *tremor cordis* at the movement of the statue moving, but we are also strangely certain that Shakespeare has been faithful
to us in this instant of potential theatrical 'infidelity' that mirrors the
potential moment of marital infidelity in Act One. Consequently, we
accept the *tremor cordis* as such. We accept it as Leontes does this time
around: as a pleasurable madness.

This act of fidelity of faithfulness entails more than one might expect.
We should recall, *pace* Agamben, that faith is not opposed to law.

When Shakespeare via Paulina says to an audience that you must 'awake your *faith*' he is not calling for a simple 'belief' in the magic of theater that temporarily suspends the laws of reality. Rather, like Paul, he is seeking to recall for an audience – more forcefully than he did in Act One by staging the *tremor cordis* – that the distinction between faith and law that emerged in history has been nullified by the messianic event of Christ's resurrection. The two are joined as one just as they are joined (almost) as one in Shakespeare's dialogue. Paulina's call to 'awake your faith' is followed immediately by the sovereign command from Leontes: 'Proceed. / No foot shall stir'. Indeed, an audience has no time to distinguish between the exhortation to faith and the lawful command. To be a bit more precise, perhaps, we could say Shakespeare is calling on an audience to commemorate with him – now, at the moment of performance – the religious and historical (to a Christian) fact that these distinctions have been nullified and that time has contracted so that even now we coincide with a pre-juridical sphere of existence where faith and law were one, bound together in a collective trust. One could say, perhaps, that a 'Eucharistic community' of messianic time is constituted in this theatrical moment when an audience awakens their faith or trust in the playwright.[19]

What makes this 'Eucharistic' community possible, however, is not the body and blood of Christ (a post-Pauline obsession), but the common affective state Shakespeare identifies as a critical feature of the experience of living in Pauline messianic time – and, of course, Shakespeare's dramaturgical skill at bringing an audience to experience and accept this common affective state. An audience experiences and accepts their own shock or 'tremor cordis' at the movement of the statue in part because they have been chastened by the experience of Leontes. Indeed, the willingness of Leontes to throw himself back into this experience of messianic time ('make me to feel so twenty years together') helps draw an audience into this experience.

For those inclined to try to graft my reading onto more traditional religious understandings of the play that focus on sin and redemption it is possible to suggest that this is the moment the audience 'forgives' Leontes for his earlier madness. But, I would insist, this is a forgiveness which is folded into the collective Eucharistic commemoration of accepting the 'resurrection' of Hermione. An audience is in no real position to forgive Leontes anything and, if I am correct, and Leontes' reaction is a manifestation of his experience of messianic time, criticism has been a bit harsh with the king, faulting him for his mad, jealous overreaction.[20] Living in Paul's messianic time, as I suggested at the outset, is something all of Christianity has struggled with up to and including the 'present' day. Ultimately implied in Shakespeare's dramaturgy, I think, is that no Christian audience should fault Leontes in Act One for not realizing what is so difficult for so many to grasp: that we all live in one moment in Christ. Conversely, entirely too much moral praise can be and has been given to Leontes for his repentance and redemption in this second part of the play, a repentance and redemption that supposedly allows him to repair in some sense an initial failing or sin. If there was no sin then there is nothing to repent. Redemption for sin was St. Augustine's concern, not Paul's. And Shakespeare, I am arguing, was reading Paul directly.

Having listened to Paulina for sixteen years, Leontes is, at best, more open to experiencing messianic time. Shakespeare is quite careful here. It is important to note, again, that at the moment of the statue's appearance Leontes does not ask for forgiveness. Nor does he ask for a magical transformation of some kind that would restore Hermione to life. He calls only for a return to the horrible affective state that opened the play. This is the playwright's focus: an acceptance of the experience of messianic time. Similarly, an audience is prepared to accept an experience of messianic time – not having listened to Paulina – but having been prompted by the dramaturgy of a Pauline Shakespeare.

The 'non' Pauline part of all this, I should note, is Shakespeare's suggestion that 'art' can somehow bring about an experience of messianic time. On this point, Shakespeare is actually closer to a supposedly non-religious materialist thinker like Deleuze. Just as Paul insisted that messianic time could not be accessed apart from chronological time, Deleuze posits that the virtual is inaccessible because of the actual and the present. Our very creatureliness inhibits our embrace of the virtual 'creating' from which we come. Still, his whole philosophy involves calling for 'counter-actualizations' to reverse as much as possible the process wherein the virtual creating becomes the actual, the creatural. The point, again, is greater access to the virtual, creativity itself. And to get this access we must lose ourselves, our very sense of identity in the oneness that is creativity.

Next to philosophy itself, art stands as the chief form of counter-actualization for Deleuze. Of course, art is an 'actual' practice and therefore as distant from the virtual as anything in creation. But because art is interested primarily in 'creating' *as such* Deleuze argues it can provide extraordinary contact with the virtual which, again, can be best described as a field of energies or pure potentialities – creativity itself. In creating a work of art, he posits, an artist liberates a *'bloc of sensations'* from the realm of the virtual, sensations that in turn animate the actual material of artistic construction. Deleuzian sensations are a 'force of creative intensity' that pulse through not just 'materials' but 'actual bodies'.[21] Sensation, Deleuze writes, is 'not the flesh but the compound of nonhuman forces of the cosmos, of man's nonhuman becomings'.[22] A Rodin sculpture, then, for example, 'lives' in a 'zone of indetermination' between the actual material used by the artist and the sensations that pass into it as it becomes a 'work' of art.[23]

I am not suggesting Hermione lives in the sense a Rodin sculpture lives in a Deleuzian framework. What I am suggesting is that when Shakespeare creates an experience of messianic time for his audience by presenting them with the 'statue' of Hermione, he exposes the ontological consistency of the sensations we experience in messianic time – indeed, sensations that constitute our very being – and the sensations that animate his art. Art can induce an experience of messianic time because, for Shakespeare, if not for Paul, the distance between art and life in messianic time is nil. For him, it seems, they are as one. Polixenes says as much in his famous speech on 'art'.

> Yet Nature is made better by no mean
> But Nature makes that mean; so over that art
> Which you say adds to Nature, is an art
> That Nature makes. ...
> This is an art
> Which does mend Nature – change it rather; but
> The art itself is Nature.
>
> (4.4.89–97)

In suggesting, then, that *The Winter's Tale* is a religious play I am some distance, I think, from the 'transcendence' so many 'materialists' fear. Shakespeare's close religious reading of Pauline messianic time ultimately produces the univocal ontology of Deleuzian philosophy. This might mean, though, that many current thinkers should at least ponder the nature of their (seemingly non-religious) desire for immanence.

Notes

1. Giorgio Agamben, *The Time that Remains: A Commentary on the Letter to the Romans* (Stanford: Stanford University Press, 2005). Agamben's book is, in part, a response to Alain Badiou's *Saint Paul: The Foundation of Universalism*, trans. Ray Brassier (Stanford: Stanford University Press, 2003). Badiou's relatively short book helped prompt a broad conversation about the relevance

of Paul to the modern world that now extends across disciplines, from historians of religion to philosophers to literary scholars. See John D. Caputo and Linda Martin Alcoff, eds, *St. Paul among the Philosophers* (Bloomington and Indianapolis: Indiana University Press, 2008) for a coherent overview. Gunther Bornkamm's *Paul*, trans. D. M. G. Stalker (Minneapolis: Fortress Press, 1995) is still particularly useful for historical context as is E. P. Sanders, *Paul: A Very Short Introduction* (Oxford: Oxford University Press, 1991). See also Jacob Taubes, *The Political Theology of Paul*, trans. Dana Hollander (Stanford: Stanford University Press, 2004); Theodore W. Jennings, Jr., *Reading Derrida/Thinking Paul* (Stanford: Stanford University Press, 2006); Garry Wills, *What Paul Meant* (New York: Viking, 2006). Julia Reinhard Lupton, *Citizen-Saints and Political Theology* (Chicago: University of Chicago Press, 2005) should be credited with joining this Pauline conversation to early modern studies. For a still very useful consideration of 'time' in the play, see Inga-Stina Ewbank, 'The Triumph of Time in *The Winter's Tale*', REL 5 (1964), 83–100.

2. See Ken Jackson, 'Tarrying with the Marxists', *Religion and Literature* (forthcoming).
3. Agamben, *Time that Remains*, 1.
4. *Geneva Bible*, 1560 (Peabody, MA: Hendrikson Publishers, 2007).
5. Agamben, *Time that Remains*, 67.
6. Henri Bergson, *Matter and Memory*, trans. Nancy Margaret Paul and W. Scott Palmer (New York: Zone Books, 198); Peter Hallward, *Out of This World: Deleuze and the Philosophy of Creation* (London and New York: Verso, 2006).
7. Hallward, *Out of this World*, 35.
8. For a lucid and provocatively interesting early modern take on this critical tendency, see Michael Whitmore, *Shakespearean Metaphysics* (London: Continuum, 2008).
9. Ibid., 23.
10. Agamben, *Time that Remains*, 114.
11. Ibid., 116.
12. Žižek's many discussions of Paul begin, perhaps, with his account in *The Ticklish Subject: The Absent Centre of Political Ontology* (London: Verso, 1999), 146.
13. Caputo and Alcoff, eds, *St Paul among the Philosophers*, 20.
14. See James A. Knapp, 'Visual and Ethical Truth in *The Winter's Tale*', *Shakespeare Quarterly* 55:3 (2004), 253–278.
15. I have argued that Shakespeare uses a comparable echo of 'Grace' in *Richard III*. See my '"All the world to nothing": Badiou, Žižek, and Pauline Subjectivity', *Shakespeare* 1:1 (2005), 29–52.
16. Jonathan Gil Harris, *Untimely Matter in the Time of Shakespeare* (Philadelphia: University of Pennsylvania Press, 2009), 2.
17. Grace echoes throughout this opening scene prior to the *tremor cordis* of Leontes. Unfortunately, Shakespeare is not clear at all about what the relationship is between this echoing grace and the *tremor cordis*. He seems to hint at some cause and effect at work but ultimately one is left to wonder how this experience is produced or even if we can say 'something' produced it. One is tempted to argue that we are meant to understand that 'Grace' echoes only as a 'resounding gong' or a 'clanging cymbal' (1 Corinthians 13:1), an ambiguous auditory image that Leontes and most audiences might hear but

cannot yet understand in full. This barely intelligible auditory image would stand, then, in illuminating contrast to the famous visual image of the resurrected 'statue' of Hermione. At one point, Paulina specifically juxtaposes the more powerful impression of this visual image to an auditory one when Polixenes asks a still silent, but moving, Hermione to explain.

> That she is living,
> Were it but told you, should be hooted at
> Like an old tale; but it appears she lives,
> Though yet she speak not.
>
> (5.3.115–18)

Based on this, one could argue the play's earlier auditory imagery was insufficient to alert Leontes and others to the state of Grace in which they live, but the striking visual image of the statue has a transformative effect on viewers. And, in fact, this sort of argument would correspond nicely with very good and interesting scholarship, most notably by Michael O'Connell in *The Idolatrous Eye: Iconoclasm and Theater in Early-Modern England* (New York: Oxford University Press, 2000) that has concentrated on this apparent hierarchy of sight over sound to argue that the emphasis on the visual suggests a residual Catholicism in Shakespeare. If we are to look for religion in Shakespeare, we need to look, rather than listen, for the drama has absorbed much of the medieval, visual Catholic culture shattered by Protestant iconoclasm. If, however, we are to trace accurately the contours of Shakespeare's dramatic religious imagination it is crucial that we be sensitive to the way in which that imagination might exceed dominant historical narratives. As Julia Lupton and Graham Hammill write in 'Sovereign, Citizens, and Saints: Political Theology and Renaissance Literature', *Religion and Literature* 38:3 (2006), 1–12:

> A cultural approach to religion ... can make little sense out of the Pauline tradition, which does not simply exemplify a religious identity ('Christianity'), but rather mobilizes a battery of geographical, historical, hermeneutic, juridical, and subjective positions and processes, as well as possible programs for their transformation and sublation. Like ghosts or viruses, religions leap across groups and epochs, practicing cultural accommodation in order to outlive rather than support the contexts that frame them. (2)

I would note that in redirecting Polixenes' efforts to extract a verbal explanation, Paulina tries first and foremost to bring attention to and preserve the initial *generalized* response all on stage and in the audience experience as Hermione's statue moves. 'Mark a little while', she says. In other words, the statue scene is not as interested in the visual as current scholarship inching toward a 'Catholic' or more medieval Shakespeare would have us believe. Rather, Shakespeare seems more interested in a certain pre-interpretive response, something that may be understood, again, in terms of materialism or immanence, as a Deleuzian 'sensation'.

18. Knapp, 'Visual and Ethical Truth', 115.

19. The term 'Eucharistic community' in this dramatic context comes to me via the scholarship of Sarah Beckwith, *Signifying God: Relation and Symbolic Act in the York Corpus Christi Plays* (Chicago: University of Chicago Press, 2001).
20. On the problems and possibilities of forgiveness see Jacques Derrida, 'To Forgive: The Unforgivable and the Imprescriptible', in John D. Caputo, Mark Dooley, and Michael J. Scanlon, eds, *Questioning God* (Bloomington: Indiana University Press, 2001), 21–51.
21. Hallward, *Out of this World*, 107.
22. Gilles Deleuze and Félix Guattari, *What is Philosophy?*, trans. Hugh Tomlinson and Graham Burchell (New York: Columbia University Press, 1994), 183.
23. Ibid., 164, 173.

9
'Loves Best Habit': Eros, Agape, and the Psychotheology of *Shakespeare's Sonnets*

Gary Kuchar

The turn to religion in Shakespeare studies has been described, perhaps somewhat surprisingly, as a return to critical theory. In Julia R. Lupton's formulation, the '"religious turn" in Renaissance studies represents the chance for a return to theory, to concepts, concerns, and modes of reading that found worlds and cross contexts, born out of specific historical situations, traumas, and debates, but not reducible to them'.[1] This characterization of the religious turn reflects a desire to move beyond the limits of thematically oriented historicisms – a desire Philip Lorenz expresses when he asserts that 'To *really* "re-turn" to religion ... would be to turn to the turns themselves: the tropes of theology that screen and animate ... Renaissance [literature].'[2] I would like to take up Lorenz's challenge by showing how Shakespeare's use of a particular Pauline figure in the sonnets reveals that the experience of love in the poems constitutes a distinctly post-Reformation species of psychotheology – a mode of subjectivity that emerges from the contradictions between Neoplatonic Eros and reformist agape.[3] In short, I want to suggest that individual poems in the sequence are often structured by converging and diverging relations between the mutuality of Neoplatonic love as theorized in the Florentine school and the asymmetry of Reformation agape as envisioned in the Protestant tradition. My hypothesis, quite simply, is that the voice in the sonnets and the non-religious wit that animates it are shaped by the speaker's variously conflicting allusions to Neoplatonic and Reformation ideas on love. The result is a poetics that remains irreducible to either context.

By pursuing this thesis I hope to move studies of the sonnets beyond a certain deadlock. The best recent readings of the sonnets have tended to keep the poems' theological features separate from their psychological interests; and they have done so at the expense of understanding

211

important dimensions of Shakespeare's wit. For example, Joel Fineman's argument that *Shakespeare's Sonnets* produces a poetic first-person who is characterized by a formally and historically determined opacity to himself remains unconcerned with how scriptural and theological allusions shape the complexity of tone in the poems.[4] On the reverse side of things, Lisa Freinkel's study of the sonnets in relation to Luther illuminates large questions of authorization and the politics of *figura* but it does not closely consider how nuances of voice and conceit are shaped in many individual poems by scriptural and theological reference.[5] Given that pre-Cartesian culture tended not to separate psychology from theology in this way, recent criticism on *Shakespeare's Sonnets* tells us as much about our own predilections as it does about the witty interplay between grace and Eros in the poems themselves. As long as criticism of the sonnets maintains this separation of the psychological from the theological, we will misconstrue nothing less than the subjectivity expressed in the poems and the ironic wit that bodies it forth.[6]

Thus, rather than considering how the sonnets anticipate modern forms of subjectivity, as Fineman has done, my aim is to resituate Shakespeare's poetry within the pneumatic-based tradition of the soul that originates with Aristotle's *De Anima* and that runs through the *Stilnovo* poets up to Scholasticism, Dante, Petrarch, and finally to Florentine Neoplatonism and then into treatises and traditions contemporary with Shakespeare such as Thomas Hoby's translation of Castiglione's *Book of the Courtier* and English Petrarchism more generally. At the heart of this variegated tradition is the assumption that the lover does not fall directly in love with the beloved, but rather becomes enamored with the lover's image or phantasm.

According to this view of love, which was popularized by Andreas Cappellanus' twelfth-century work *De amore*, the lover becomes obsessed with the mental-image of the beloved, attaching himself to it and becoming, cognitively and affectively, one with it. In several sonnets, this phantasmatic theory of Eros in which the lover becomes one with the beloved crosses with the different, even opposing, view of agape as theorized in Protestant accounts of justification. Whereas the Neoplatonic view of Eros presumes a genuine reciprocity between lover and beloved, effected through the operations of the phantasm, Protestant notions of agape presume an asymmetrical relation in which God forensically imputes grace to an undeserving beloved. By exploiting the tensions between mutuality and asymmetry inherent in traditions of Eros and agape, Shakespeare generates a psychotheology that is marked by ironically indecorous allusions to competing views of love.

The competition between Neoplatonic and Protestant ideas of love in post-Reformation Europe turned, in part, on interpretations of St. Paul. It is thus not entirely surprising that in each of the sonnets that I consider, Shakespeare impresses St. Paul's figure that divine grace involves putting Christ on as though he were a garment into non-religious use. In what follows I trace how the voice in *Shakespeare's Sonnets* and the homoerotic drama it expresses are constructed through references to Romans 13:14, Corinthians 15:53, and Galatians 3:27, each of which expresses love as a putting on of the beloved as a garment.[7]

Although Paul's image of putting on Christ shapes Shakespeare's psychotheology in several sonnets, it appears to have gone unobserved in critical and editorial commentary on the poems. This gap is particularly notable because Shakespeare's allusions to Paul's figure may also illuminate some connections between the sonnets and the early modern theater. As sonnet 15 intimates, the idea that grace can be figured as a change of clothing gets interwoven in Shakespeare's poems with the Elizabethan system of livery and its view of investiture in which changes of clothing produce near-miraculous translations of identity. The Elizabethan conviction that clothing is determinative rather than peripheral to identity is crucial to the kinds of conversions one sees on the early modern stage, just as it is crucial to the kinds of regeneration imagined by the speaker of the sonnets.[8]

Sonnet 62: 'loves best habit'

One of Shakespeare's most explicit engagements with religious tropes occurs in sonnet 62. In this poem, the phantasmatic conception of love as *immoderata cogitatio* crosses with a mock-Pauline conversion scene. What results is a good example of how Shakespeare exploits the indecorum inherent in secular love poetry's impressing of religious tropes into its own service. In particular, the poem is indicative of Shakespeare's rather bold misapplications of Paul's figure that the old man becomes a new man by putting on Christ.

The opening quatrain draws on the psalmic language of devotional poetry, making it appear that the speaker is praying for conversion:

> Sinne of selfe-love possesseth al mine eie,
> And all my soule, and al my every part;
> And for this sinne there is no remedie,
> It is so grounded inward in my heart.
>
> (1–4)[9]

Parodying the language of religious despair, the speaker 'resolves' his crisis by engaging in a conversion scene that puts the young man in the place of Christ and physical beauty in place of spiritual justification:

> Me thinkes no face so gratious is as mine,
> No shape so true, no truth of such account,
> And for my selfe mine owne worth do define,
> As I all other in all worths surmount.
> But when my glasse shewes me my selfe indeed
> Beated and chopt with tand antiquitie,
> Mine own selfe love quite contrary I read
> Self, so selfe loving were iniquity
>> T'is thee (my selfe) that for my selfe I praise,
>> Painting my age with beauty of thy daies.
>>
>> (5–14)

The poem's concluding irony partly rests on the speaker's indecorously strained allusion to passages such as Romans 13:14: 'But put ye on the Lord JESUS CHRIST, and take no thought for the flesh.'[10] Rather than making no provision for the flesh through a putting on of Christ, the speaker imagines himself as beautiful by painting himself with the young man's youth. The sonnet's wit thus rests on the way it parodies the Protestant logic of forensic justification; rather than praising his own merit *per se*, he praises himself as the unworthy possessor of the young man's beauty and value. The poem's indecorum lies in the speaker's ironic self-assurance that he is physically beautiful rather than spiritually justified. The logic here parallels, even as the scene itself distortedly strains, the Protestant view that the regenerate soul is *simul justus et peccator* – both justified and a sinner. Just as elect Protestant souls are 'sinners in fact but righteous in hope',[11] so the speaker of sonnet 62 experiences himself as old in (empirical) truth but young in (loving) faith.

Protestant exegetes routinely cite Paul's trope of being covered by Christ as evidence that grace is imputed not imparted, that it is an unmerited not an earned gift. Calvin, for example, cites Paul's figure more than six times in his discussion of justification in Book III chapter 11 of the *Institutes*. Drawing on this familiar figure, the speaker imputes the beloved's beauty to himself – rather than having it imputed to him as the logic of forensic justification would demand. What makes such presumptive imputation meaningful within the context of secular love, however, is the way the phantasm takes possession of the lover so that everything he sees is, to paraphrase sonnet 113, shaped to the beloved's feature – including, in this case, his own face.

The poem is thus structured by the ironic interplay between two economies of love: the Eros of the phantasm in which the lover actively becomes one with the beloved (seeing everything through his eyes) and the agape of forensic justification in which the soul is passively imputed to be one with Christ but is, in itself, still an alienated sinner. This tension between the symmetrical reciprocity of pneumatic-psychology and the asymmetry of Protestant agape is characteristic of how the young man sonnets generally express the oscillations of union with and separation from the young man.

Although the speaker of sonnet 62 may be deluded, his delusions are those of a genuine lover obsessed with the image of the beloved. His 'contrary reading' of himself as painted with the beauty of the beloved is something other than an act of meaningless fancy; it is an act of folly not altogether unlike the wisely foolish acts committed in sonnet 138 'When my love sweares that she is made of truth'. In sonnet 138, 'loves best habit' consists of knowing that love has the structure of a fiction: one gracefully imputes things to the beloved that are known to be empirically untrue, so that what is unjust becomes, through love, just: 'But wherefore sayes she not she is unjust? / And wherefore say not I that I am old? / O loves best habit is in seeming trust, / And age in love, loves not t'have yeares told' (9–12). Throughout the sonnets, the drama of love unfolds through similarly ironic and indecorous misapplications of Pauline figures for witty and, more often than not, homoerotic ends.[12]

Sonnet 22: of grace and ecstasis

Sonnet 22 is a provocative case in point. This sonnet is a good example of how intersections between Protestant agape and Neoplatonic Eros produce witty effects of dramatic irony. In the course of the poem, the speaker's claim that he and the beloved are of one flesh collapses under the pressure of his contradictory theological investments. On the one hand, he marshals a Neoplatonically inflected allusion to ideal mutuality of lover and beloved; on the other, he appeals to the Calvinist idea of irresistible grace, thereby introducing competing views of love, the soul, and ecstasis. As a result, the precise tone of the poem, with its combination of tenderness and aggressiveness, dependence and self-possession, is an effect of how competing views of love converge and diverge in the sonnet.

The speaker's major premise unfolds in the first two quatrains and consists of the proposition that he and his beloved have literally, rather than just metonymically, exchanged hearts and that they are thus truly one flesh:

> My glasse shall not perswade me I am ould
> So long as youth and thou are of one date,
> But when in thee times forrwes I behould,
> Then look I death my daies should expiate.
> For all that beauty that doth cover thee,
> Is but the seemly rayment of my heart,
> Which in thy brest doth live, as thine in me,
> How can I then be elder then thou art?
>
> (1–8)

As Stephen Booth has observed, the wit of Shakespeare's treatment of the Petrarchan heart-conceit 'derives from seeming to take a metonymy literally'.[13] This indecorously literal treatment of a trope enhances the indecorum of figuring their erotic union through scriptural allusion. Like sonnet 62, the speaker's assertion that he shares his beloved's beauty because his heart is clothed by the young man's skin is animated by its parallels with Paul's assertions that the old man becomes a new man and that corruption puts on incorruption, by putting on Christ – thereby becoming one with him.

Henry Smith interprets Paul's figure of putting on Christ in a sermon titled *The Wedding Garment* (printed seven times between 1590 and 1592) by harmonizing the passages Shakespeare condenses in his major conceit. According to Smith, Paul's 'straunge speeche' in Romans 13:14, 'But put ye on the Lord IESUS CHRIST', indicates that

> wee must put on Christ, for the word signifyeth so to put him on, as if thou wouldest put him in, that hee may bee one with thee, and thou with him, as it were in a body toherher [sic] ... we must put on all his graces, not halfe on, but all on, and claspe him to us, and girde him about us, and weare him even as wee eare our skinne, which is alway about us.[14]

By taking a conventional trope literally, sonnet 22 has the effect of turning the spiritually regenerative process of becoming a new man by being incorporated in Christ's body into an indecorously, because quite literally, fleshy affair between himself and his beloved. As Smith writes of Paul's figure, 'You put on Christ … when you cast off the old man, which is corrupt … and put on the new man' (43). So while the allusions to Paul rest on the conventionally indecorous Petrarchan practice of depicting the beloved as quasi-divine, its hyper-literal adaptation of spiritual union and regeneration take such indecorum one step further than the normative conventions.

As the title of Smith's sermon indicates, Protestant exegetes often harmonized Paul's trope about the soul and Christ in Romans 13:14 with the nuptial union of man and wife in Ephesians 5:29–30: 'For no man yet ever hated his owne flesh, but nourisheth and cherisheth it, even as the Lord *doeth* The Church.' Appearing in the matrimonial service dictated by the Book of Common Prayer, the allusion to Ephesians that Stephen Booth notes in quatrain three of sonnet 22 would have been readily audible to Shakespeare's audience: 'O therefore love, be of thy selfe so wary / As I not for my selfe, but for thee will, / Bearing thy heart which I will keep so chary' (9–11).[15] Just as Smith connects the union of soul and Christ in Ephesians 5 with the garment metaphor of Romans 13, so too Shakespeare's Ephesian image of the lovers being one flesh achieves its full indecorous effect when heard resonating within the larger Pauline context of becoming a new man by putting on Christ. Shakespeare's speaker thus grounds an argument for same-sex love by alluding to a matrimonial rite dedicated to husband and wife. Given Shakespeare's homoerotic appropriations of Christianity's most authoritative anti-sodomite, is it any wonder an early annotater of the sonnets thought the poems 'a heap of wretched Infidel Stuff'?[16]

The indecorousness of this allusion to the Book of Common Prayer is redoubled by the poem's evocation of related scriptural passages, most importantly Galatians 2:20 in which Paul declares that 'I live *yet*, not I now, but Christ liveth in me.' Luther glosses this passage by saying that 'This attachment to Him causes me to be … pulled out of my own skin, and transferred into Christ.'[17] Sonnet 22 is characterized by the way its commonplace Petrarchan conceit crosses a Protestant idiom with which it is at odds. For the speaker's image of ecstatic mutuality in sonnet 22 is not designed to express the kind of total dependence on the beloved that reformers see expressed in Paul's depiction of Christ as a garment. Rather it is designed to express the ideal reciprocity envisaged by Neoplatonic ecstasis. As M. A. Screech observes, references to Galatians 2:20 are found throughout Christian Neoplatonic philosophy as a way to verify the philosophical commonplace 'that lovers live not in themselves but in the one whom they love, their souls departing from their bodies and entering that of the beloved, in a mutual exchange (*Phaedrus*, 244C–265B)'.[18] For Christian Neoplatonists writing in the wake of Pseudo-Dionysius, such as Ficino, 'the love of Christ and man is a reciprocal ecstasy. Both are outside themselves: Christ lives in man and man lives in Christ as lovers who are both "outside themselves": for [Pseudo-Dionysius] that is what St Paul meant by "Christ liveth in me".'[19] The poem's irony thus turns on how the speaker rests his case for ideal reciprocity on an overly literal application of a Petrarchan conceit that is buttressed by contradictory versions of ecstatic love. In the process, the speaker's hyperbole betrays an increasing degree of insecurity.

Even a Catholic reformer such as Erasmus makes clear that Neoplatonic and Pauline ecstasy are incompatible with one another. As Screech explains, Erasmus' depiction of Christian ecstasy in the conclusion to the *Praise of Folly* suggests that he 'finds abhorrent' the Pseudo-Dionysian reading of Galatians 2:20 so central to Neoplatonic views of ecstasy in the Renaissance.[20] While Erasmian ecstasy 'remains in some ways Platonic ... it is conceived as a one-way enrapturing of the human soul into the Godhead of Christ'.[21] Erasmus' resistance to the Neoplatonic interpretation of Paul thus anticipates, even as it is radicalized by, Protestantism's insistence that the soul is wholly dependent upon Christ for its spiritual well-being, thereby precluding the kind of reciprocal ecstasy Neoplatonism imagines.

For Neoplatonists, ecstasy is a mutual exchange of identities – a form of reciprocity that is achieved not through a gift that is passively received but as the result of active contemplation.[22] For reformers, on the other hand, the soul is entirely dependent upon Christ in a way that precludes reciprocal ecstasy achieved through the work of contemplation. As I have indicated, the figure of being clothed by Christ is adduced by reformers to make precisely this point. Henry Smith, for example, emphasizes the soul's dependence on Christ when he draws on Paul's figure to explain how 'Christe doeth cover us like a Garment, and defende us like an armoure. Hee hideth our unrighteousnesse with his righteousnesse, he covereth our disobedience with his obedience' (11). The oppositional language here conveys a fundamentally non-symmetrical relation of soul to Christ. While such asymmetry defines Reformation views of ecstasis, it is antithetical to Neoplatonic versions of the experience.[23]

Shakespeare's poem exploits this crucial distinction between types of ecstasis.[24] By putting both Calvinist and Neoplatonic interpretations of Paul into play in the same conceit, Shakespeare's speaker undermines the ideal reciprocity that he tries to imagine. Although he wants his relation with the beloved to have the perfect reciprocity of a Neoplatonic ecstasis, the relation slowly reveals itself to have the kind of non-symmetrical dependence associated with reformed views of Pauline ecstasis. Figuring his relation with the young man in the same language that reformers use to figure the soul's relation to Christ thus comes at the cost of effectively undermining the reciprocity dreamed of in the Neoplatonic image of the exchanged hearts that grounds the whole argument. This play of competing contexts suggests that Shakespeare's speaker stacks his argumentative deck. Just as the unconventionally literal application of the exchange of hearts betrays a certain desperate insecurity, so too do the hyperbolic theological echoes. The overall effect generated here is a kind of dramatic irony; the speaker's ostensibly confident claims of ecstatic union and ongoing youth betray a pained awareness of differentiation and age.

This irony is capped in the third quatrain by the tension between the speaker's claim, on the one hand, that he and his beloved are of the same age and his assertion, on the other, that he will care for the beloved the way a tender nurse cares for an infant: 'O therefore love, be of thy selfe so wary / As I not for my selfe, but for thee will, / Bearing thy heart which I will keep so chary / As tender nurse her babe from faring ill' (9–12). In the very act of asserting full mutuality, the speaker betrays his own feeling of dissension between himself and the young man. Moreover, a certain desperate selfishness might be audible in a potential pun on Shakespeare's own name in the line: 'As I not for my selfe, but for thee will' (10). In any case, tension between imagery and argument can now be added to the speaker's investment in competing notions of ecstatic union – all of which paints a picture of emotional insecurity rather than assured union.

These ironic tensions help account for the desperate, even somewhat insidious, quality of the final couplet. The combination of tender love and appropriative desire unfolding in the poem comes into relief when we bear in mind that sonnet 22 rewrites Petrarch's sonnet 21 by literalizing its figures, thereby generating a different ending. Unlike Petrarch's poem, or Wyatt's imitation of it, the exchange of hearts in Shakespeare's sonnet is presupposed from the beginning. No less importantly, this exchange of hearts is presented as a literal sharing of flesh rather than serving as a figure for reciprocal love. Petrarch begins his poem, 'A thousand times, O my sweet warrior, in order to have peace with your lovely eyes, I have offered you my heart; but it does not please you to gaze so low with your lofty mind.'[25] Thus where Petrarch's problem is the absence of a figurative exchange of hearts, the problem for Shakespeare's speaker is how to maintain a presumed literal exchange.

Petrarch tries to resolve his problem by telling Laura that without her love, his heart will die: 'Now if I drive him [my heart] away, and he does not find in you any help in his sad exile, nor can stay alone, nor go where some other calls him, his natural course of life might fail, which would be a grave fault in both of us, and so much the more yours as he loves you the more' (lines 9–14 in Italian). Shakespeare's speaker ups the ante of Petrarch's strategy. Instead of arguing that his own heart will die should the beloved abandon him, Shakespeare's speaker implies that the beloved's heart will fail too. This surprising turn is achieved in the final couplet: 'Presume not on thy heart when mine is slaine, / Thou gav'st me thine not to give back againe' (13–14). The speaker intimates that it is not just his life that may be 'expiated' due to an inordinate love – a key-word whose etymological meaning is to appease by means of sacrifice – but the beloved's as well.

Once the ending of Petrarch's sonnet 21 has been reversed this way, the speaker's argument for graceful reciprocity has been effectively undermined. The result is that the speaker ends by betraying how his love conforms to a mode of desire that Thomas Aquinas defines as *amor concupiscentia* – a form of affection in which 'the lover is not satisfied with any external possession or enjoyment of the beloved; but seeks to possess the beloved entirely, by penetrating into his most inward parts, as it were'.[26] This mode of love opposes the genuinely reciprocal, even altruistic, form of filial love or *amor amicitiae* the speaker seeks to show. In such a love-of-friendship, the lover loves the beloved in the beloved's own person, without any appropriative dimensions. Perhaps most interestingly, the speaker's stark assertion that the lover's heart was given not to be returned further invests the argument in the post-Reformation Pauline background evoked in quatrain two. For insofar as the exchange of hearts results in the speaker wearing the beloved's flesh in a way that parallels the soul's wearing of Christ, it follows, from an orthodox Reformation standpoint, that the heart cannot be returned. The irony of the final couplet thus rests in its oblique evocation of the Calvinist doctrine of irresistible grace: the idea that once given, prevenient grace will not be taken away.[27]

As I have suggested, it is the indecorous difference between Paul's Christological contexts and the sonnet's homoerotic context that betrays the speaker's conceit in quatrain two as more optative than indicative, more possessive than altruistic. Although this indecorous grafting of psychological and Christological discourses is crucial to what Fineman calls the subjectivity-effect produced by the sonnets – their ability to generate a poetic voice that is characterized by self-difference – it is a strategy that has not been adequately explained in criticism of the poems.

If sonnet 22 is characterized by the dramatic ironies I have just claimed, then it is partly because the Pauline figure of putting on Christ activates Elizabethan religious controversies over figurative and literal meanings. By presupposing that his heart is really and truly present in his beloved's breast, the speaker deploys the Pauline allusion with exactly the kind of literalness Elizabethan Protestants often warn against with reference to Romans 13:14 and Galatians 3:27. In Reformation prose works of the period, such as John Foxe's *The pope confuted* (1580) or Smith's *The Wedding Garment,* Paul's figure of putting on Christ raises questions about figurative versus literal meaning. In his anti-Catholic tract, for example, Foxe attacks the doctrine of transubstantiation by asking, 'O why shoulde you bee more squemishe at a figurative speeche where we are saide to eate Christ, then where we are commanded by the scriptures to put on Christe?'[28] Smith makes a parallel point when he says that 'we must put on Christ, as if wee did eate him, not as the Papists doe in their Masse, but as the meate is turned into the substance of the body' (37). The irony of Shakespeare's poem is deepened not just by hearing allusions to Paul's trope, but by understanding how such allusions tended to signify in Elizabethan culture as proof that one should avoid exactly the kind of literalism exploited by the speaker.[29]

Taken together, the intersections between Pauline agape and Neoplatonic Eros in sonnet 22 bespeak a lover immoderately and insecurely in love rather than one enjoying anything like perfect mutuality. The poem is powerful not because it presents a persuasive or even beautiful argument but because it reveals the pathos of a dramatically ironic persona.

Sonnet 26: naked thought

Just as the interplay between Reformation and Neoplatonic contexts consistently thwarts the desire for ideal reciprocity in sonnet 22, so the intersections between theology and faculty psychology in sonnet 26 work to signify dissension between lovers in the very midst of their apparent unity. The poem begins with the speaker declaring that he writes his Lord 'to witness duty, not to show [his] wit' (4). Ostensibly confessing an impoverishment of invention, the speaker asks that his

Lord bestow upon him the good favor of what he calls an 'all naked' conceit. This phrase appears in parentheses in the original Quarto and demarcates the precise point where theology and phantasmology, as well as speaker and addressee, are simultaneously united and set asunder. The second and third quatrains read:

> Duty so great, which wit so poore as mine
> May make seeme bare, in wanting words to shew it;
> But that I hope some good conceipt of thine
> In thy soules thought (all naked) will bestow it:
> Til whatsoever star that guides my moving,
> Points on me gratiously with faire aspect,
> And puts apparrell on my tottered loving,
> To show me worthy of thy sweet respect.

$$(5-12)^{30}$$

The phrase 'all naked' describes both the young man's conceit and the speaker's bare wit, making it function grammatically as well as conceptually as a copulative – a lexical and erotic point of union between the two men.

While the parentheses embrace both men together as naked in thought, the phrase itself operates in a surprisingly technical way. When read as a reference to the Lord's conceit, 'all naked' refers to the principle that an intelligible idea is formed in the intellect through a process of denuding or stripping the material qualities of a sense impression out of which an idea is formed. Sir John Davies summarizes this process in *Nosce Teipsum*, noting how the mind 'abstracts the *formes*' from the 'grosse *matter*' of sensual images.[31] Shakespeare describes the kind of thought resulting from this process as 'naked' because the imagination works by a 'progressive "disrobing" (*denudatio*) of the phantasm from its material accidents'.[32] The wit of sonnet 22 turns on the speaker's demand that his bare-invention be clothed by such a fully denuded conceit. The sameness between the Lord's naked soul and the speaker's bare-wit ironically denotes the absence of mutuality that is more explicitly bemoaned in the sonnet beginning 'A Womans face, with natures owne hand painted'. Indeed, the incongruity of mixed metaphor voices a promised but somehow inconceivable unity between lovers of the same gender.

This incongruous crossing of mixed metaphors derives its power by the way it grafts phantasmology onto Pauline agape. Helen Vendler helps shed light on this play of contexts when she observes that the 'saintlike action [of] "clothing the naked," [which is] one of the seven corporeal works of mercy ... is predicated so equally of the beloved and the guiding star as to make them by succession one'.[33] Yet as the Pauline allusions put into play by sonnet 22 indicate, this action is Christ-like rather than merely saint-like: to be clothed in the beloved's gracious thought is to achieve the greatest gift in the romance of erotic soteriology. The Christological allusions implicit in this image of being clothed in the good favor of the Lord's naked conceit thus coincide with the phantasmological view that the highest form of thought consists of a denuded phantasm. The figure is ingenious precisely inso-far as it works along two distinct registers – one theological, the other phantasmological.

The intersections between grace and Eros are not only thematic, how-ever, but also formal. The figure of being clothed in the grace of a naked thought confirms Freinkel's thesis that catachresis, or abusive metaphor, informs the poetics of *Shakespeare's Sonnets*, just as it structures Luther's theology. The real brilliance of Freinkel's argument lies in how she takes the standard reading of Luther's relation to medieval theology one step further than it is normally developed by historians of the Reformation. Freinkel reveals that once Luther demystifies medieval commitments to analogical approaches to the relations between human and divine, what he is left with is not symbol but catachresis – an abusive yoking of mixed metaphors. Although the play of catachresis in *Shakespeare's Sonnets* does not confirm any direct causal relation between Luther's thought and Shakespeare's poetry, it does remind us that the sonnets are written in the wake of a general dismantling of patristic and scholastic configurations of the human–divine relation along an axis of simili-tude.[34] Just as Luther insists that the bond between the soul and God needs to be configured according to the category of relation rather than the category of substance (through contiguous interactions rather than through mutually shared essential properties) so Shakespeare figures the promised unity between the young man and the speaker through a mixed metaphor that substantially disjoins them while simultaneously supplicating for their identity.[35]

If anything like an actual union between lover and beloved is prom-ised in the parentheses of a naked thought, the very hope of such a mir-acle comes at the price of metrical uniformity as the line stutters at the crucial moment of supplication only to veer off into an extra half-foot: 'But that I hope some good conceipt of thine / In thy soules thought (all naked) will bestow it' (7–8). It is almost as though the mere promise of a gracefully reciprocal love were too much to bear for the outworn conventions of English Petrarchism. While *Shakespeare's Sonnets* may be one of the first collections of poems to express this post-Reformation sense that the operations of grace are better expressed through incongruities of form than through the symmetries of analogy, this ethos will come to shape the more orthodox forms of Protestant poetics associated with poems such as George Herbert's 'Grief' where poetic symmetry are similarly at odds with the unifying force of agape.[36]

But *pace* Freinkel's emphasis on the Lutheran background of Shakespeare's poems, sonnet 26 is not dominated by Reformation contexts any more than is sonnet 22. Multiple theological codes are put into play in the poem: medieval ideas about the corporeal acts of mercy and merit circulate within the same textual space as post-Reformation forms of grace. Like Donne in the *Holy Sonnets*, Shakespeare thrives on mixing pre- and post-Reformation vocabularies as a way of generating specific literary effects.[37] Though *Shakespeare's Sonnets* is not concerned with how the coexistence of competing theological systems generates anxiety about salvation, as Donne and *Hamlet* are, it does draw on competing ideas of merit, will, and grace. Pulling at contradictory theological threads, sonnet 26 finds a powerful way of expressing the speaker's desire for a young man who is at best distant, at worst unattainable.

Perhaps even more importantly, Shakespeare's appropriation of theological figures for poetic ends is itself a key pleasure offered by the sonnets. If not directly blasphemous in all instances, such mistranslations of religious tropes nonetheless offer the sort of libertine pleasures more often associated with Marlowe than with Shakespeare.

In our current climate of reassessing Shakespeare's place in Paul's legacy, it's important that we not forget the pleasures of 'infidel' Will. If Shakespeare's *The Winter's Tale* reimagines medieval drama through as earnest an engagement with Paul's conception of time as Ken Jackson suggests in the previous essay, the sonnets nevertheless further ironize the appropriation of religious figures for erotic ends inherent to Renaissance love poetry. Yet any contradictions between Jackson's argument and mine are more apparent than real as the motivating force in both instances is genre; Shakespeare is playing two entirely different games in the sonnets and the romances. In the first instance, wit rules over all other concerns. Poetic form and the rich history embedded within it controls the thoroughly human (which is to say 'secular') play of psychotheology in the sonnets. In *The Winter's Tale*, however, life-altering questions of forgiveness and repentance are dealt with in the context of a dramatic form that aspires toward the status of wisdom literature: hence Jackson's contextualization of the play in relation to a more sincere reception of St. Paul than that which occurs in the sonnets. Despite the differences between *The Winter's Tale* and *Shakespeare's Sonnets*, both works reveal how sensitive Shakespeare was to the Pauline nature of his culture, particularly to how the cultural fault lines of early modern England turned on competing readings of Paul.

Sonnet 15: of grace, grafting, and investiture

The idea that the transformative power of grace might be expressed by changing one's tattered weeds for a new garment had particular power for a readership caught up in the Elizabethan system of livery and its belief in the transmuting powers of investiture – or so sonnet 15 should lead us to think.

In sonnet 15, the speaker's promise to engraft the young man anew unfolds alongside a parallel promise to remake him by re-clothing him. This secondary promise resonates against the background of the Elizabethan investiture system – the idea that a person's clothing gave them, in Ann Rosalind Jones and Peter Stallybrass' words, 'a form, a shape, a social function, a "depth"'.[38] The Elizabethan conviction that clothing imprints an identity on a person, producing an interiority rather than expressing a preexistent inwardness, testifies to the social importance of livery: 'the social and sartorial system by which lords dressed retainers, masters dressed apprentices, husbands dressed their wives'.[39] According to Jones and Stallybrass, 'Livery acted as the medium through which the social system marked bodies so as to associate them with particular institutions. The power to give that marking to subordinates affirmed social hierarchy.'[40] The theological and social audacity of sonnet 15's wit lies in the way it crosses sartorial and Pauline contexts. For re-clothing the young man has the effect of reversing the class differences that many of the first 126 poems seem to presume. But what is more, this class-defying power is made possible by the claim that the lasting force of poetry is analogous to the power of the divine Word to clothe the soul. In this way, the poem de-cathects the Christological dimensions of Paul's figure, only to re-cathect them to the poet himself.

The speaker's claim to re-clothe the young man as a way to renew him unfolds along converging and diverging lines of Neoplatonic, Pauline, theatrical, and poetic modes of reference. The poem begins with the speaker setting the image of a decaying world against the youthful image of his beloved's beauty. In the first two quatrains, the world is figured as a huge stage in which men 'wear their brave state' or best finery out of memory:

> When I consider every thing that growes
> Holds in perfection but a little moment.
> That this huge stage presenteth nought but showes
> Whereon the Stars in secret influence comment.
> When I perceive that men as plants increase,
> Cheared and checkt even by the selfe-same skie:
> Vaunt in their youthfull sap, at height decrease,
> And were their brave state of out memory.

(1–8)

As Ingram and Redpath note, the image of wearing a brave state 'out of memory' suggests the way actors wore clothing that had been handed down to them by nobles even after it had 'lost all gloss and fashion'.[41] Yet the idiom here is biblical as well as theatrical, alluding to passages in Psalms 102:25–26 and Isaiah 51:6 where the aging world is construed as being like an outworn garment.[42] The biblical idiom activated in the octave establishes the context for the concluding idea that a figurative change of clothing represents a shift from temporal to eternal orders, as it does in Psalm 102:26. For images of grafting and clothing are paralleled in the poem precisely insofar as both involve incorporating the young man into a new body, be it the body of a new wife or a body of writing.

When the speaker perceives his beloved youth in the context of a world as outworn as a tattered garment, the young man appears all the more beautiful. Inspired by this juxtaposition of images, the poet promises that he will engraft the young man anew after his day has changed into sullied night:

> Then the conceit of this inconstant stay,
> Sets you most rich in youth before my sight,
> Where wastefull time debateth with decay
> To change your day of youth to sullied night,
>   And all in war with Time for love of you
>   As he takes from you, I engraft you new.

(9–14)

The variously intersecting contexts at work in these lines pivot around the verbs 'Sets' and 'engraft' and through the implied parallel between men and plants in line 5: 'men as plants increase'. The verb 'Sets' means 'to place' or 'to evoke' indicating that the beloved's image is placed before the speaker's inward eye, even as it implies 'to plant' or 'to engraft' suggesting that this process of evoking the young man's image is not ephemeral but lasting. The parallel between humans and plants thus seems at least as biblical as it is classical in idiom, resting not only in Plato's description of man in the *Timaeus* as a '"heavenly plant" whose "head and root" are suspended from above' but also in the many scriptural passages in which God tends to the garden of the soul.[43]

Read in these ways, 'Sets' activates, among other things, a psychological context in which the young man's image is grafted in the speaker's mind according to the principles of phantasmatic psychology. The Renaissance lover loves insofar as he grafts himself onto the image of the beloved, becoming one with it. Ioan Couliano explains this principle when he observes that the

> true object ... of Eros is the phantasm, which has taken permanent possession of the spiritual mirror [or fancy] ... The phantasm that monopolizes the soul is the image of an object. Now, since man is soul, and since soul is totally occupied by a phantasm, the phantasm *is* henceforth the soul. It follows that the subject, bereft of his soul, is no longer a subject: the phantasmic vampire has devoured it internally. But it also follows that the subject has now grafted itself onto the phantasm which is the image of the other, of the beloved. Metaphorically, then, it can be said that the subject has been changed into the object of his love.[44]

Read in this phantasmatic context, the line means not only, as Booth glosses, 'I give you new life (by writing about you)',[45] but also 'I make myself into your image, which has been set in my mind by this inconstant stay' or 'I am made into your image, which has been set in my mind by this inconstant stay'.

Ficino describes a related process of erotic grafting that further indicates how the speaker's renewing of the young man also signifies a uniting of higher and lower classes, of older and younger lovers, as well as of phantasm and soul. Citing Pseudo-Dionysius, Ficino writes:

> Love ... we understand to be a certain grafted and mixing virtue [a certain instinct for joining and uniting] which certainly moves superior things to the care of inferior things, also reconciles equals to social intercourse with each other, and lastly, urges any inferior things to turn toward greater and higher things.[46]

Viewed in these terms, the act of engrafting the young man anew indicates not only the act of immortalizing him by clothing him in verse, nor simply convincing him to marry, but also remaking him in the soul's eye through the act of phantasmatic love. In each case, Shakespeare's speaker assumes the kind of power that Pico della Mirandola attributes to the Renaissance magus: 'And as the farmer weds his elms to the vines, so the "magus" unites earth to heaven, that is, lower orders to the endowments and the powers of the higher.'[47]

What makes these Neoplatonic claims to poetic and imaginative power unusual is the way they are made to intersect with the regenerative force of Pauline agape. Freinkel comments on the theological dimensions behind this poem's grafting imagery when she argues that 'Shakespeare, like Luther, envisions a new "art of ingrafting": one that only binds flesh to spirit, inconstant youth to eternal constancy, insofar as it remarks their ineradicable difference.'[48] In other words, just as Luther suggests that Christ overcomes sin not by resolving flesh into spirit but by transcending it in a movement that further reiterates the very difference between temporal and eternal orders, so Shakespeare generates a language of union that is forever marked by difference: hence the importance of catachresis to both Luther's theology and Shakespeare's poetics; in catachresis vehicle and tenor are often characterized by incongruity and difference rather than similitude or analogy.

Though Freinkel focuses primarily on the Lutheran contexts at work in the sonnets, grafting is a common Reformation metaphor for how the human soul is adopted into Christ, as Donne indicates when he declares that 'the spirit of adoption hath ingraffed us' into God's covenant.[49] This Reformation vocabulary of being engrafted new is paralleled in the poem with the biblical idiom that an aging world is a tattered garment in need of being replaced. The poem's closing lines thus suggest that the process of renewing the young man involves changing him in both sartorial and horticultural senses.

The image of being clothed by grace is similar to that of being grafted by it insofar as both images presume that one is renewed by incorporation into a new body. As Jones and Stallybrass indicate, this Pauline image had particular force for early modern readers because in 'the livery companies, "translation" meant the incorporation of a member through investiture with livery into the guild'.[50]

Donne, for example, connects the process of adoption into Christ's body with the process of being incorporated into a livery system, when he preaches that God 'gives us outward distinctions, liveries, badges, names, *visible markes in Baptisme:* yea he incorporates us more inseparably to himself'.[51] In Reformation exegesis, to be grafted or clothed by grace means being incorporated into a new, higher, body. The act of re-clothing the young man in sonnet 22 involves a quasi-soteriological process of suturing him into another body; both the body of a future wife as indicated by previous sonnets; as well as a body of writing, as intimated by subsequent poems; and also by the etymological roots from *graphein* 'to write' and *graphis* meaning stylus.[52] Shakespeare's elision of Paul's representation of grace through images of grafting and clothing is remarkable because it shows the speaker presuming a quasi-divine power, making him into a kind of *vates* even Sidney didn't envision. Sonnet 15 thus anticipates the speaker's blasphemous appropriation of God's tautological self-denomination in Exodus: 'I am that I am' later in sonnet 121 (line 9).

As I have indicated, though, the poet's offer to remake the beloved by setting his image in mind unfolds not only in the terms of Pauline regeneration, but also in something like the regenerative terms of Ficino:

> The soul ... puts the beloved's visual image beside its own interior image, and if anything is lacking in the former ... the soul restores it by reforming it. Then the soul loves that reformed image as its own work ... This is how it happens that lovers are so deceived that they think the beloved more beautiful than he is. For in the course of time they do not see the beloved in the real image of him received through the senses, but in an image already reformed by the lover's soul, in the likeness of its own innate idea, an image which is more beautiful than the body itself.[53]

The remaking of the beloved is thus highly overdetermined in the poem, occurring along phantasmatic, soteriological and poetic lines.

Indeed, when the speaker claims to engraft his beloved anew he refers simultaneously to an erotic-epistemological process of birthing a reformed image of the beloved and to a quasi-soteriological process of incorporating him into a new body of verse. This act of incorporation involves an imaginative appareling of the young man in clothes that are not subject to decay, an appareling that is a poetic version of the anagogic garmenting promised in Psalm 102: 'Thou hast aforetime laid the fundation of the earth, and the heavens *are* ye worke of thine hands. Thei shal perish, but thou shalt endure: even thei all shal waxe olde as doeth a garment: as a vesture shalt thou change them, and thei shalbe changed.'

Shakespeare's sonnet resolves by promising to replace the tattered weeds that have been worn 'out of memory' with a new garment, one evocative of the kind of Christological garment central to Paul's theology. In this way, sonnet 15 evokes the two contradictory associations of clothing that Jones and Stallybrass identify as characteristic of Elizabethan culture: fashion as change and inconstancy and fashion as demarking one's incorporation into a particular social body. While the poem's depiction of the worn-out world rests on the former view of clothing as inconstancy, its solution rests on the latter one of clothing as lasting identification. In the final analysis, clothing the young man, like engrafting him, means incorporating him into something larger than himself. By performing this feat the poet not only reverses class hierarchies, he recreates the beloved in a loving act of poetic and phantasmatic regeneration. In the process, he compares himself to God.

Conclusion

As Jones and Stallybrass help us see, Paul's idea of being dressed by the power of God's salvation made particular sense to a culture in which clothing was a form of 'material memory' and a mode of identification. The image of putting on Christ signaled not only a translation of one's identity, but it also served as a reminder of one's mortal sinfulness and thus of one's indebtedness to Christ.[54] The Pauline figure that circulates in *Shakespeare's Sonnets* denotes the sequence's investment in the work of memory, its desire to sustain both Shakespeare's will and the youth of the young man. Shakespeare thus puts Paul to work for poetic and homoerotic ends.

While the sonnets have long been viewed as being aloof from post-Reformation conflicts, they clearly are not. Quite the contrary. Shakespeare draws a good deal of literary power from the tensions between Pauline agape as construed by reformers and Neoplatonic Eros as construed by Ficino and others. In particular, such tensions help produce some powerfully ironic, strikingly indecorous, effects. As I have argued, understanding Shakespeare's psychotheology involves understanding how he grafts the pneumatic idea that Eros consists of having an image cover one's mind onto the Pauline idea that grace involves putting on new skin. This intersecting figure is but one example of how theology and psychology are interwoven in the sonnets in ways that unsettle our post-Cartesian tendency to rigorously separate such discourses. Further analysis of how Shakespeare manipulates the contradictions between Reformation theology and pneumatic psychology will result in better comprehensions of the speaker's voice and wit. For it is through the play of such cross-contexts that Shakespeare does much more than 'keep invention in a noted weed'. Through it, he shows that his 'best [lies in] dressing old words new' (sonnet 76, ll. 6, 11).

Notes

This essay was made possible by funding from the Social Sciences and Humanities Council of Canada.

1. Julia Reinhard Lupton, 'The Religious Turn (To Theory) in Shakespeare Studies', *English Language Notes* 44:1 (2006), 145–149 (146).
2. Philip Lorenz, 'Notes on the "Religious Turn": Mystery, Metaphor, Medium', *English Language Notes* 44:1 (2006), 163–172 (164).
3. Eric L. Santner coined the term 'psychotheology' to denote the way human subjectivity must cope with the excessive over-presence of others, with the uncanniness of our neighbors. As I explain below, I use the term to express how subjectivity in the sonnets, as in Petrarchan discourse more generally, is mediated by theological conceptions of love and desire. For Santner's use of the term, see *On the Psychotheology of Everyday Life: Reflections on Freud and Rosenzweig* (Chicago: University of Chicago Press, 2001).
4. See Joel Fineman, *Shakespeare's Perjured Eye: The Invention of Poetic Subjectivity in the Sonnets* (Berkeley: University of California Press, 1986).
5. See Lisa Freinkel, *Reading Shakespeare's Will: The Theology of Figure from Augustine to the Sonnets* (New York: Columbia University Press, 2002). See also Murray Krieger, *A Window to Criticism: Shakespeare's Sonnets and Modern Poetics* (Princeton: Princeton University Press, 1964) which anticipates Freinkel's thesis that *Shakespeare's Sonnets* are written in the wake of a decline in medieval traditions of typology.
6. Margaret Healy's examination of alchemical figures in the sonnets, '"Making the quadrangle round": Alchemy's Protean Forms in Shakespeare's Sonnets and *A Lover's Complaint*', in Michael Schoenfeldt, ed., *A Companion to Shakespeare's Sonnets* (Oxford: Blackwell, 2007), 405–425, is one recent effort to connect psychological and religious features in the sonnets. Unlike Healy, though, I do not see the sonnets as earnestly concerned with questions of spiritual improvement, even if those questions are posed in the unorthodox language of alchemy. Rather I see the sonnets as misapplying Pauline tropes for homoerotic ends. It should, of course, go without saying that my argument says nothing of Shakespeare's private beliefs – a matter I, like you, know nothing about.
7. As far as I have been able to tell, Shakespeare is the only Elizabethan Petrarchanist to use this Pauline figure. A more straightforward instance of it does appear, however, in Guido Cavalcanti's 'Veggio negli occhi de la donna mia'. For an English translation of this poem see Giorgio Agamben, *Stanzas: Word and Phantasm in Western Culture*, trans. Ronald L. Martinez (Minneapolis and London: University of Minnesota Press, 1993), 106.
8. As one might expect, *The Comedy of Errors* seems most preoccupied with this Pauline figure.

9. Unless otherwise noted, references to *Shakespeare's Sonnets* are from the 1609 Quarto as reproduced in *Shakespeare's Sonnets*, ed. Stephen Booth (New Haven: Yale University Press, 1977).

10. References to Scripture are from *The Geneva Bible: A Facsimile of the 1560 Edition*. I modernize u and v.

11. Cited in Brian Cummings, *The Literary Culture of the Reformation: Grammar and Grace* (Oxford: Oxford University Press, 2002), p. 98. For Luther's account of justification by faith see 'A Commentary on St. Paul's Epistle to the Galatians', in Martin Luther, *Selections From His Writings*, ed. John Dillenberger (New York: Anchor Books, 1962), 99–165, especially 126–127. For Calvin's views on the same matter see *Institutes of the Christian Religion*, trans. Henry Beveridge (Grand Rapids, MI: Eerdmans, 1989), Book 3, Chapter 11.

12. Although Heather Dubrow has challenged the tradition of reading *Shakespeare's Sonnets* as consisting of two main sequences, the so-called young-man sequence (sonnets 1–126) and the so-called dark-lady sequence (127–154), my reading appears to verify the critical tradition. See Dubrow, '"Dressing old words new"? Re-evaluating the "Delian Structure"', in Schoenfeldt, ed., *A Companion*, 90–103.

13. *Shakespeare's Sonnets*, ed. Booth, 170.

14. Henry Smith, *The Wedding Garment* (1590), 10 and 38. Subsequent references are cited parenthetically in the text.

15. See *Shakespeare's Sonnets*, ed. Booth, 170.

16. Cited in Katherine-Duncan Jones, *Shakespeare's Sonnets*, Arden 3rd Series (London: Thomson Learning, 1997), 49.

17. Martin Luther, *Luther's Works: Lectures on Galatians 1535 Chapters 1–4*, ed. Jaroslav Pelikan (Saint Louis: Concordia, 1963), 167.

18. M. A. Screech, *Ecstasy and the Praise of Folly* (London: Duckworth, 1980), 56.

19. Ibid., 56.

20. Ibid., 56. The place of *Praise of Folly* in the sonnets has yet to be studied. Similarly, Shakespeare's allusions to and parodies of ecstasis in the sonnets has not been analyzed. Sonnet 22 is only one example of how Shakespeare engages this tradition, one central to Petrarchism.

21. Ibid., 57.

22. For Neoplatonic views of ecstasy, see Ioan P. Couliano, *Eros and Magic in the Renaissance*, trans. Margaret Cook with foreword by Mircea Eliade (Chicago: University of Chicago Press, 1987: 'But with what kind of grace are we concerned? It is not a gift passively awaited and received but rather the result of active contemplation', 67.

23. The controversies surrounding Paul's trope of putting on Christ were heightened in the Reformation because Romans 13:13–14 was the passage that occasioned Augustine's conversion which served as a model for subsequent Christian conversion experiences and narratives. See Saint Augustine, *Confessions*, Book 8.

24. Shakespeare's interest in Erasmian and other views of ecstasy is evinced by the character of Bottom and to a lesser extent that of Falstaff. See Leonard Barkan, 'Diana and Acteon: The Myth as Synthesis', *English Literary Renaissance* 10 (1980), 317–359 and Walter Kaiser, *Praisers of Folly* (Cambridge, MA: Harvard University Press, 1963).

25. *Petrarch's Lyric Poems: The Rime sparse and Other Lyrics*, trans. Robert M. Durling (Cambridge, MA: Harvard University Press, 1976), sonnet 21, lines 1–4 in the Italian.

26. St. Thomas Aquinas, *Summa Theologiae* (London: Blackfriars 1963), part 2, question 28, article 2, page 95. 'Amor namque concupiscentiae non requiescit in quacumque extrinseca aut superficiali adeptione vel fruitione amati; sed quaerit amatum perfecte habere, quasi ad intima illius perveniens.'

27. For Calvin's views on grace see *Institutes*, Book 3.

28. John Foxe, *The pope confuted* (London, 1580), 99.

29. A similar play on over-literal readings of Pauline figures occurs in *The Comedy of Errors* 2.2.140–144. For an astute discussion of this sequence see Edward Berry, *Shakespeare's Comic Rites* (Cambridge: Cambridge University Press, 1984), 73–74.

30. I follow Booth in amending 'their' in line 12 to 'thy' as it appears to be a typographical error.

31. Sir John Davies, *The Poems of Sir John Davies*, ed. Robert Krueger (Oxford: Clarendon Press, 1975), line 541.

32. Agamben, *Stanzas*, 79.

33. Helen Vendler, *The Art of Shakespeare's Sonnets* (Cambridge, MA: Belknap Press of Harvard University), 148–149.

34. See Freinkel, *Reading Shakespeare's Will*, chapters 3 and 4.

35. For Luther's discussion of the category of relation as the proper means of expressing continuities between human and divine orders see *Luther's Works*, ed. Pelikan, Volume 3, 122.

36. For a discussion of 'Grief' along these lines see Mark Taylor, *The Soul in Paraphrase: George Herbert's Poetics* (The Hague: Mouton, 1974), 46 and Gary Kuchar, *The Poetry of Religious Sorrow in Early Modern England* (Cambridge: Cambridge University Press, 2008), 87–89.

37. For the most influential discussion of how multiple theological codes circulate in Donne's *Holy Sonnets* see Richard Strier, 'John Donne Awry and Squint: The "Holy Sonnets"', *Modern Philology* 86 (1989), 357–384.

38. Ann Rosalind Jones and Peter Stallybrass, *Renaissance Clothing and the Materials of Memory* (Cambridge: Cambridge University Press, 2000), 5.

39. Ibid.

40. Ibid.

41. Cited in *Shakespeare's Sonnets*, ed. Booth, 157.

42. Psalm 102:25–26: 'Thou hast aforetime laid the fundation of the earth, and the heavens *are* ye worke of thine hands. Thei shal perish, but thou shalt endure: even thei all shal waxe olde as doeth a garment: as a vesture shalt thou change them, and thei shalbe changed.' Isaiah 51:6: 'the heavens shal vanish away like smoke, and the earth shal waxe olde like a garmcnt, and thei that dwell therein, shall perish in like maner: but my salvation shalbe for ever'. See also Calvin, *Institutes*, Book 2, Chapter 10, paragraph 15.

43. The reference to Plato is from Lyndy Abraham, *Marvell and Alchemy* (Aldershot: Scholar Press, 1990), 181.

44. Couliano, *Eros and Magic*, 31.

45. *Shakespeare's Sonnets*, ed. Booth, 158.

46. Marsilio Ficino, *Commentary on Plato's Symposium on Love*, trans. Sears Jayne (Dallas, TX: Spring Publications, 1985), Speech 3, Chapter 1, page 64. Jayne indicates in a footnote that the phrase 'grafted and mixing virtue' 'means "a certain instinct for joining and uniting"'.

47. Giovanni Pico Della Mirandola, *Oration on the Dignity of Man*, trans. A. Robert Caponigri (Chicago: Henry Regnery Company, 1967), 57.

48. Freinkel, *Reading Shakespeare's Will*, 203.

49. John Donne, *Sermons*, ed. George R. Potter and Evelyn M. Simpson (Berkeley: University of California Press, 1959), Volume 5, 102. See also, for example, John Calvin, *Commentaries on the Epistles of Paul to the Galatians and Ephesians*, trans. Rev. William Pringle (Grand Rapids, MI: Eerdman, 1957), 224.

50. Jones and Stallybrass, *Renaissance Clothing*, 220.

51. Donne, *Sermons*, Volume 5, 107.

52. Booth notes these etymologies in *Shakespeare's Sonnets*, 158.

53. Ficino, *Commentary*, 114.

54. See Jones and Stallybrass, *Renaissance Clothing*, 270.

Part III
Rematerialisms

10
Against Materialism in Literary Theory

David Hawkes*

I

'Materialism' has been a shibboleth in cultural analysis for three hundred years. If we discard the word itself and concentrate instead on its various, shifting significances, it has been a major bone of contention in philosophy and aesthetics for three thousand. In fact, materialism predates theoretical thought altogether, and there seems a distinct possibility that it will also postdate such thought. Because of its invariably pivotal position in humanistic discourse, a survey of materialism's historical vicissitudes illuminates the progress taken by abstract thought in general. It also suggests some explanations for materialism's increasing prominence today in the field of literary theory. For although that field is as riven by contention as it has ever been, it is close to unanimous in one regard. The vast majority of today's literary theorists, like the overwhelming majority of Western intellectuals as a whole, share a methodological commitment to materialism. In fact, this commitment is often so deep as to be unconscious.

The firm roots that materialism has sunk within the contemporary intellectual psyche are discernible from the fact that a methodological materialism is practiced even by critics who abandon the rhetorical commitment to materialism that was practically *de rigueur* during the closing decades of the last century. Gabriel Egan's elegant and effective contribution to this volume, 'Shakespeare, Idealism and Universals', correctly notes that such commitment was frequently automatic and reflexive:

> Since the 1980s idealism, essentialism, and universals have become dirty words as the New Historicism and Cultural Materialism

237

popularized an unthinking association between these philosophical principles and political conservatism. In these new and related schools, the alleged antidote to all three evils was said to be materialism, which meant paying more attention to the physical (often the economic) realities of a system under consideration than to the ideas in it.

Egan rightly objects to the 'unthinking' nature of this tendency, and proposes instead to 'argue that essentialism and Platonic idealism are reasonable ways to think about the various manifestations of a play'. The argument he actually makes, however, immediately concedes the main materialist position concerning human subjectivity. Egan blithely announces that the Cartesian notion of an autonomous, non-material core of subjectivity is manifestly false: 'most people when pushed will accede that it cannot actually be true. It is certainly difficult to see how there could be an interface between the body and an immaterial spirit such that the latter could control the former.' We are not informed why this should be difficult. Despite the claim that 'most people' find it impossible to conceive of an autonomous non-material subject, the vast majority of people throughout human history have found no difficulty whatsoever believing in such a phenomenon. More immediately problematic, it is hard to see how a Platonic argument of any sort can be constructed from a position that rejects the existence of an autonomous non-material subject. Any such position is materialist by definition.

And indeed Egan proceeds to offer an impeccably materialist explanation for belief in the autonomous non-material subject. He contends that 'it has served an evolutionary purpose'. Far from defending Platonism, in fact, the essay bases its case on the most dogmatically materialist of all methodologies: evolutionary psychology. This approach to the human subject, popularized by thinkers such as Daniel Dennett and Richard Dawkins, claims that human behavior can be explained by reference to the physical structure of the brain, and that this structure has been formed according to the principles of Darwinian evolution. Although he concedes that Dawkins himself 'only half-intended' his theory of 'memetics', which suggests that culture operates in an evolutionary fashion analogous to biology, Egan readily employs it in his analysis of literary texts. He finds it confirmed by cognitive neuroscience, and he argues for example that the 'mirror neurons' of our brain account for the empathy we feel for Hamlet or Lear:

> These neurons fire not only when we perform an activity but also when we watch someone else performing that same activity. They appear to be the reason that it is difficult to watch someone

yawning or laughing without joining in, and equally why it is difficult to watch Lear's agony at the death of his daughter without sharing in the emotion. Our mirror neurons make us feel his pain even though we know we are watching only an imitation.

Such arguments reduce ideas to matter: they are materialist. The tautological reasoning, biological reductionism, and ideological function of evolutionary psychology are discussed below.[1] My point here is to note that an essay that begins by announcing its intention to defend Platonism against 'unthinking' materialism should have recourse to the most uncompromising form of materialism in pursuit of that end. This is eloquent testimony to the almost instinctual hold that materialist assumptions currently exert on the Western intelligentsia. Today's debates are not usually over whether materialism is a desirable theoretical orientation, but over which approximation is most faithful to materialism's authentic nature. To understand how this situation has arisen, a glance at materialism's ancient and illustrious lineage will be useful.[2]

Materialism is an instinctive response to the world, initially based on sense-perception alone. To the entirely unreflective eye, it appears that matter is all that exists, for only matter is perceptible. Thus materialism was the first position that the Greeks arrived at when they began to consider their situation in conceptual terms. Such a reaction to experience shows a failure to distinguish between appearance and essence. Primitive materialism assumes that the way things appear to be is the way they really are. Once it is accepted that all existence shares the single characteristic of being material, the natural next step is to identify an *arche*, a single element within all matter, which would provide it with a definitive characteristic and a unifying principle. Thus in the early 'Ionian' school of philosophy, materialism takes the form of 'monism', the attempt to impose unity on the multifarious, to insist that apparent difference is in fact identity. Such a principle was a theoretical rather than an empirical necessity. The earliest known Western philosophers, the 'pre-Socratic naturalists', include Thales, who held that everything was composed of water, and Anaximenes, who believed that everything was made up of air. Empirical observation was beside the point of such assertions; the point was to establish a unifying principle within all matter.

This seminal materialism was expanded and elaborated by Democritus, who recognized that all matter was composed of atoms. But Democritus also departed from his predecessors by elaborating the distinction between matter and ideas, and by producing an account of how material circumstances influence the mind. He suggested that objects transmit images, or *eidola*, that impact the organs of our senses to produce our impressions of the world. Epicurus further refined this atomism, arguing that matter emanated physical images, called *lamina* or *simulacra*, which were shaped like matter itself, and whose impression on the eye gave rise to our perception of images. In the Roman poet Lucretius, this kind of materialism is used to refute the existence of the gods, and as an antidote to superstition in general. Democritus, Epicurus, and Lucretius move beyond the eliminative materialism of Thales and Anaximenes by acknowledging the objectivity of ideas. They remain materialists, however, because they trace the development of ideas to an origin in material stimulation of the senses.

These early forms of materialism eventually faced formidable, concerted opposition from Platonic idealism. Reversing the approach of the materialists, Plato believed that the realm of ideas creates the realm of matter. For him, human experience is always mediated through ideas, or concepts. It is impossible for a human being to have a merely sensory experience, for we inevitably impose concepts on the data that we receive through our senses. Although these ideas do not have any material existence, they nevertheless determine the way human beings experience their surroundings, and in this sense they create those surroundings for us. This reasoning led Plato beyond the contention that ideas determine our material experience, to the conclusion that ideas constitute the only objective reality, and that the material world is a mere illusion. Platonic idealism is thus the mirror-image of materialism. Both approaches reduce the relation between ideas and matter to one of its poles, assuming that only one side of the dichotomy is authentic, and claiming to explain how it creates the illusion of the other.

Platonic idealism became an extremely important influence on Christianity, and the institutional power of Christianity ensured that for almost two thousand years materialism was relegated to a minority opinion among Western thinkers. It did not disappear altogether, it always survived as an oppositional undercurrent, but generally speaking philosophers and theologians assumed not just the ontological but also the ethical priority of ideas over matter. Materialism was not only mistaken, it also showed a morally reprehensible orientation toward carnality. In fact the modern reaction in favor of materialism drew much of its impetus from a revulsion against the religious imposition of idealism, which often took the form of a puritanical denial of fleshly pleasure.

Platonic thought also provided a much-needed rationalization of slavery, for which it drew on an alliance with Aristotelian teleology.

Aristotle claimed that the proper purpose, or *telos*, of a human being was 'an activity of the soul in accordance with virtue'. This claim coalesced with Platonic idealism, by elevating intellectual speculation above physical labor, associating the former with nobility and the latter with servility. A slave was by definition not a fully human being, because he served the purposes of his master rather than his own. He was a 'property' of his master. According to Aristotle, the majority of the human race were 'natural slaves' because they reversed the Platonic hierarchy between ideas and matter within their own souls. To be a natural slave was to prefer the pleasures of the flesh over those of the spirit, and to act in accordance with the demands of the body rather than those of the soul. The ideological utility of this argument, as well as its philosophical coherence, helped to bolster idealism's long-term intellectual dominance over materialism.

Over the course of the seventeenth century, however, materialism enjoyed a dramatic resurgence. The scientific empiricism of Francis Bacon depends upon the basic materialist proposition that our knowledge of the world comes from sensory experience. This led Bacon to advocate a highly successful 'instrumentalist' view of science, which would pursue and evaluate theories according to the practical achievements they made possible. At the same time, the atomist physics of Pierre Gassendi was giving rise to a revived materialist approach to scientific theory. The dramatic scientific advances facilitated by this pragmatic materialism gave it an unassailable advantage over the abstract speculations of idealism.

The literature of the Renaissance frequently intervenes in this debate, generally denouncing materialism under the rubric of 'worldliness' or 'carnality'. As Ian Munro observes in his contribution to this volume, 'Theater and the Scriptural Economy in *Doctor Faustus*', Marlowe's Faustus evinces unmistakable tendencies toward materialism. He espouses an empiricist ontology in such remarks as 'I think Hell's a fable' (1.5.126). The play's message, however, is that such beliefs are absurd: Faustus' confident opinion is voiced to Mephistopheles, a manifest inhabitant of the place in which he claims not to believe. The devil mockingly notes the self-refuting nature of Faustus' skeptical empiricism: 'Ay, think so still, till experience change thy mind' (1.5.127). He assures Faustus that 'I am damned, and now in Hell' (1.5.131), but the magician's refusal to admit the objective existence of what is non-material prevents him from accepting the devil's assertion: 'How! Now in Hell! Nay, an' this be Hell, I'll willingly be damned here' (1.5.132). The literary texts of the early modern period regularly link philosophical materialism with personal villainy. When *The Tempest*'s murderer Antonio is asked about his conscience, he gives the archetypal materialist response: 'Ay, sir: where lies that? If it were a kibe, / 'Twould put me to my slipper; but I feel not / This deity in my bosom' (2.1.273–275).

Antonio is unambiguously evil, but his depraved opinions were gradually growing more common. In the mid-seventeenth century the philosophical implications of Bacon and Gassendi's scientific theories were developed by Thomas Hobbes, whose epistemology led him into a skeptical form of empiricism. Believing that only matter existed, he claimed that experience of matter produced consciousness. However, Hobbes also claimed that there was no reason to assume that our sensory experience gave an accurate impression of its objects. It followed that human knowledge must inevitably be imperfect and provisional. Since absolute knowledge was inaccessible, it made sense to regard human thought as guided by the material interests of its advocates rather than by the search for objective truth. The Darwinian identification of self-interested market behavior with human nature is among materialism's first fruits. This kind of materialism molded the skeptical relativism that predominates in the post-humanistic discourses of the twenty-first century.

Modern materialism implies a relativist morality, in which human beings are guided by the pursuit of economic self-interest. Hobbes suggested that, because the appetites were natural, they must be accommodated. In the eighteenth century this philosophical assumption became the basis of the new science called 'political economy'. Showing the influence of the Calvinist notion of 'total depravity', according to which human nature was completely and inescapably corrupt, thinkers like Bernard de Mandeville made the case that, left to themselves, people will always pursue their selfish desires. This proposition is still the fundamental assumption of mainstream economic theory. Early political economists like David Ricardo and Adam Smith gave an optimistic gloss to this bleak view of humanity by asserting that if every individual sought to maximize his material self-interest, the cumulative result would be beneficial for society as a whole.

By the nineteenth century, political economy was under attack by early socialist theorists, but most of these shared the basic assumption that human ideas were rooted in material self-interest, if not of the individual, then of a particular social class. The kind of socialism advocated by Karl Marx became especially closely associated with philosophical materialism. Marx was challenging the idealist thought of G. W. F. Hegel, who had refined Plato's idealism into a dialectical historicism.

Like Plato, Hegel believed that ideas determine people's experience, but unlike Plato he conceived of ideas as changing and developing in the course of human history. Marx agreed with Hegel's historicism, but he emphasized the role of material circumstances, especially economic circumstances, in determining historical developments.

However, it is misleading to think of Marx as a 'materialist', just as it is to conceive of Hegel as an 'idealist'. Both Hegel and Marx were 'dialecticians', which means that they conceived such paired contradictions as the one between ideas and matter as mutually determining. They thought that each pole of the dichotomy brought the other into existence, a doctrine known as 'the interpenetration of opposites'. It would be impossible to conceive of 'matter' unless we also held the opposite conception of 'idea'. It is thus a 'reductionist' fallacy to claim either pole of the dichotomy determines or creates the other. This vital insight has frequently been obscured in subsequent philosophy. The followers of Marx, led first by his friend Engels and later by his most successful disciple Lenin, emphasized the materialist elements in Marx's argument, and the institutional communism of the twentieth century insisted on a dogmatic and unsophisticated form of doctrinaire materialism. In this philosophy, the 'economy' was conceived as material, and as giving rise to the 'ideologies' in which social classes understood and advanced their collective interests.

Although communism and capitalism are opposed modes of thought in many ways, they share some core materialist assumptions in common. In particular they both ascribe determining power to the 'economy'. Over the course of the twentieth century the 'dialectical materialism' which was the official ideology of the communist world converged with the materialism fostered by the capitalist marketplace, which depends upon the notion that the acquisition of wealth is the natural purpose of human life. The notion that the 'economy' is 'material' hardly stands up to close analysis, however, and as the last century drew to a close it became progressively harder to maintain. Economic developments were increasingly driven by consumption rather than production, and thus by psychological decisions instead of material activity. The very concept of the 'economy' as a discrete field of human behavior began to break down, and with it the materialist determinisms that had dominated twentieth-century philosophy.

But this certainly did not mean the end of materialism. The fall of the wall dividing the 'economy' from other areas of life enabled the view of human nature that was first developed by political economists to colonize every part of experience. The view that the pursuit of material self-interest is natural, and thus inevitable, appeared to be corroborated by Darwin's evolutionary theory, and today it is solidly entrenched in both popular culture and the academy. By the end of the twentieth century, eliminative materialism had become the dominant approach to the study of the mind. Just as astronomy had eliminated astrology, just as chemistry had superseded alchemy, it was claimed that the insights of cognitive neuroscience, which equates ideas with the neurological patterns of the brain, could and should abolish the 'folk psychology' which conceived of ideas as occupying a separate sphere from matter. Ideas were no more real than the elves and fairies of popular mythology.[3]

A majority of today's literary critics take the basic assumptions of materialism for granted, although their application of these tenets varies considerably. 'Cultural materialism', a movement largely inspired by the work of Althusser's disciple Michel Foucault, continues to thrive within literary studies. Some critics use the term 'materialism' to designate a field of interest rather than a theoretical approach. They include 'historians of the book', who analyze the development of printing and the physical shape of books, and those critics with a particular interest in the way objects such as furniture or clothing function within literary texts. More recently, such forms of materialist criticism as 'cognitive' or 'evolutionary' theory have gained significant followings. Above all, the word 'materialism' is frequently used for polemical effect, to indicate the practitioner's opposition to essentialist or idealist approaches to the literary artifact. Having considered how this situation has arisen, we can now examine some of the current forms taken by materialist literary theory.

II

Raymond Williams' *Culture and Society* (1958) is often cited as the first application of cultural materialism to literary studies. However, the movement's intellectual roots lie further back, in Antonio Gramsci's *Prison Notebooks*, composed in the late 1920s and early 1930s, which mark an important departure from the materialist determinism espoused by institutional communist dogma. Gramsci argued for a 'relative autonomy of the superstructure', meaning that the realm of culture could change and develop in a manner distinct from the economic 'base'. Culture and aesthetics could thus become venues for political action, and the cultural materialist critics conceived of their work as interventions in broader power struggles. It was this sense of political engagement, more than any inherent philosophical bias, which led them to call their work 'materialist'. Their focus on the influence of social forces and power relations on literary texts enabled the cultural materialists to break new interpretive ground, and by the 1980s, critics like Stephen Greenblatt, Jonathan Dollimore, and Stephen Orgel had forged a formidable body of innovative materialist work.

The contention that the superstructure is autonomous of the base, even 'relatively' so, is a retreat from dogmatic materialism. In fact the cultural materialists inhabit something of an oxymoron, for they are generally quite prepared to admit the existence of ideas, and also the influence of ideas on people's material activities. Their brand of materialism is concerned to emphasize the historical circumstances in which an aesthetic work is produced, but it does not necessarily prioritize 'economic' factors. Cultural materialists can abandon economic determinism and still call themselves 'materialists' because they contend that all culture is material, not just the economy.

Toward the end of the twentieth century, the term 'materialism' became code for leftist or liberal political commitment. A materialist critic would tend to pay attention to relations of class, gender, and sexuality in literary texts, although he or she would not necessarily trace these back to foundations in the economy. In fact, cultural materialists would be more likely to point out the way such relations are organized by and through systems of signification. The title of 'materialist' could then be justified by pointing out that such systems inevitably express themselves in material form. Thus poststructuralism and philosophical neo-pragmatism, which lay heavy stress on the role of language in determining ideas, are frequently classed as species of materialism. The term 'materialist' thus grew increasingly capacious as the twentieth century wore on.

By the turn of the millennium, in fact, it was no longer clear who the materialists were opposing. They had achieved a philosophical predominance not seen since the pre-Socratics. Seventeenth-century materialists had been an avant-garde minority who often made it their business to scandalize respectable society. In his 'Satyr against Reason and Mankind', the Earl of Rochester made a libertine case for materialism, pouring scorn on what he views as humanity's vain faith in abstract reason:

> The senses are too gross, and he'll contrive
> A sixth, to contradict the other five,
> And before certain instinct, will prefer

> Reason, which fifty times for one does err;
> Reason, an *ignis fatuus* of the mind,
> Which, leaving light of nature, sense, behind,
> Pathless and dangerous wand'ring ways it takes
> Through error's fenny bogs and thorny brakes

$$(8-15)^4$$

Unlike such seventeenth-century cavaliers, today's materialism shocks nobody. Eighteenth-century materialists fought bravely against the political and intellectual power of religion, but that battle is long over in the West. Nineteenth-century materialists were often committed to socialist or communist political causes that seem impossible or undesirable today. In the first half of the twentieth century, literary studies was still dominated by post-Romantic individualists, who luxuriated in the subjective affect and emotions generated by the text, but the materialists have long routed such feeble opposition. At the beginning of the twenty-first century, it seems that there are very few literary critics who are *not* materialists.

Galloping onto this wide open field, the materialists have broken off into clusters. One version of materialism, found in works like Jean Howard's *Theater of a City* (2006) and Stephan Mullaney's *The Place of the Stage* (1988), directs our attention to the influence exerted by the physical locations of early modern theaters on the plays performed there. Another materialist tendency devotes itself to the analysis of objects. Historians of the book like Peter Stallybrass and Elizabeth Eisenstein study the physical shapes and textures in which semiotic significances have been transmitted to the minds of readers. Other critics, like Margreta de Grazia and Natasha Korda, concentrate on the ways that physical objects are represented in literary texts, often to brilliantly illuminating effect. Occasionally, however, this kind of materialism can degenerate into what Douglas Bruster has unkindly called 'tchotchke criticism'[5]: an interest in objects for their own sake which, ironically enough, frequently eschews the cultural contexts in which objects acquire meaning. In order to distinguish it from both cultural materialism and Marxism, this object-centered approach is sometimes referred to as the 'new materialism'.

Object-centered criticism is often exciting and informative. For example, Jonathan Gil Harris' *Untimely Matter in the Time of Shakespeare* undertakes the task of 'recasting matter as an actor-network'[6] in a manner that yields important new insights into Renaissance drama. Drawing on the work of philosophers like Bruno Latour and Paul Virillo, as it has been expanded into an 'object oriented ontology' by younger phi-losophers like Graham Harman and Levi Bryant, Harris points out that what we identify as a unitary and coherent 'object' is in reality a formal construct, made up of many other 'objects', and finally of subatomic particles that are not material at all. Objects are constantly changing, becoming larger or smaller, cleaner or dirtier, harder or softer. Objects have their own conditions of possibility, circumstances that must be in place before they can come into being and that determine their nature, such as particular relations of gravity, air pressure, internal chemistry.

Objects, in short, are historical – and this has nothing to do with our subjective experiences of them, but is an inherent property of the objects themselves. To take this fact seriously is to experience all objects as palimpsests, containing different levels of historical significance. Like other object-oriented critics, Harris perceptively notes that the market's attribution of independent agency to commodities is reflected in much of the period's literature. He demonstrates, for example, how the mate-rial body of Desdemona's handkerchief becomes a subjective force in Shakespeare's *Othello*, and thus 'enters into a diverse array of actor net-works'.[7] The handkerchief plot is recapitulated in Ben Jonson's *Volpone*, when Corvino tells his wife: 'you were an actor with your handkerchief' (2.3.40). Although he notes that the kind of power bestowed upon the handkerchief is fetishistic and magical, Harris' account of the acquisi-tion of subjective agency by objects is by and largely ethically neutral. For the people of sixteenth- and seventeenth-century Britain, how-ever, magic was quite literally a satanic activity. They protested loudly against the quasi-magical attribution of agency to objects. In John Bale's *Comedy Concernynge Thre Lawes* (1538), a character named 'Idololatria' boasts of her ability to animate objects by 'charmes of sorcerye: I can make stoles to daunce / And eaerthen pottes to praunce …'[8] Bale's con-cern is to establish the satanic source of such illusory animation, and the moralistic horror with which such agency was generally portrayed in early modern Europe contrasts with the ethically neutral position adopted by materialist criticism.

The usury controversy provided a particularly congenial venue for protests against the magical illusion that objects can act independ-ently, for in usury money takes on an artificial agency that displaces the human activity it originally represents. Roger Bieston's *The Bait and Snare of Fortune* (1559) offers an allegorical dialogue between Man and Money, in which the pair bicker over which of them is responsible for usury. 'Man' has forgotten that 'Money' is nothing more than his own alienated activity, and 'Money' is forced to remind him that: 'I as of my selfe can nothing doe nay say / In thee lieth al the dede ...'[9] Thomas Floyd's *The Picture of a Parfit Commonwealth* (1600) is typical in its outraged declaration that: '[u]sury is an actiue element that consummeth all the fewell that is laid upon it, gnawing the detters to the bones, and sucketh out the blood and marrow from them ...'[10] When Marlowe's villainous Barabas declares 'I hope our credit in the Custome-house / Will serve as well as I were present there' (1.1.57–58), he alludes to the false subjectivity that money was achieving before the audience's eyes.

In our own time, such category confusion between subject and object has led thinkers like Paul and Patricia Churchland and Daniel Dennett into an eliminative materialism which denies the existence of subjective experiences, or 'qualia' altogether. For them the Kantian dichotomy between the 'for us' and the 'in itself' can be resolved by the simple abolition of the former. Phenomena such as pain or desire amount to nothing more than specific configurations of neurotransmitters in the brain.[11] This philosophical 'eliminative materialism' is beginning to influence some literary critics. For example 'evolutionary criticism' takes its inspiration from the allegedly scientific discipline of evolutionary psychology, and purports to show how the material operations of the human brain are reflected in literary texts.

In the words of Mary Thomas Crane: 'literary theory derived from cognitive science ... offers new ways to locate in texts signs of their origin in a materially embodied mind/brain'.[12] In *Shakespeare's Brain*, Crane aims to show how '[s]everal of Shakespeare's plays experiment with different forms of polysemy and prototype effects in ways that leave traces of cognitive as well as ideological processes in the text'.[13] In *Toward a Theory of Cognitive Poetics* (1992), Reuven Tsur claims that responses to literary texts are 'constrained and shaped by human information processing'.[14] On the basis of this claim, Tsur builds many more specific assertions, such as the discovery of a 'definite spatial setting'[15] for emotion produced by poetry which, he believes, works by channeling language through the right hemisphere of the brain, which is not usually used in linguistic comprehension. Although the instruments they use and the detail in which they can analyze the brain are vastly more sophisticated, cognitive neuroscientists share the essential assumptions of nineteenth-century phrenologists, who sought to study behavior and character by analyzing the shape of the skull. In the *Phenomenology of Mind* Hegel famously satirized phrenology for its assumption that 'the spirit is a bone'.[16] This materialist tradition assumes that human beings are identical with their bodies so that, in the words of Stephen Kosslyn and Oliver Koenig: 'the mind is what the brain does'.[17]

This kind of materialism has moved away from historicism, since it declines to take account of culture or society as formative influences on the personality. Evolutionary psychology, which was formerly known as 'sociobiology', is extrapolated from the theories of ultra-Darwinist biologists like Richard Dawkins. Its essential presupposition is that, since the human brain has evolved according to evolutionary requirements, and since human behavior is caused by the brain, all social and cultural relations ought, in theory, to be explicable according to evolutionary principles. The dogmatic claims of this school's adherents accrue confidence from its allegedly scientific basis. In *Consilience: The Unity of Knowledge* (1998), E. O. Wilson declares that the assumptions and method of Darwin, being objectively accurate and true, ought to be imported wholesale into the humanities. Wilson explicitly connects this desire to unify all knowledge to the efforts of Thales and the other pre-Socratic materialists to find a single *arche* underlying all existence: he refers to his endeavor as 'the Ionian enchantment'.[18]

According to Joesph Carroll, 'the study of literature should be included within the larger field of evolutionary theory'.[19] This is because Darwinism 'necessarily provides the basis for any adequate account of culture and literature. If a theory of culture and literature is true, it can be assimilated to the Darwinian paradigm; and if it cannot be reconciled with the Darwinian paradigm, it is not true.'[20] Evolutionary critics note that narrative storytelling is a feature of all human societies, and that it therefore presumably predates civilization and must have played a role in the evolution of the human race. Paul Hernadi argues that 'the protoliterary experiences of some early humans could, other things being equal, enable them to outdo their less imaginative rivals in the biological competition for becoming the ancestors of later men and women'.[21] Michelle Sugiyama speculates that 'those individuals who were able (or better able) to tell and process stories enjoyed a reproductive advantage over those who were less skilled or incapable of doing so, thereby passing on this ability to subsequent generations'.[22] William Flesch suggests that narrative fictions trained human beings in 'social scanning', and that realism in literature enables evolutionary advantage: '[e]ffective narratives are therefore likely to be accurate representations of human interactions, just because genuine human interactions are what we are so attuned to monitor'.[23]

The focus of evolutionary critics on the modular structure of the physical brain takes a slightly different form in 'cognitive criticism'. This approach departs from the notion that ideas are embodied in material form, and examines instead the ways in which they are determined by informational structures. It deserves to be called 'materialist', however, because it shares materialism's skeptical attitude toward autonomous consciousness and subjectivity. Cognitive critics are influenced by evolutionary theorists such as Stephen Pinker, the 'posthuman' approach of anthropologists such as Donna Haraway, and the poststructuralist philosophy of Jacques Derrida. As summarized by Katherine Hayles:

> Among the characteristics associated with the posthuman are a privileging of informational pattern over material instantiation; a construction of consciousness that sees it as an epiphenomenon rather than the seat of identity; a view of the body as an originary prosthesis that we all learn to operate at birth and that is supplemented later in life by other prostheses; and above all, a configuration of the human so that it can be seamlessly articulated with intelligent machines. The posthuman can be understood as an extension of postmodernism into subjectivity, carrying the projects of fragmentation and deconstruction into the intimate territory of nerve and bone, mind and body.[24]

Like evolutionary critics, cognitive theorists are committed to the reduction of subjectivity to the functions of the brain, but they depart from their colleagues in conceiving of those functions as informational rather than physical. Unlike evolutionary critics, cognitive theorists can therefore gain access to the insights of Derridean deconstruction. Thus Ellen Spolsky describes her approach as 'based on an analogy between some elementary facts about the human evolved brain and the post-structuralist view of the situatedness of meaning and of its consequent vulnerability to the displacements and reversals that deconstructionist criticism reveals'.[25] Despite their differences, however, all of the materialisms currently prominent in literary studies share one fundamental assumption. They all believe that the human subject, mind or soul is an illusion. Object-oriented critics neglect subjectivity in favor of analyzing the representation of physical things; evolutionary critics view subjectivity as merely an advantageous adaptation produced by the development of the brain; cognitive critics consider subjectivity an epiphenomenon produced by patterns of information. Any evaluation of materialism's benefits for literary analysis must therefore focus on this core shared assumption. Is it true that human beings have no soul?

III

The dominance of materialism in today's literary studies is in large part a result of the colonization of the human by the natural sciences. This in turn results from the enormous boost in self-confidence that the natural sciences have received since the Second World War, from the 'new synthesis' between Darwinian evolution and genetics. Many scientists believe that this synthesis provides a universal explanatory key that can account for *everything* – animal, mineral, vegetable – in the entire universe, the arts and humanities very much included. This assertion is made with ever-increasing confidence, and it involves a blanket dismissal of the non-material subject as an outmoded, superstitious fantasy. The geneticist Walter Gilbert is apparently fond of beginning his lectures by brandishing a compact disk containing genomic information and announcing to the audience: 'this is you'.[26]

That is a provocative gesture, but many natural scientists believe that it is nothing but the truth. DNA, we frequently hear, is destiny. Not only our bodies but our character, our essence, is contained in our genes. Genes are eternal, while the bodies that contain them are mortal. It is no exaggeration to say that in the work of evolutionary geneticists like Richard Dawkins, genes have replaced both the soul and its traditional source: God Himself. Dawkins makes no secret of his colonialist ambitions toward the humanities, and has proposed that units of signification that he calls 'memes' function in culture in a fashion analogous to the operation of genes in nature. And it is certainly true that evolutionary and genetic determinism have profound implications for literary texts. That is one reason for the almost undisputed reign of materialism in today's departments of literature. However, one of materialism's traditional tenets is that truth-claims can never be considered in isolation from the wider social context in which they arise. It therefore seems appropriate to consider how the various claims of contemporary materialism fit into the current political, social, and economic context.

There is no doubt that materialism can serve politically progressive causes. By showing how human society and the human subject are formed by external circumstances, it provides a potent antidote to essentialist arguments that portray particular social formations or subjectivities as natural. Materialism can thus be enlisted in the service of identity politics, and used to show, for example, that femininity or homosexuality are not immutable conditions that must occupy particular roles. It can challenge the contention that a certain race or gender is naturally or inevitably dominant; indeed it can call into question the very existence of races or genders. By demonstrating the contingency of culture, it can encourage us to envisage different and more equitable social arrangements from those that currently pertain. By emphasizing the importance of practical engagement in political affairs, materialism can facilitate ameliorative engagement with oppressive hierarchies that were once assumed to be unavoidable. When applied to literary texts, materialist criticism can enhance our understanding of why particular forms, themes, and effects come into being and pass out of common usage.

However, I would argue that these benefits actually derive from historicism rather than from materialism. The contention that there is no fixed human nature or natural mode of social organization is best advanced by locating the objects of study within their contingent historical circumstances, but it is not necessary to claim that those circumstances are solely material. There seems no reason why an idealist historicism should not achieve the same ends. In fact it seems likely that the association of materialism with leftist or liberal political agendas is a historical accident. It originally arose out of the need to challenge the political power of organized religion. It was consolidated by the imperative to improve the living conditions of the industrial proletariat. During the eighteenth, nineteenth, and twentieth centuries, conservatives often took refuge in idealism, identifying their own power with the state of nature, and appealing to invariant human nature as a bulwark against progress.

Today, in contrast, arguments for materialism emanate largely from advocates of capitalism and the market economy. From its inception Darwinian evolutionary theory has been twisted into rationalizations of the competitive marketplace. Indeed Darwin himself was aware of the economic implications of his theory, as the famous passage from his *Autobiography* reveals:

> In October 1838, that is, fifteen months after I had begun my systematic inquiry, I happened to read for amusement Malthus on *Population*, and being well prepared to appreciate the struggle for existence which everywhere goes on from long-continued observation of the habits of animals and plants, it at once struck me that under these circumstances favourable variations would tend to be preserved, and unfavourable ones to be destroyed. The results of this would be the formation of a new species. Here, then I had at last got a theory by which to work.[27]

Theories of evolution are ancient, dating back to Anaximander, and no serious thinker disputes that organisms evolve through interaction with their environments. However, that does not mean that Darwinism can be regarded in isolation from its own environment as an absolute, axiomatic set of truths. Darwinist biology is often invoked to suggest that the competitive marketplace is the natural mode of social organization, and several materialist critics are now applying this connection to the literary canon. In *Fiction Sets You Free* (2007), Russell Berman argues that by its very nature, literature 'contribute[s] to the value structure and virtues of a capitalist economy', and to 'the dissemination of capitalist behavior', because all fictional writing 'cultivates the imaginative prowess of entrepreneurial vision'.[28] It does this, Berman suggests, simply because it is not true. By describing situations other than those that actually pertain, 'literature imposes an economic choice on the reader'. All fictional texts are thus 'indispensable sources for capitalist psychology' because they address themselves 'to entrepreneurial risk takers who have the will to imagine'.

Berman returns to the roots of modern philosophical materialism, laying heavy stress on the Hobbesian *bellum omnium contra omnes*. He claims that a work of literature comes into existence surrounded by antecedent texts 'which threaten to crush it', and that it is immediately forced to 'assert itself against its competitors and predecessors'. The deployment of evolution as a universal explanatory key has already spread beyond biology into sociology, psychology, and, above all, economics. Evolutionary theory imports it into literary studies. Berman uses 'a Darwinian axiom' and an 'evolution-theoretical claim' to support his contentions, and displays the influence of Richard Dawkins' theory of memes in his speculations on 'literature's genomic character'. Dawkins' 'memetics' is an attempt to extend the synthesis of Darwin and genetics into the realm of signification. He suggests that, rather than being driven by the power of reason or the motor of history, ideas replicate autonomously, in a manner analogous to genes. Such attempts to colonize the humanities with the materialist assumptions of economics and the natural sciences are proving quite successful: it seems they strike a chord in the *Zeitgeist*.

This is a paradoxical consequence of materialism in the humanities. It was precisely in order to assert the influence of culture and society on literary texts that leftist critics of the 1980s insisted that these were material phenomena. They desired to show how art reflects the real power struggles of society, and so they claimed that the sphere of culture was material. But if that is true, if art and literature really are material, then they must be susceptible to study by the same meth-ods as the materials of the natural sciences. And such methods will frequently contain political implications that are the reverse of those desired by the cultural materialists when they initiated their project twenty and thirty years ago. For the reduction of the human self to matter, the objectification of the subject is the prime ideological effect of capitalism. The present prominence of materialism in literary theory reflects the bleak conclusion of Theodor Adorno in *Negative Dialectics*: 'The subjective consciousness of men is socially too enfeebled to burst the invariants it is imprisoned in. Instead, it adapts itself to them while mourning their absence. The reified consciousness is a moment in the totality of the reified world.'[29]

The facts that philosophical materialism began its modern rise to prominence at the same time as capitalism, that it has blossomed and flourished to the same degree as capitalism, and that its current virtually undisputed power coincides with the global triumph of capitalism may all be coincidental. But there are also theoretical reasons for suspecting collusion between materialist philosophy and capitalist economics. A capitalist economy is a vast machine that seems almost consciously designed to reduce people to the status of objects. It universalizes the condition that Aristotle described as slavery, whereby a human being is not free to pursue his own ends but must serve the ends of his mas-ter. By this definition, everyone who works for a wage is a slave, and it is reasonable to suppose that the psychological objectification that Aristotle associated with slavery has spread and solidified along with slavery itself.

But even if we leave the subjective effects of wage-labor aside, the capitalist economy objectively transforms its participants into com-modities. Virtually everyone in such an economy must sell his or her time for money. Time is indistinguishable from life, so that everyone in capitalist society must constantly translate his or her life, his or her self, into the objective form of financial representation. It is not surprising that the idea that human beings are purely material objects should gain credibility in such a system. Nor is objectification limited to the sphere of production. Today's consumer societies erase the distinction between production and consumption, and the job of consuming commodities is at least as economically important as the job of producing them. A vast array of ideological apparatuses is therefore devoted to convinc-ing people that their identities can be constructed through the com-modities they consume, and postmodern capitalism encourages us to equate identity with image. It is not difficult to see how this kind of society would produce philosophies arguing that subjectivity springs out of material representation.

Eliminative materialism is not content to derive ideas from matter but proceeds to the conclusion that only matter exists. We have seen how ancient thinkers such as Epicurus accounted for the undeniable fact that we experience ourselves as having ideas by describing them as material *simulacra* that are cast upon our sense-organs by the action of material objects. In the postmodern world, thinkers like Jean Baudrillard make a similar argument using identical terminology. For Baudrillard the whole of experience is made up of *simulacra* that combine to produce a 'hyper-reality' in which the distinction between sign and referent is obsolete.[30] As with the pre-Socratics, appearance is equated with essence: what seems to be is identified with what is. This is the logical terminus of eliminative materialism and, once again, it is demonstrably an effect of the market economy.

An exchange-based society will systematically replace the inherent use-values of objects, which are inseparable from their physical bodies, with symbolic exchange-values, which are grafted onto physical bodies by the human mind. Because use-value is implanted into objects through productive labor and manifested in the use of objects by human beings, use-value is inseparable from human activity. It follows that, since exchange-value represents use-value in symbolic form, it is ultimately a representation of human labor-power, or finally of human life itself. The objective form of exchange-value is money, and a fully developed capitalist economy allows money to breed and reproduce independently of any human intervention. Such societies bestow absolute power on money, allowing it to rule the entire world, and money is nothing but the objective representation of human life. Once again, it seems clear that a money-based society will give rise to eliminative materialism as the inevitable theoretical expression of its practical activity.

This is not the place to ask whether capitalism is a good or a bad thing. Perhaps a phenomenon of capitalism's scale and scope is in any case not susceptible to a straightforward ethical evaluation. But there seems no doubt that materialism, in its twenty-first-century manifestations, is the ideological form of capitalism. It may be possible to be both a materialist and a political progressive, if identity politics are regarded as progressive causes, as I think they should be. Materialism is not, however, compatible with anti-capitalism. On the contrary, materialism *is* capitalism in philosophical form. I suspect that, while some literary theorists like Berman are well aware of this, the majority of critics who consider themselves materialist are not. I think that if they can be convinced of this connection, they are likely to reconsider their commitment to materialism, which I believe is now largely sentimental and rhetorical in any case. I hope that this essay will provide some impetus toward that process.

Notes

*From the editors: for productive engagement with this essay, please see the special issue of *Early Modern Culture* (Issue 9, 2011) devoted to the subject.

1. For a brilliant critique of evolutionary psychology, see John Dupré, *Human Nature and the Limits of Science* (Oxford: Oxford University Press, 2002), and my appreciative review in the *Times Literary Supplement* (11 January 2002), 5–6.
2. The most comprehensive and authoritative history of materialism remains Friedrich Lange's magisterial *The History of Materialism and Criticism of its Present Importance* ([1925] London: Routledge & Kegan Paul, 1957).
3. See Stephen Stich's summary of eliminative materialism's 'conclusion that beliefs, desires, and other posits of folk psychology do not exist'. *Deconstructing the Mind* (Oxford: Oxford University Press, 1996), 4.
4. John Wilmot, second Earl of Rochester, 'Satyr against Reason and Mankind'.
5. Douglas Bruster, *Shakespeare and the Question of Culture* (New York: Palgrave Macmillan, 2003), 203.
6. Jonathan Gil Harris, *Untimely Matter in the Time of Shakespeare* (Philadelphia: University of Pennsylvania Press, 2009), 25.
7. Ibid., 181.
8. John Bale, *Comedy Concernynge Thre Lawes* ([1538] London and Edinburgh: Tudor Facsimile Series, 1908).
9. Roger Bieston, *The Bait and Snare of Fortune* [1556], no page numbers. Early English Books Online, Arizona State University Library, 14 February 2010.
10. Thomas Floyd, *The Picture of a Parfit Commonwealth* [1600], 276–277. Early English Books Online, Arizona State University Library, 14 February 2010.
11. See in particular P. S. Churchland, *Neurophilosophy: Toward a Unified Science of the Mind/Brain* (Cambridge, MA: MIT Press, 1986).
12. Mary Thomas Crane, *Shakespeare's Brain: Reading with Cognitive Theory* (Princeton: Princeton University Press, 2001), 4.
13. Ibid., 4.
14. Reuven Tsur, *Toward a Theory of Cognitive Poetics* (Amsterdam: Elsevier, 1992), 1.
15. Ibid., 360.
16. G. W. F. Hegel, *The Phenomenology of Mind*, trans. A.V. Miller (Oxford: Oxford University Press, 1977), 208.
17. Cited in Crane, *Shakespeare's Brain*, 10.
18. E. O. Wilson, *Consilience: The Unity of Knowledge* (New York: Alfred A. Knopf, 1998), 4–5.
19. Joseph Carroll, 'Evolution and Literary Theory', *Human Nature* 6 (1995), 119.

20. Joseph Carroll, 'Post-structuralism, Cultural Constructivism and Evolutionary Biology', *Symploke* 4 (Winter/Summer 1996), 214.
21. Paul Hernadi, 'Literature and Evolution', *SubStance* 30:1/2 (2001), 56.
22. Michelle Scalise Sugiyama, 'Narrative Theory and Function: Why Evolution Matters', *Philosophy and Literature* 25:2 (October 2001), 233.
23. William Flesch, *Comeuppance: Costly Signaling, Altruistic Punishment and Other Biological Components of Fiction* (Cambridge, MA: Harvard University Press, 2008), 72.
24. Katherine Hayles, 'Desiring Agency: Limiting Metaphors and Enabling Constraints in Dawkins and Deleuze/Guattari', *SubStance* 30:1/2 (2001), 146.
25. Ellen Spolsky, 'Darwin and Derrida: Cognitive Literary Theory as a Species of Post-structuralism', *Poetics Today* 23:1 (2002), 43–62, quote from 44.
26. Dorothy Nelkin, 'Less Selfish than Sacred? Genes and the Religious Impulse in Evolutionary Psychology', in Hilary and Stephen Rose, eds, *Alas, Poor Darwin: Arguments against Evolutionary Psychology* (New York: Harmony Books, 2000), 22.
27. Charles Darwin, *The Autobiography* (New York: W. W. Norton, 1993), 34–5.
28. Russell Berman, *Fiction Sets You Free: Literature, Liberty, and Western Culture* (Iowa City: University of Iowa Press, 2007). Quotations are cited from my review of Berman's book in the *Times Literary Supplement* (24 October 2008), 24–25.
29. Theodor W. Adorno, *Negative Dialectics* ([1966] New York: Continuum, 1983), 95.
30. See Jean Baudrillard, *Simulacra and Simulation*, trans. Sheila Faria Glaser (Ann Arbor: University of Michigan Press, 1995).

11
Performativity of the Court: Stuart Masque as Postdramatic Theater

Jerzy Limon

In times when theory as such has become suspect, and attempts to define or demarcate literary and artistic works are rejected on ideological grounds as ultraconservative, if not reactionary, it is not easy to find conditions suitable for a serious intellectual exchange of thought. An ideological minefield, created in the last quarter of a century, has resulted in an unprecedented gap between the humanist intellectual tradition, with its basic concepts about aesthetics and art in general, and the theory and practice of what today is labeled art. Inevitably, this leads to extreme polarization of opinions, incompatibility of basic concepts and categories, and to mutual misunderstandings of all the sides involved (and there are many more than just two, the radical and the conservative). Art history and criticism have been the first conspicuous battlefield, with many losses and few gains, it seems; now time has come for the theater to take an active role in the ongoing controversy. What is being proposed now is a totally new way of understanding theater, with concomitant rejection of the tradition, both as a theater practice, criticism, and theory. One trend that has played an important role in the dispute is that of the so-called postdramatic theater, which undermines many of the traditional criteria, basic notions, and categories. It is therefore worthwhile to look at what the new line of thought has to offer to theater theory and performance analysis, and to determine in what ways the new is really novel.

According to Hans-Thies Lehmann, the well-known if not the chief proponent of the new, one of the basic features of postdramatic theater is the predominance of the non-verbal components of the performance. Lehmann even speaks of 'visual dramaturgy' or 'theatre of scenography'.[1] He also mentions a number of other features distinguishing postdramatic theater from all that came before, which include, among others, the merger of scenic 'genres' (dance, performance, or narrative theater). The disbelief in the power of language to describe reality adequately makes one look for some other means of expression: action (if it is retained) is 'depsychologized'; attention is focused on the human body in a manner analogical to body art; the aesthetics of beauty is transgressed as this body is frequently deformed.[2] An important feature of the theatrical trend discussed is its departure from *mimesis*, the blurring and effacement of the boundary between the fictional and the real, the rupture with traditional acting based on 'pretending' and creation, not on 'action' and presence. All this is connected with the new understanding of theater not as a performance but as a social situation, where dialogue is replaced by monologue or soliloquy, and theater is seen not as a finished work of art but as a 'process' taking place between the performers and the addressees,[3] set in spaces chosen for a given 'staging' (*site-specific theater*), although the latter feature is not new in theater history.

These phenomena are obviously present in current theatrical trends, yet – as I intend to show – they are no novelty in the history of the medium. I am not opposed to the new, as I consider this movement artistically interesting and intellectually often refreshing and stimulating, but I think it necessary to temper the enthusiasm of its proponents, who – setting postdramatic theater against all that came before – sometimes too hastily voice judgments that are not entirely justified historically. One cannot discuss here all the similarities and differences between postdramatic theater and other trends from the past. Still, we may note that even the selection of a Christian church space – the more so at Easter time – was a *sine qua non* choice of medieval liturgical theater, which is not only *site specific* but also *time specific*: its staging in a space other than a church and at a different time deprives it of meaning. Such transposed staging can be at most a reconstruction, which is not capable of transferring the meanings generated by a liturgical play. Medieval drama, also in its more developed forms, displays the tendency to blur the boundaries between fiction and reality, with numerous addresses *ad spectatores*, drawing the spectators into the 'action' and engaging them in all sorts of interaction; in such cases one can hardly speak of psychologization or spatio-temporal barriers established between the stage and the auditorium. It is well known that medieval theater is not yet fully 'dramatic': it is dominated by sequences of monologues, and not by any animated, fully drawn exchange of dialogues; drama will fully emerge in the later centuries. Much has also been written on the medieval fascination with corporeality, excretions, or physiognomic deformation.[4] These, again, are often seen as attributes of the new trends in art and theater, but in fact they continue the tradition of many centuries.

Still, the Middle Ages are not an isolated period in this respect. What particularly strikes the attention is the surprising similarity between what is defined as postdramatic theater and the court theater in the sixteenth and seventeenth century.[5] It may sound paradoxical to many, but it is precisely the theater created by aristocrats that demonstrates almost all the features of postdramatic theater, so enthusiastically described by Lehmann, who seems to neglect what actually occurred in theater a few hundred years ago. This should not come as a surprise: with theater studies and criticism being ideologized to a considerable degree, both the historical period and the theatrical trend in question are often left out by the ideologues of the 'new', probably because all the weighty artistic, aesthetic, and technical achievements of court theater are ideologically alien to them, as well as indicating the derivative nature of many seemingly revolutionary accomplishments of postmodern theater artists.[6] The very fact that such a revolutionary and revitalizing theatrical trend could have been created by the class of exploiters and idle rich seems difficult to accept. Moreover, even when, as today, the criteria used in judging art are blurred, the heralds of the new – contradicting themselves – are quite certain about one thing: bourgeois theater is the expression of lack of taste and kitsch. However, one should not forget that kitsch has briskly entered contemporary art (and theater), where it immediately made itself at home, though of course it now has a different function. The aforementioned contradiction results primarily from the persistently voiced yet, in my opinion, totally absurd theory that the boundaries of a work of art are not valid any longer, which is why one cannot distinguish an artwork from other human creations, behaviors, or social situations. This explains the popularity and influence of the writings of such authors as Erving Goffman or Richard Schechner, with their all-inclusive understanding of theater.[7] In reality the editorial, expository, and theater practice clearly indicates what is and what is not art, even though the aesthetic criteria are (seemingly) rejected or their validity suspended. For some reason, even the strongest opponents of 'academic aesthetics', hostile to any attempt to define or demarcate art, know very well what art is, distinguish good art from not so good art, select works for exhibitions, and yet create a situation as if this knowledge and ability to distinguish art from non-art was God-given, an illumination of some sort rather than an immediate result of aesthetic (or any other) evaluation. And evaluation (selection) by definition implies the existence of criteria. Moreover, if we cannot mark the boundaries of the phenomenon discussed, the more so are we unable to relate it to other phenomena hitherto considered artistic.[8] If we do so, however, we undoubtedly make use of some specific criteria making it possible to distinguish the rules of the 'new' from all that is classified as a relic of the past.

To return to Lehmann's book, the vision of the past supposedly rejected by postdramatic theater is open to much objection. It is not only oversimplified but also in many respects simply false. It is not true that the whole past of European theater has been dominated by the word. It is not true that the entire tradition is limited to illusionistic theater: illusion in theater, at least in the sense of the material shape of the staging, is a relatively recent stylistic trend, and just one of many simultaneously present on European stages. Illusion was to be found neither in ancient theater nor in medieval or Elizabethan. Psychologism was alien to burlesque or commedia dell'arte. Synaesthesia, which Lehamnn considers the chief feature of postdramatic theater, is a trait typical of theater as a whole; it has always characterized the manner in which meanings were created onstage. Similarly, the development of temporal 'disfiguration' techniques is precisely the basic feature of theater as an art (this problem is discussed in detail in my book *Piąty wymiar teatru* [*Theater's Fifth Dimension*][9]); the dissolution of the temporal frame of the performance is a scenic trick employed from the Renaissance, particularly favored by the Baroque. Notable scenic transgressions of the aesthetics of beauty, as well as fascination with the body, are to be found in the Middle Ages, the Renaissance, and the Baroque. To sum up, what Lehmann understands as the dramatic theater is in fact just one of many theatrical trends, which did not predominate in all periods. In fact, some cultures did not develop it at all.

The wealth and diversity of theatrical heritage is demonstrated, among others, by Renaissance and Baroque court theater, which in England became known as the masque.[10] That is why in this essay I intend to describe a few of this theater's chief features, which will straightforwardly reveal some significant convergences with what is now called postdramatic theater. These analogies are the following: the rejection of the predominance of the word, the creation of the theater of scenography, the increased role of dance and music, as well as of special effects (fireworks!), the new function of the body, often provocatively naked and sexually ambiguous,[11] the move beyond spaces of institutionalized theater into urban space, onto the river, or into tournament ground, etc., the substitution of monologues for dialogues, the breaking of the barrier separating performers from the audience, their physical integration (common dancing and feasting, even direct communication between the performers and the audience[12]), the departure from fiction in favor of the 'real', the rejection of the story and verbal coherence, the departure from communication in favor of participation, and so on. What we are dealing with here is thus a far-reaching phenomenon of 're-theatricalization', discussed by Lehmann as the chief feature of postdramatic theater (50–52), as well as some characteristics of the 'solo theatre', that is the tendency to monologize the verbal element (125–129). These performances were also ascribed to the spaces where they were held, establishing diverse relationships with the iconography, ornamentation, and ideological program of the interiors, thus generating meanings. They sometimes referred to contemporary events, such as a royal wedding or the creation of a prince. They were thus to a large extent *site specific*, sometimes also *time specific* (contemporaries talked of the *present occasion*).[13] It is worth pointing out that theater as such is also *substance specific*,[14] and the court theater was additionally *audience specific*, i.e., it created meanings in relation to the particular members of the audience, the royals in particular. It seems that the more specificity (of the kind mentioned) there is in a particular production, the more inseparable it becomes from the time, space, signaling matter, and the participating audience. The blend or merger is thorough and complete, creating a very complex semiotic structure. The complex relationships established are intrinsically connected with the occasion and its timing, even with the particular audience, and cannot be shown or reproduced elsewhere. These relationships not only merge all the spaces, time, people, and various phenomena into one inseparable whole, but also generate meanings, which would not surface if any of the participating elements were missing. In this way, in the masque the court does not really present a fictional fable or plot, but through the merger described, it stages itself and its audience, incorporating both in the propagandist program of Stuart dynasty, in which art and theater played a significant role. Thus, indeed, the space of the court at Whitehall signals its real contiguity with the metaphysical realms created on the stage, establishing an indexical link with the outer world. The performativity of the court creates a blend, in which both the spheres become inseparable and are not demarcated by any visible or invisible boundary. In this way, the court presents itself as a gateway to divinity.

Despite the aforementioned considerable degree of dislike or even open hostility toward court theater, many historians of theater and drama do admit that the majority of technological novelties have their source in princely entertainments. Court theater was the first to introduce new machinery which made it possible to change stage design in a veritably miraculous way, with chariots riding on clouds, mountains opening or submarine worlds being discovered. It was there that the innovations in artificial lighting were introduced, creating truly fairy-tale worlds in the eyes of the contemporary audience. This technological revolution may only be compared to the introduction of, say, holography or the new media on to theater stages at the end of the twentieth and the beginning of the twenty-first century. However, behind the continuously improved technology there was an ideology which in a spectacular way announced to the world the beginning of a new era, the golden age. This ideology disclosed the eternal and unchanging laws governing the world. It showed humanity the road to salvation; it elevated monarchs and princes, who were to lead humanity to a luminous future, having first trampled down the heresy, falsehood, and evil rampant in the world. Also today, much of recent ideologically engaged theater shows social and political awareness to the point that it offers ready solutions that might lead humanity to liberation from whatever or whoever the oppressor is, a road being an equivalent of Christian salvation. Let us notice, though, that these works contained no developed action based on propaganda texts; they also lacked explicit indoctrination. As today, meanings were predominantly created by means of images,[15] 'speaking pictures' – to use Ben Jonson's phrase, iconographic references being made to emblems, codes, and texts external in relation to the theater, all of which would make up the ideology of the ruling household or the doctrine of the state.

Court theater's principal feature was its departure from the drama dominating the public stage, or – more generally speaking – its abandonment of the word in favor of the image, stage design, costume, music, special effects, dance, and light. In this respect the court theater is 'postdramatic' from the very beginning. In the literary component of these works – which is very limited indeed – direct references (but just references and echoes) are made to rural comedy of carnival provenance (this is especially the case with the so-called anti-masque), as well as to allegorical, mythological, and pastoral drama, or even to what we would today call *fantasy* or *science fiction*. Penned by the best writers of the epoch, monologues, dialogues, and songs in court spectacles also refer to allegorical, bucolic, satirical, and panegyric poetry, and finally to the much earlier medieval romances, dialogues and disputes, even to riddles. Word and image refer to the tradition of emblematic literature, constituting the international language of the educated elites of the time.

Still, all these elements do not make up a coherent dramatic structure; the predominance of the word declines, as it is subordinated to the stage picture and scenic movement. The word supplements and clarifies, it complements the image as poetic ekphrasis, but in itself it usually does not create any action, plot, or story. It follows that the verbal component of the works discussed was basically non-dramatic, actionless, and far from psychologism, while dialogue often gave place to sequences of monologues, especially in the masque proper (in the so-called anti-masque there is more 'true dialogue', as the form of this 'induction' is more theatrical than the masque proper). What is more, when it comes to court theater, frequently enough we are dealing with forms vividly resembling contemporary narrative theater or what Lehmann calls scenic parataxis (86–87). Changeable stage design – often truly wondrous – presented various transformations to the spectators' eyes, proving that such effects are possible, and, furthermore, showing that the true laws that govern reality are different from the ones our mind or science would suggest; analogically, the force and divine power emanating from the monarch are capable of retransforming the world. It is here that allegory, metaphor, and the topos of *theatrum mundi* gain additional meaning: they show the world as a mechanism capable of retransformation just as stage design is changed by means of the invisible scenic machinery. The changeable spaces of the palace find their prolongation in the changeable stage design; it is here that the space of the palace meets the supernatural world. This process is underlined by the iconography of the stage, costumes, and ornamentation, which frequently refer to the iconographic program of the palace interior, creating an iconographic blend. The inventions of the poet and the 'architect of the stage' (i.e., stage designer) gave insight into the true nature of this world; they provided the audience with a detailed description of the reality that – as was believed – escapes strictly scientific description. This invention broke with fiction and provided the scenic world with a trace of reality: in the space of a palace theater the real world of the palace and royal power linked up with the metaphysical world. It may be said that similarly to today's postdramatic theater, the court spectacle wanted to be perceived not as fiction, but as reality.[16]

One may say that, generally speaking, these spectacles generate meanings on the basis of various codes and rules, belonging to diverse mediums, systems, and traditions, which include: theater, literature, mythology, history, music and dance, hermetic studies and alchemy, and finally the fine arts, especially graphic arts, architecture, and emblematic.[17] Characteristically enough, as today, court theater does not provide for a variety of possible readings, just a single one is correct. Hence, this is a world closer to allegory than symbol, which is semantically unclear in its very essence. Even if allegorical senses are blurred on the textual plane, court spectacles provide us with a metatextual explication. Many verbal utterances are of such character: once again, this is a feature shared with postdramatic theater. The texts employed explicate what appears on the stage. There is a remarkable similarity with contemporary art, which also does not create meanings in an entirely independent way and frequently enough requires verbal explication. The same can often be heard of postdramatic theater, which frequently alludes to texts that refer to what is today called art.

The rules governing the world in court spectacles are already known and unchanging; in order to understand them one just has to acquaint oneself with the 'languages' of their artistic notation. This language is a specific one, being composed of miscellaneous signs, as mysterious as hieroglyphics: signs of diverse substance, which belong to numerous systems and mediums, yet nevertheless combine on the stage to make up complex theatrical signs. The stage language discussed requires interdisciplinary knowledge because of the nature of its substance and of its encoding rules, which originate in the borders of the broadly understood literature, architecture, music, and the fine arts. That is why theater becomes a truly ideal site for the creation of a model of the world which is in itself highly codified and theatrical on an everyday basis (i.e., the court). The mechanism of transformations undergone by people and worlds is presented as analogical to theater. Things which might have been overlooked by some in the space of the court or palace are here scenically visualized and foregrounded. Laws of physics are suspended, the flow of time is stopped, palace interiors come to life, and the fluorescence of the monarch becomes a fact (some figures are blinded by the light emanating from the king). In other words, scenic synaesthesia became the principal feature of the program of palace space, where visitors could notice the fictitiousness of ideology, allegorization, or mythologization of the interior and human comportment, where for many there was an obvious dissonance between meaning and substance. Court performances were thus aimed at bridging this gap; they served to demonstrate, strengthen, and visualize the invisible laws governing this world. This is the reason why they did not want to be perceived as fiction. In order to achieve that, the court staged itself, blending the two ontologies and two times of the two spheres, the superhuman (i.e., the court) and the divine (i.e., the stage).

Deciphering signs of this type of theater resembles reading a three-dimensional book of emblems, which – as is well known – were considered a sort of illustrated encyclopedia of knowledge about the world. The book of the universe in court theater is thus shown as a changing sequence of three-dimensional emblematic scenes, complemented with word, song, dance, and the surrounding space. Owing to the thus created context of reading, the sometimes unclear allegory finds its explication. Still, this sequence is usually not related to the development of any action, and temporal relationships between separate scenes resemble temporal relationships in multi-scene paintings, creating visual polyptics, where we usually get simultaneous presentation instead of historical chronology. That is why the performances discussed rarely take on the shape of a scenic parable, and their meanings are created through analogical transference rather than through pointing the story (Solomon is related to his temple in the same way as, for instance, the English king James I is to the palace of Whitehall). Figures' actions and utterances usually take on the shape of allegorical paintings created pictorially or kinetically on the stage. It seems that such selected ordering of three-dimensional stage pictures constitutes the crucial, dominant combination and stream on the performance's syntagmatic axis. Establishing various interrelationships, signs belonging to other systems and mediums are subjected to this axis. The monarch's encounter with mythological deities or allegories is not related to any specific historical moment or action; it draws the king into the seemingly fictional realm and can make him become the object of allegoresis; pictures, sculptures, or theatrical performances which depict this meeting are in no way interested in capturing a historical moment, as they present no story, and the more so no fictional action. What they show instead – most frequently through allegorical and emblematic language – is the essence of power and the laws governing the human world. These laws are unchanging and they are not rooted in time, neither do they result from any historical events which people are involved in. Whenever historical events are invoked – even if presented simultaneously – in the individual reception we can put them in the right order if need be owing to our knowledge of the codes employed. Yet, generally in court theater it is completely unimportant whether, for example, Merlin announces his prophecies prior to the king's meeting with Minerva, or does so before – or after – the king's feast with the gods, as well as whether this happens at exactly the same time as Hercules rips off the Hydra's head or Prometheus steals fire from the gods. Other issues similarly deprived of significance include the psychology of the figures, the probability of the presented events, or their relation to 'real history'. Indeed, it is the monarch himself that is history, and history is created by him; in fact, it follows that truth about the world may only be grasped through the presentation of the monarch and the origins of his power over the world, which is included in the iconosphere of the monarchy, and not through the combination of various past events that do not directly result from his reign.

Instead of concentrating on effects, court art focuses on the cause. Hence, as was believed, its intellectual superiority and depth. The world created in court spectacles is to a large extent closed, yet it defines itself in its successive variants. It is a cyclical world, multiplied in concentric images like the scrolls of a Baroque volute, as Walter Benjamin would have it. And if it tells a story, it does so not only in the space of stage decorations, but also in the space of the entire palace, which – as opposed to stage design – remains unchanging and timeless. That is why it breaks with fictional time of the anti-masque and presents figures and events which neither pretend nor imply that what they signify is different from what can be seen or heard. They are not just the substance of signs, but the meaning. Instead of representation, we witness the presence. What we are watching is not a reconstruction of past events but something taking place in the here and now of the spectators. Thus, as it creates artistic events that resemble mobile installations, court theater turns away from action, fictionality, scenic semiosis, and *mimesis* or 'representation'. Instead of separating the fictional world of the stage from the real world of the auditorium, it leads to a merger of both, by which the court becomes part of the show and, as I said, stages itself. Once again the resemblance to postdramatic theater is noticeable. Indeed, it is the return of the real.

However, one should also stress that the departure from tradition as well as the transformation of theater into another type of scenic event is more interestingly and more accurately rendered by court theater, where a hermetic work of art is transformed into a social situation. Hence, an important structural element of each entertainment of the type discussed was the so-called 'anti-masque', a kind of prelude to the spectacle proper. However, this was not an introduction or exposition in the sense of an action, for the masque basically neither continues nor develops the plots of the anti-masque; it does not even refer to these in any way. What is more, the anti-masque has the features of a typical theatrical one-act play.[18] It is performed by professional actors, who take the parts of fictional figures (the masque proper is acted out by aristocrats who usually play themselves).[19] The anti-masque develops around a 'story' which is most frequently connected with the figures' desire to reach the king, to whom they have something to communicate, be this in the form of a court play, a letter, or a petition. As the action is taking place on the stage inside the royal palace and the figures constantly signal to us their existence in a different time stream and in a different place, it is obvious to the spectators that we are dealing with pure theater. Only at the end of the anti-masque do they manage to get to the royal court; it is at this point that the fictional space overlaps with the real one, and the fictional time 'catches up' with the spectators' real time in the manner so typical of theater. However, at this precise moment the theatrical play ends, and the figures retire at the sight of the monarch, whom they usually recognize by the emanating light. The superhuman space is inaccessible to common mortals.

At this point the performance proper, that is the court masque, begins. The curtain is drawn back, exposing spaces contiguous with the royal palace and figures who usually do not pretend to be somewhere else and at a different time, but who are in the 'here and now' of the spectators. Their utterances unmistakably indicate that they do notice the king and other spectators, so there is no spatio-temporal gap so typical of theater.[20] Visitors from the metaphysical sphere, usually as messengers from the gods, or the gods themselves, appear in person and communicate messages directly to the king, and the masked aristocrats who are discovered in some cave or in a cloud, descend onto the stage, and first 'make obeisance' to his Majesty before they start to dance.[21] Thus he appears not only as the main spectator of the masque, to whose location everything is subjected, but also as the source of magical power, by which the world represented on the stage is governed by laws different from those that govern the human realm. The figures do not create fictional time or spaces; they do not pretend, nor do they signal that we are dealing with a game, an agreement, or a convention. They want to be perceived as something real, denoting what can be seen and heard. Thus, as in postdramatic theater, the actors who play and 'pretend' are replaced by performers, who neither pretend nor play, because they are 'themselves', sharing the time and space with the spectators.[22] It so remains until the end, when one of the successive scene changes reveals the masquers seated on a cloud or in a cave, who speechlessly ride down onto the stage in a chariot and descend among the spectators in order to ask the ladies out for a dance, which becomes the sign of common merrymaking that will last a few hours more. In this way the world of the stage merges with the auditorium and the two can hardly be distinguished (according to Lehmann, the shared space of spectators and performers is one of the principal features of postdramatic theater, 122–125). In court entertainments of the type discussed one can thus notice a clearly marked transformation of the represented worlds, as well as of the ways of describing them: what is shown is the transformation of dramatic theater, which is based on theater-specific rules of creating temporal and spatial structures, into a social situation, or into a non-dramatic – or postdramatic – spectacle. As it breaks with drama and limits the dominance of the word in favor of non-verbal means of expression, such as picture, music, and dance, the spectacle becomes a performance set in the time and space of the addressee, which creates meanings in a way alien to theater. It draws spectators into its boundaries and creates a court-specific type of ritual, where the spatio-temporal unification of stage and audience is accompanied by the complete integration of performers and spectators. This has indeed much in common with ritual or a game that courtiers loved to play, and not with theater in the traditional meaning of the word.

Moreover, some of the masques are preceded by other forms of spectacles and spectacular shows, such as fireworks, shows of arms, sea battles, or progresses through the streets of London, that are inseparably linked to the masque proper. The same fictional figures that appeared in these outdoor shows, made their way to the royal court and the indoor stage. Through this, the space of the masque was expanded and incorporated, metonymically, the City of London.[23] It seems the city was often a sign of the whole country, or of its inhabitants, as indicated literally in one of the masques by Inigo Jones and William Davenant, *Britannia Triumphans* (1637): '... and afar off a prospect of the city of London and the river Thames [was seen]; which, being a principal part, might be taken for all of Great Britain'.[24] Even if the stage-set did not provide a visual representation of the city, the real progresses through the city, involving hundreds of participants, preceded several masques, for example, *The Memorable Masque* (1613), *The Masque of the Inner Temple* (1613), and *The Triumph of Peace* (1633). In these instances, the city was incorporated into the spectacle and thus transformed into a stage and a scenographic sign that created meaning in relation to other components of the macro text. Again, the use of spaces traditionally not treated as theater is a typical feature of the postdramatic (avantgarde, postmodern, and the like). The space of the stage proper finds its continuation outside the theater building, and once again a given production becomes site-specific, for it cannot be shown in any other combination of spaces without a significant loss of meaning. Again, the theatricalization of the city was not aimed at creating fiction; on the contrary, the show was presented as factual. So, when the costumed actors from the *Memorable Masque* are paraded through the streets of London, they are given an official welcome, as if they really were ambassadors from abroad.

The whole show begins with a procession through the city, which departed from the house of Sir Edward Philips, Master of the Rolls, and consisted of the mixture of the fictional and the real. It included, among others, fifty gentlemen with footmen, a mock-masque of baboons, two chariots with musicians disguised as Phoebades, Virginian priests worshiping the sun, and chief masquers dressed as Virginian princes: in Indian habits, on horseback, each sided by two Moors, with torchbearers mounted before them. As radiance spreads from the noble performers onto the spectators, in the course of the progress, the city is allegorized. The two spaces and the two times merge, creating a blend of an illusion of reality. The masquers were followed by a chariot with Capriccio/Eunomia and Phemis/Honour and Plutus. In his description of the procession, George Chapman noted that 'These, thus particularly and with propriety adorned, were strongly attended with a full guard of two hundred halberdiers; two marshals (being choice gentlemen of either House) commander-like attired, to and fro coursing to keep all in their orders.'[25] Note that the fictional figures do not pretend to be somewhere else and at a different time than the spectators gathered along their path. This means that there is no spatial or temporal division separating the two realms.

Let us also draw attention to the fact that in court entertainments we are frequently dealing with the human body which ceases to signify some fictional body, as it wants to be perceived as a body in its own right.[26] Strikingly similar was the function of the body in court performances to its use on today's stage. Often deformed or almost naked (which must have shocked contemporaries), parading its ugliness, nakedness, and deformation, transgressing the categories of beauty or aesthetics, underlining the fluidity of gender, under the influence of the monarch's presence it is sometimes retransformed, gaining what it had previously contradicted, that is harmony. The entire court entertainment may be regarded as a transformative process where chaos changes into harmony. Let us notice that while chaos was the domain of theater, harmony belonged to anti-theater or court ritual, which, instead of pretense, costume, or fiction, presented the truth about the laws governing the universe. Theater remains the domain of people and their frequently repulsive physicality; court ritual takes place in the superhuman space close to the divine harmony of body, poetry, music, dance, and geometry (of a painting). One should again notice a strik-ing similarity with postdramatic theater, in its aesthetics of ugliness, deformation, vagueness of anatomy, and sex (men playing half-naked women – with false bare breasts – were also to be found in court enter-tainments), and with the aesthetics of indefiniteness (Roman Ingarden's 'spots of indeterminacy') described by Lehmann (100–101). Thus, as in modern art, nudity – including the nakedness of female aristocrats taking part in these spectacles[27] – becomes a challenge, a transgression of human limits and culture. What is signified by nakedness is not so much erotic challenge or licentiousness as divinity.

One more thing: I must stand up for theater as art also when Lehmann presents postdramatic theater as frequently deprived of 'meaning'. The careful reader of his book will notice with surprise that the distinctive feature of the new theater, which the author considers crucial, is supposed to be its lack of coherence. One cannot accept this view. Even if they are unclear or 'fluid', meanings will nevertheless result from the rules that make it possible to create a given staging. Even Lehmann himself notices that the director, like a poet, creates associa-tive fields between words, sounds, bodies, movement, light, space, and objects (111). The emergent structure is constructed on the basis of the system of rules incorporated by the work (or, if we must, process). These principles can be recreated by analyzing the work in question, even when they transgress the rules of another trend in theater. Even if the work rejects these rules in practice, as well as in the aesthetic or ideological declarations, its relationship with what it departs from gen-erates meaning. One cannot reject the past without foregrounding the fact of rejection; the same applies for transgressing the tradition or any rules. The very fact of rejecting and choosing what and how to reject is meaningful and creates a dialogic situation and all sorts of relation-ships between the rejected and the rejecting form. That is why one can-not accept Lehmann's thesis that postdramatic theater 'does not mean anything', or may not mean anything. It creates meaning exactly by rejecting or undermining the traditional ways in which meaning was created in theater. One should express this differently: theater can be non-semantic and 'adducent', it can even signify exclusively itself, but this does not mean it is devoid of meaning. If an artistic system is non-semantic (it does not refer to the external world), it creates meanings in relation to rules that make its emergence possible and justify it; then it is an instance of what Roman Jakobson calls introversive semiosis, the best examples of which are music and dance. We are hence dealing with the absolute predominance of the aesthetic function, and consequently with the phenomenon one may call 'true art' (or 'art for art's sake'). Just like music or dance, such theater may mean nothing in the referential sense, but it will nevertheless expose its systemic features, it will be 'pure aesthetics', it will itself amaze.[28] Yet theater of this kind cannot engage in social or political affairs, as in doing so it ceases to denote exclusively itself and the referential function starts to prevail. What is more, the predominance of the aesthetic function distinguishes artistic texts from non-artistic ones, which – contrary to some currently fashionable claims – makes it possible to delineate the boundary separating art from non-art, and the more so art from life.

It seems that theater of the new type has found a way to solve this dilemma: it has created a set of rules related iconographically with the ideology of modern art. What has been ideologically described by the art of recent years is metonymically invoked by the elements of costume, make-up, prop, motion (dance), or music.[29] One can hence escape the need to define social or political affairs verbally, with its inevitable smack of propaganda or didacticism, using the system of iconographic attributes, codes, and rules understandable for some recipients to create certain polysystemic compositional structures which invoke artistic and ideological texts external to theater. Owing to this, postdramatic theater (if this term is indeed to be retained) creates two interrelated semiotic schemes: a scenic and a contextual – or intertextual – one. The emergent structure clearly refers to the discussed tradition of court theater. That is why the claims about the supposed lack of meaning in theater of this type are grossly unjust and lead to an oversimplified vision of this theater's artistic and intellectual values. Equally invalid is the claim that postdramatic theater is a new phenomenon, a reaction to its immediate predecessors. As I have tried to show, it evokes and echoes a number of features from theaters of the past. Ironically, its foremost predecessor is the court masque.

Notes

An earlier version of this essay appeared in Polish as 'Barok postdramatyczny – reaktywacja', *Teatr* 7–8 (2007), 94–100; this version has been translated into English by Agnieszka Żukowska and revised and expanded by the author for the present publication.

1. Hans-Thies Lehmann, *Postdramatic Theatre*, trans. Karen Jürs-Munby (London and New York: Routledge, 2006), 93–94. Subsequent references are cited parenthetically in the text.
2. For me (partly following Roman Jakobson), 'aesthetic' simply means a form of human communication, in which part of the message is oriented toward

itself. In other words, the text partly talks about itself, about the rules that enabled its composition and appearance in the form or modeling and substance given (theater is not only site specific, but it is also substance specific). An artistic text (object, whatever) justifies its appearance in the space, time, and substance given; non-artistic texts do not reveal that feature. Quite contrary to literature, where in most cases the actual shape and color of fonts or letters bears no relation to the semantics of the text, and these fonts and letters usually become transparent during the process of reading, in theater the substance of signs is of great importance: not only does it draw attention to itself, to its otherness, by which it ceases being transparent, but it creates all sorts of relationships with other substances of a given production and with the elements of the implied denoted world (and, of course, directly or indirectly with the elements of our empirical reality). This in itself is a proof that the artistic aim of a production is not the description of the fictional world only, but also to establish meaningful relationships between its particular components or signifiers.

3. Theater can of course be regarded as a social situation, but in the context of the traditional understanding of art it will not be a work of art or artistic text. Still, the term 'art' is nowadays repeatedly used to describe diverse phenomena which up till now have not been regarded as art. As claimed by the defenders of the new, the boundaries between life and art have been obliterated, with the effect that they themselves find it hard to determine what exactly they are talking or writing about. This is a much wider phenomenon, which takes on an absurd dimension, because authors writing about art declare in advance that what they are writing about cannot be defined (Lehmann, for instance, does this on page 180, as does Kerstin Mey in a recently published work, *Art & Obscenity* [London and New York: Palgrave Macmillan, 2007]). In such cases there is no ground for understanding those who think that artistic texts are definable. This, of course, generates all sorts of controversial issues, bringing the discussion to an ideological minefield. For instance, we may ask why do we need to define theater at all? Art history and theory has to a large extent rejected that futile and reactionary approach (thus becoming new art history). As I see it, we are interested in identifying the medium, because depending on its nature, we adjust our reading, hence interpretation, to the rules generating meaning. In theater, we do not watch humans interacting, but actors at work, who can only be signs of humans interacting. If we are not able to distinguish between humans interacting and actors impersonating humans interacting, we shall put ourselves in the position of naïve spectators or, rather, witnesses of an event which we treat at face value, and, hence, the event witnessed is not capable of generating meanings that go beyond the literal or free connotations. Of course, much of today's theater imitates life and lifelike utterances and behavior, but we must not be misled: this is just another convention, which may be analytically distinguished and described. Naturally, the case is different in performance art, which is not theater (but may contain a large dose of theatricality). Thus, defining the theater as a medium is a necessary procedure not to define its demarcation, but to define the rules that enable a justified selection, combination, and modeling of substances and other stage phenomena.

4. In radical writings, the Middle Ages are looked upon with some degree of tolerance: even though we are dealing with religious art and theater, it is still pre-capitalist, hence easier to digest in the ideological sense. Of the numerous works devoted to the study of the body in postmodern art, see, for instance, Francesca Alfano Miglietti, *Extreme Bodies: The Use and Abuse of the Body in Art* (Milan: Skira, 2003).

5. See the last chapter of my book *Między niebem a sceną* [Between Heaven and Stage] (Gdańsk: Słowo/Obraz terytoria, 2002), where I describe the phenomenon of English court theater in the early Stuart period, as well as the earlier *The Masque of Stuart Culture* (Newark and London: University of Delaware Press, 1991).

6. Perhaps they not so much attest to the derivative character of these accomplishments as prove that what we are dealing with is reference to tradition rather than any revolutionary parting with it, and that theater – if it remains an art – despite all the declarations is still making use of the wealth of its heritage.

7. For a convincing refutation of these concepts, see Eli Rozik's *Generating Meaning in Theatre: A Theory and Methodology of Performance Analysis* (Brighton: Sussex Academic Press, 2008), 86–88.

8. We must not forget that in theater we are dealing with temporal and spatial boundaries that are not lifelike. Every performance has a beginning and an end, clearly demarcates its space, and within these limits everything enters into closer or more distant relationships, which means that meaning is generated in ways totally different than in real life. Thus, the seeming similarity of art and life on the stage is just a convention, often employed in the past. Once we recognize theater as a specific communiqué, we understand that we have to select the ingredients that take part in the stage reactions (semiosis, if you will) from those that do not. And we, the spectators, are expected to switch from one way of perceiving reality to another one, just as if we were treating the stage signals as a sort of an unusual text, a communiqué. Consequently, we activate cognitive receptors different from the ones which we normally use to take in perceiving the empirical world. Thus, marking the boundaries between theater and non-theater goes far beyond ideology: it concerns cognition, the different ways we perceive reality (and a theatrical performance is part of that reality, modeled in a very unusual way).

9. This was published in Gdańsk in 2006 by Słowo/Obraz terytoria. I have discussed the issue in some of my articles published in English; see, for example, 'The Fifth Wall: Words of Silence in Shakespeare's Soliloquies and Asides', *Shakespeare Jahrbuch* 144 (2008), 47–65.

10. For bibliography see my *The Masque of Stuart Culture* and David Lindley, ed., *Court Masques: Jacobean and Caroline Entertainments, 1605–1640* (Oxford and New York: Oxford University Press, 1995).

11. Speaking parts in court theater were taken by professional male actors. Thus, when, for instance, a half-naked Diana appeared on the stage, she was played by an appropriately disguised man with the attached sexual attributes. Gender fluidity is obviously one of the obsessions of contemporary art, theater included. Let us not forget that noblewomen acting on court stage, who took mute parts, frequently had bare breasts, as well as – what

equally shocked their contemporaries – naked arms. We are thus dealing with an obvious transgression of morals imposed by the patriarchal society.

12. Notice that similar phenomena sometimes occur in public theaters: Beaumont and Fletcher's *The Knight of the Burning Pestle* (1607/1608) provides us with what Lehmann considers the distinctive feature of postdramatic theater (120–123), that is, the negotiations between spectators and actors resulting in the disruption of the performance, in place of which a new play is staged by the actors with the participation of some chosen spectators. Obviously enough, in this case the spectators are also actors playing the audience. Similarly, when it comes to contemporary theater, one may wonder to what extent the authors of the performance actually supervise issues such as integration with spectators, improvisation, or even negotiations. Lack of supervision may lead to unforeseeable and undesired consequences.

13. To read more on this subject, see *The Masque of Stuart Culture*.

14. The theater as an artistic medium is always oriented toward itself. Theater wants to draw our attention to its internal beauty, to the rules and materials with which it is constructed, and which justify its appearance in this particular shape and not in any other. The signifier denotes something in the world of fiction, but it also denotes itself; it is self-referential. This is where the aesthetic function reveals its existence. This also means that the medium becomes (part of) the message. It is substance specific. As is always the case with artistic texts, theater, as an exhibitionist and a narcissist combined, uncovers its anatomy, and allows us to contemplate its internal substance and composition, always unique, sometimes strikingly strange and surprising, but capable of conveying an abundance of information.

15. See my article 'Kto trzyma kalejdoskop, czyli o zalążkach terroru' [Who's Holding the Kaleidoscope, that is, on the Origins of Terror], *Dialog* 10 (2006), 100–111.

16. See my essay 'The Monarch as the Solo Performer in Stuart Masque', in Ute Berns, ed., *Solo Performances: Staging the Early Modern Self* (Amsterdam: Rodopi, 2009), 229–248.

17. The close relationship between emblematics and court culture, theater in particular, is discussed by – among others – Jane Farnsworth in '"An equall, and a mutuall flame": George Wither's *A Collection of Emblemes* 1635 and Caroline Court Culture', in Michael Bath and Daniel Russell, eds, *Deviceful Settings: The English Renaissance Emblem and its Contexts* (New York: AMS Press, 1999), 83–96.

18. In the anti-masque, professional actors and dancers enact fictional figures, inhabiting fictional worlds, being representations of the real ones, and are engaged in dialogic exchanges and plot sequences, and they are separated from the audience by time and space. The actors, through their gestures and utterances, signal to us that the figures they impersonate are set at a different time and a different space, and they also signal the figures' model of perceiving reality, which is conspicuously different from that of the spectators. Thus, what we have in the anti-masque is typical for theater: at least two streams of time, two present times, two spaces, and two different models of perception are contrasted, that of the fictional figures and of the spectators. What we see and hear on the stage is not what the stage figures see and hear,

and the two present times, two spaces, and the two models of perceiving reality, that of the fictional figures and of the spectators, are constantly at play. It is only in some instances that the figures enter the space of the court and their fictional time catches up with the present time of the audience, in which case they can pay homage to the king and comment on the glory of the court. In many other cases when that happens, the figures are blinded by the light emanating from the monarch and simply run away.

19. I think we should abandon the tradition of using the notion of a 'character' to denote a fictional being created on the stage. 'Character' implies an independent psychological construct, whereas the fictional figure does not have any individuality originating in its own psyche: it is a joint construct, resulting from the signals provided by the dramatist, the actor, the director, and the costume designer (choreographer, etc.) that affect the mind of the recipient. It may also be noted that it is rather difficult for an amateur actor to create a 'convincing' fictional figure. What usually happens is that we see an actor 'in action', but what we do not see is the result of that action that would be capable of creating elements of fictive realms. What we see is a human being attempting to create an image of someone else, without the result being obvious or convincing. Again, postdramatic theater often attempts to break the boundary dividing the actor and the figure; consequently, the temporal and spatial split or hiatus is annulled. That may, of course, be a temporary feature of any production, but if it becomes dominant, it undermines the basic qualities of theater as art, which inevitably thus turns into another system, such as performance or happening. Lehmann gives a number of examples of spectacles in which the recipient finds it impossible to distinguish between the actor and the figure. Moreover, Lehmann insists that the performers do not create fiction: they remain themselves throughout the show. This of course undermines our basic criteria for distinguishing theater from non-theater. The actor may, of course, remain him- or herself, as a human being enacting itself, but he/she has to create a temporal hiatus between his/her appearance on the stage, and the audience. In other words, the actor may remain him- or herself, but it has to be him- or herself from time passed: an hour, a week, or two years. Otherwise, we are dealing with the art of performance, which is not theater, or with some other type of live show. If we did not make these distinctions, we would not be able to differentiate a 'postdramatic situation' as defined by Lehmann from, say, a situation in which a grandmother is reading a fairytale to her grandchild ('narrative theater'), or a political rally, or any other communicative situation that occurs in real life.

20. The surviving texts of Jacobean masques leave no doubt: the king is the source of everything that happens in the masque proper, which is more like a ritual or a game than theater (as opposed to the anti-masque, which may be seen as a one-act play governed by the rules of theater rather than anything else), and the royal presence in the Banqueting Hall surpasses the role of a mere spectator. For the miraculous events shown on the stage, and the unveiled direct communications channeled with the metaphysical sphere, are all the result of the king's presence and his superhuman qualities. He is even the source of light, which metaphorically means also life. Thus, the fluorescent monarch becomes the life-generating agent.

21. As described, for instance, by Horatio Busino, chaplain to the Venetian Ambassador: 'After they had made an obeissance to his Majesty, they began to dance in very good time, preserving for a while the same pyramidacal figure, but with a variety of steps ... When this was over, each took his lady ... all making obeissance to his Majesty first and then to each other' (*Calendar of State Papers, Venetian, 1617–1619*, ed. Robert Ashton, 242).

22. Characteristically, in his book Lehmann scarcely ever talks of acting and actors in the theatrical trend he discusses. He almost always uses the word 'performer'. See the discussion on 134–144.

23. Not much has been written on the subject; for a perfunctory treatment see Irène Mamcarz, 'The Representation of Cities in the Baroque Opera and the Development of "Italian Style" Scenography', *Medieval English Theatre* 16 (1995), 142–164.

24. Inigo Jones and William Davenant, *Britannia Triumphans* (1637).

25. Quoted from Lindley, *Court Masques*, 77, ll. 111–114.

26. On the various, often provocative, uses of the human body in art, see, among others, Francesca Alfano Miglietti (FAM), *Extreme Bodies: The Use and Abuse of the Body in Art* (Milan: Skira Editore, 2003); R. Betterton, ed., *Looking On: Images of Femininity in the Visual Arts and Media* (London: Routledge, 1987); Rebecca Schneider, *The Explicit Body in Performance* (London and New York: Routledge, 1997); and Helen McDonald, *Erotic Ambiguities: The Female Nude in Art* (London and New York: Routledge, 2001). Interestingly, most scholarship is devoted to the study of the female body, with a conspicuous evasion of male bodies.

27. As noted above, what shocked contemporaries were not only the naked breasts of women, but also their naked arms.

28. Obviously enough, the aesthetic function has nothing in common with the category of beauty, as is frequently tendentiously claimed by the heralds of the 'new': the aesthetic function distinguishing artistic from non-artistic texts denotes the work's focus on itself, that is on the rules making its creation possible in this shape and not another. By rejecting this function as a distinctive feature of art, we run into irreconcilable contradictions.

29. Lehmann himself admits several times that only now has theater begun to tackle issues which have long been exploited by the fine arts, in the course of which it takes on their aesthetics (106–107, 110–111, 164–165).

12
Shakespeare, Idealism, and Universals: The Significance of Recent Work on the Mind

Gabriel Egan

For most the twentieth century, the serious study of Shakespeare's works was founded on a general acceptance of the principles of Platonic idealism and essentialism, and a belief in the existence of universals. In textual criticism, the New Bibliography that emerged from the work of A. J. Pollard, W. W. Greg, and R. B. McKerrow in the first decades of the century assumed a relatively unproblematic application of Platonic idealism for the relationship between the play as conceived in the mind of the dramatist and the play as performed or written down. Since the 1980s idealism, essentialism, and universals have become dirty words as the New Historicism and Cultural Materialism popularized an unthinking association between these philosophical principles and political conservatism. In these new and related schools, the alleged antidote to all three evils was said to be materialism, which meant paying more attention to the physical (often the economic) realities of a system under consideration than to the ideas in it. In respect of the textual condition of Shakespeare's plays, this meant attending to the material particularities of the early quarto and Folio texts rather than seeking to extrapolate back from those to a lost authorial manuscript that preceded them or, worse still, to whatever it is Shakespeare had 'in mind' when he composed them.

Under this new intellectual dispensation, materialism was displayed as a badge of pride, a way of appearing to be tough and pragmatic by talking about the degree to which literature was (albeit covertly) really about money, matter, and production rather than love, ideas, and self-reflection. This was a curious development, since the particular kind of materialism the New Historicists and Cultural Materialists professed to practice was the nineteenth-century philosophical materialism best known through the works of Karl Marx. Ironically, far from rejecting ideas and valorizing harsh realities, Marx's materialism was an abiding concern for how ideas arise *from* material circumstances. To assert, as Marx did, that social being does not arise from consciousness but rather consciousness arises from social being is not to leave ideas out of the matter but to insist that they are the mysterious phenomena that are in need of an explanation rather than simply popping into existence to account for the way the world is. Finding a plausible account of the relationship between the hard facts of existence and the nebulous realm of ideas was a problem that dogged the twentieth-century refiners of Marxist determinism, notably Georg Lukács, Antonio Gramsci, Theodor Adorno, Herbert Marcuse, and Louis Althusser (whose approaches are admirably summarized by Eagleton[1]), which tradition continues most exhilaratingly, although as yet without a solution, in the work of Slavoj Žižek.[2]

In analyzing cultures from the past (especially the early modern period), New Historicism in particular drew upon a fresh approach from outside the Marxist tradition, articulated by the anthropologist Clifford Geertz who influentially characterized culture as essentially a literary phenomenon. According to Geertz, symbolic forms are the key to social behaviors – we must always ask 'what does this activity mean to these people?' – and the tools for reading culture are essentially literary-critical, not scientific. In an early and much anthologized essay, Geertz argued that Enlightenment science's approach to investigating the nature of humanity was to assume that like other phenomena (planetary motions, chemical reactions) this one would have universal principles underlying its apparent heterogeneity, and that finding these universals, this human nature, was the project for anthropology. Rejecting this, Geertz insisted that culture itself made people what they are, for it was 'a set of control mechanisms – plans, recipes, rules, instructions ... for the governing of behavior', needed because humans are uniquely lost without such an 'extragenetic, outside-the-skin control mechanism'.[3] In his most famous articulation of what follows from this, Geertz wrote that 'there is no such thing as a human nature independent of culture'.[4]

Geertz's primary evidence for this claim was that modern humans differ from their predecessor primates markedly in brain capacity and little else and that until a great deal of learning has been done a newborn human is not ready to survive unaided. Moreover, it was not that humans emerged and then became clever enough to have culture, but that culture (tool-making, collaborative hunting, and so on) slowly emerged among the primates and gradually turned *Australopithecus* into us. This account is itself not greatly controversial, but Geertz had a habit of drawing pithy conclusions that could be misread as overstatements of his case. A typical example is his claim that by 'submitting' to culture 'man determined, if unwittingly, the culminating stages of his own biological destiny. Quite literally, though quite inadvertently, he created himself.'[5] Clearly, Geertz had difficulty assigning agency here. Humankind is said to 'submit' to culture as an outside force and yet to be made up by it. Geertz conceived culture and biology as mutually interactive and presumably mutually determining, since no amount of creative invention could produce tools that proto-human biology could not wield.

When, at the birth of New Historicism, Stephen Greenblatt repeated Geertz's assertion that there is no human nature independent of culture,[6] academics whose subject was literary culture found themselves thrillingly located in a central position after years of marginality within social science. If human nature were not a given but a product of culture then existing critical methodologies concerned with how literature illuminates or reflects the human condition had missed the point. Might not literature itself be constructing our sense of what it is to be human while seeming only to represent it? If so, it was performing the work of ideology as theorized in Marxism. An enlarged role for literary studies as a branch of social science and politics coincided with an ascent to orthodoxy of cultural relativism that offered the possibility of partially undoing the harm of European ethnocentrism. The cost of these developments for the study of literature has been high. Literary students most commonly encounter linguistics via the outmoded structuralism of Ferdinand de Saussure, which gives priority to language itself as an outside-the-skin system of signification, and empirical developments in the second half of the century (such as Noam Chomsky's widely accepted view that humans have an innate skill for a universal grammar of which particular languages are only instances) are generally underappreciated in English departments. The dogma of anti-essentialism has driven a wedge between English studies and most scientific approaches to the human condition and has blinded literary scholars to what is being discovered about the qualities that humans really do share simply by virtue of being human. Moreover, there are startling insights about just how much of what we call culture (including such things as ethics, politics, and play) is common also among the more intellectually advanced animals.

While Geertz may be sympathetically read as indicating that culture has embedded itself deeply in human existence, he is more commonly used to assert simply that human nature itself does not exist and that there are no universals. As Bertrand Russell pointed out, we all believe in universals of some sort else we could not accept the truth of a geometrical proof. The opening premise 'Let ABC be a rectilinear triangle' rests on our acceptance that although any triangle we might draw to illustrate the proof will necessarily be imperfect (we cannot draw absolutely straight and infinitely thin lines), the principles of how imaginarily perfect objects behave are easily understood and moreover are objectively true.[7] We do not allow our experience of measuring a hand-drawn triangle's interior angles and summing them to 179.5 degrees to deflect us from the principle that with perfectly straight lines the angles would (indeed, must) sum to 180 degrees. Rather than being a common feature of a collection of disparate objects (as with the human universal of smiling to present a non-threatening demeanor), this kind of universal is more like the Ideal or Form described by Plato in *The Republic* and *The Timaeus* in having no material manifestation and yet embodying a truth. Despite this benign meaning, many materialists habitually reject Platonic thinking on principle.

Both kinds of universal are unfashionable in modern literary studies, being widely suspected as merely camouflages for conservative and illiberal values. To suggest that all human beings share discoverable common traits can seem reminiscent of the ethnocentrism of much nineteenth-century anthropology in which non-European cultures were chauvinistically measured by the degree to which they had acquired European attributes that were deemed to represent the universally achievable ideal. To suggest that Platonic idealism might be a useful way to approach the relationship between a play and its physical embodiment as scripts and performances was for a long time usual in textual studies (it was the standard model in New Bibliography), but is now routinely dismissed as essentialism. Indeed, it *is* an essentialist model of textuality, but this only seems problematic when all forms of essentialism are rejected out of hand without reasonable cause. A strict anti-essentialist would have to argue that Shakespeare's plays (or indeed anybody else's) cannot be translated into another language, since translation is predicated on the assumption that there exists an essence of the play that stands apart not only from its manuscript and printed texts but also from the particular words written by the dramatist. This essay will argue that essentialism and Platonic idealism are reasonable ways to think about the various manifestations of a play, as performance, as written script, and as originating ideas in the mind of the dramatist. As any materialist must accept, ideas have a basis in the organization of matter in the human mind – ideas are to that extent physically real – and the new scientific studies of consciousness (especially memetics and various models of cognition) show that the Platonic analogy provides a good way to conceptualize the distinction between the play as abstract (but nonetheless material and somewhat embodied) thought and its various further embodiments in manuscript and print textualizations and as performances.

Dualism and the intentional stance

In his *Meditationes de Prima Philosophia* (Meditations on First Philosophy), René Descartes responded to the radical skepticism of those such as Michel de Montaigne by asserting that while one could doubt the accuracy of one's sensory experiences, and hence doubt the existence of the world (including one's body) known from those experiences, one could not doubt the existence of the thinking mind that was, in that moment, doing the doubting.[8] The thinking mind, at least, had to exist, and for Descartes this special status set the mind off from the rest of the body and the wider universe, all of which was made from matter; the mind, for Descartes, had to be immortal and (in his terminology) unextended and impossible to fragment. This is the essence of Cartesian dualism, dividing the material body from the immaterial (and unified) mind. Most people report that their mind does indeed feel unitary and immaterial (quite distinct from the body it inhabits), and when asked to consider its physical location in the brain many of us respond that it seems as though the sensory perceptions of the body were being delivered to a single, central location – a command center – where the visual images are played on a screen and the sounds delivered through loudspeakers for the benefit of the real self operating the controls. The Cartesian model suffers the distinct problem of infinite regression: how does the inner self watching the screen and listening to the speakers gain its consciousness, unless we posit yet another smaller homuncule inside the first, and so on? The dualism of the Cartesian mind/body distinction seems to exert a powerful grasp on the way we think about ourselves, even though most people when pushed will accede that it cannot actually be true. It is certainly difficult to see how there could be an interface between the body and an immaterial spirit such that the latter could control the former. On reflection, materialists tend to agree with Gilbert Ryle that there can be no 'ghost in the machine' and that the mind is the name we give for the collection of activities that the brain, and the body of which it is a part, collectively perform.[9]

It may be that the dualist habit is hard to discard because it has served an evolutionary purpose, for it allows us to treat the environment around us as full of intentions. An efficient way to make sense of how other creatures behave and to predict their future actions is to treat them as having purposes: a predator exists 'for' chasing, prey exists 'for' being chased, caves exist 'for' sleeping in, and so on.[10] Children are highly prone to imputing purposes to objects, but adults do it too. In a famous experiment, Fritz Heider and Mary-Ann Simmel showed a simple animated film of a pair of triangles and a circle moving around and inside a box and asked viewers to report what they saw.[11] Overwhelmingly, viewers constructed from the movements of these images a narrative of rivalry, pursuit, and flight. It is as though the rational mind's acceptance that these are only shapes, not creatures with purposes, is overwhelmed by an innate predisposition to interpret certain movements as purposeful and thus to endow each shape with its own internal spirit. This kind of (possibly innate) benign dualism suits the dramatic arts. In an after-show discussion of a work-in-progress airing of his puppet-masque adaptation of Shakespeare's *Venus and Adonis*, Gregory Doran discussed the ease with which the mind treats a wooden puppet as though it were a living being. So convincing was the puppet Venus, Doran reported, that during rehearsals he found himself giving instructions directly to 'her' rather than the puppeteers.[12] Theater with human performers also promotes the habit of dualism as a mind inferred from a character's word is temporarily put in command of the actor's body.

In this we might see an inherent contradiction between emerging knowledge about the mind and the habits of thought that enable dramatic impersonation. We know that the mind is not a distinct entity apart from the body and also that it is not unitary. Sigmund Freud's division of the mind into conscious and unconscious parts remains popular in literary studies although recent research into the operations of the brain posits further empirically verifiable divisions, such as that between left and right hemispheres and between smaller specialized units. According to Daniel Dennett's recent description of the mind, distinct modules in the human brain have worked more or less independently and unconsciously for millennia doing the important tasks involved in keeping us out of danger, while the illusion of a singular, conscious self arose only quite recently after the invention of language.[13] Dennett's account is controversial among philosophers, and it puts language at the center of its explanation of consciousness. Just what language itself is for remains a tricky anthropological question. The obvious answer might seem to be that it aids social cooperation, and so arose as an evolutionary innovation that enabled our ancestors to out-compete rival animals for the control of resources. However, our oversized brains (presumably needed for big thinking) are a physical disadvantage, consuming considerable energy even when apparently doing nothing useful for us and making childbirth considerably more dangerous than it would otherwise be. One of the best reasons for taking seriously Richard Dawkins' only half-intended invention of the meme – the cultural equivalent of the gene – is that it gives an explanation for this peculiar fact about our bodies.[14]

As Susan Blackmore argued, a memetic pressure to increase brain size could have overwhelmed the genetic pressure to keep the brain small.[15] Once imitation became useful to *Homo habilus*, about 2.5 million years ago, good imitators (that is, those whose brains happened to be good at copying others' behavior) benefited from a genetic-selection pressure in reproduction. The memes that were imitated might be genuinely useful (the making of sharp tools or the fashioning of clothes) but since, once culture took off, it made good sense from a survival point of view to mate with good imitators in preference to poor imitators, non-useful memes (say, for singing or decorating caves) could flourish too. At this point, according to Blackmore, the memes took over and were able to drive up brain size as if for their own ends. In fact, this view of human development had been somewhat anticipated by Geertz:

> The slow, steady, almost glacial growth of culture through the Ice Age altered the balance of selection pressures for the evolving *Homo* in such a way as to play a major directive role in his evolution .... [The effects of culture] all created for man a new environment to which he was then obliged to adapt. As culture, step by infinitesimal step, accumulated and developed, a selective advantage was given to those individuals in the population most able to take advantage of it – the effective hunter, the persistent gatherer, the adept toolmaker, the resourceful leader – until what had been a small-brained, proto-human *Homo australopithecus* became the large-brained fully human *Homo sapiens*.[16]

Geertz's account lacks only Blackmore's insight that once being a good copyist conferred a genetic advantage, the things copied need not be exclusively practical, since unproductive memes too could flourish in this new intellectual substrate. That Geertz dates this process to about 3.5 million years ago and Blackmore to about 2.5 million years ago is irrelevant to the larger point about culture.

The claims that memetics makes about the origin of human physiology and culture have gained support from neurological science with the discovery of the mirror neurons, first in monkeys[17] and then in humans.[18] These neurons fire not only when we perform an activity but also when we watch someone else performing that same activity. They appear to be the reason that it is difficult to watch someone yawning or laughing without joining in, and equally why it is difficult to watch Lear's agony at the death of his daughter without sharing in the emotion. Our mirror neurons make us feel his pain even though we know we are watching only an imitation, and the principle seems anticipated in Hamlet's conviction that watching a representation of a crime is all the more acutely painful for those who have committed the like action. The memeticians' claim that we are merely the conduits for self-replicating practices and ideas may seem difficult to accept, but analogues of it have been common in the humanities and social sciences for some time. A correlative of the structuralist view of language and literature was that these extra-personal corpora of meaning-bearing distinctions speak, as it were, through us in our utterances and texts. The structuralist anthropologist Claude Lévi-Strauss identified his aim as not 'to show how men think in myths but how myths think in men, unbeknownst to them'.[19] Where Lévi-Strauss took up the process from the perspective of the myth instead of the teller, W. D. Hamilton and his followers such as Dawkins took up the perspective of the gene instead of the organism, and showed that the replicator fashions the organism to get itself copied. Equally, one might take up the view of the object you are holding and say that from the book's point of view the scholar (who researches in a library) is just a library's way of getting more books made.

Recent work on the mind and on language should be of special interest to those who engage with drama, since theater is to a large extent concerned with the artificial construction of what seem like human minds in dialogue. We do not know just how realistic the theater of Shakespeare's time aimed to be, and must not make the mistake of anachronistically applying twenty-first-century assumptions about psychological plausibility to early modern drama. Psychological approaches to acting such as those taught by Constantin Stanislavsky, Stella Adler, and Lee Strasberg might well seem suitable for the portrayal of a character like Hamlet, whose unseen inner mental life is explicitly a part of the role. But they are scarcely useful for a brainless role like Rumour in *2 Henry 4* or a minimally motivated one such as the easily overlooked Adrian in *The Tempest*. But for characters that seem intended to be realistically human, certain insights about how the mind works throw light on theatrical impersonation and the construction of characters. According to F. Elizabeth Hart, we can see this at work when the characters themselves are constructing fictional realities for one another.[20] In *Othello*, Iago has a 'theory of mind', a term coined by David Premack and Guy Woodruff and meaning the ability to impute mental states to oneself and others.[21] Children over the age of about four years are generally able to comprehend that they might know things that another does not and that another might hold beliefs that are actually false, say because that person is being tricked into believing falsehoods. With this 'theory of mind', a child is able accurately to predict others' behavior even when deceptions are being practiced. Children under the age of about four, or with conditions such as Downs syndrome and autism, are generally unable to make this distinction between the way things really are and the way that another might think they are.[22]

Iago not only assumes that Othello has a mind and predicts how it will react, he also assumes that Othello himself has a 'theory of mind' and is trying to make sense of Iago's reactions. As Hart pointed out, humans are 'compulsive mind-readers' and can be made to misread a mind by someone like Iago who is able to simulate the self-editing of a mind trying not to reveal itself. Othello constructs, from scraps of evidence, a version of Iago's mind that is self-editing to avoid being read:

> [OTHELLO]
> 'Think, my lord?' By heaven, thou echo'st me
> As if there were some monster in thy thought
> Too hideous to be shown! ...
> ...
> [Thou] weigh'st thy words before thou giv'st them breath,
> Therefore these stops of thine fright me the more;
>
> (*Othello* 3.3.110–125)[23]

Iago only rarely lies, since he seldom needs to. Iago knows that habitually we read others' minds and he lets Othello misread his by dropping scraps of evidence from which Othello can construct the central falsehoods of the play for himself.

Early modern cognition and textual transmission

The art of creating plausible impersonations of fictional humans includes, of course, the creation of dialogue from which the audience may infer the existence of a mind thinking up the lines being spoken. Indeed, the entire process of early modern theater (as opposed to merely the problem of how to act) is illuminated by recent work on the mind, for certain of its practices required feats of mental activity that seem to us prodigious. It is clear from the diary of theater impresario Philip Henslowe, which lists dates and titles of performances at his playhouses, that early modern actors performed as many as six different plays in a week, with a new play entering the repertory about every two weeks. We might suppose, as Tiffany Stern did, that they managed this by sticking to formulas and having each man play the same kind of character in each play.[24] But Evelyn Tribble has suggested that in fact the physical and documentary mechanisms of early theater took some of the burden from the individual acting mind. According to Tribble, the 'parts', the 'plot', the conventions of movement, the company structures, and the theater building itself together comprised a cognitive system that enabled a playing company to maintain its seemingly miraculous high-turnover repertory. In a cognitive approach the tools we use are not strictly distinct from the minds that use them, but rather the tools are part of what we think with: they are 'cognitive prostheses'.[25] Actors did not receive the whole script of a play, only the 'part' containing each of one character's speeches, topped by a 'cue', the last two or three words of the previous speaker's speech. Thus the 'plot', a single folio sheet listing the scenes and who is in them, was an actor's only chance to see the whole play represented all-at-once, like a map, and so to see what parts he was doubling and how the scenes related one to another. Another prosthetic was verse speaking itself, since it not only helps remembering but it also conceals forgetting: a good actor can stay in verse even if he uses the wrong words.[26] Simple rules about stage movement, such as Andrew Gurr and Mariko Ichikawa's refinement of Bernard Beckerman's 'restaurant kitchen' rule (one door for getting 'on' and one for getting 'off'), make the least cognitive demands on the performer and hence are the most plausible of the various theories regarding staging conventions.[27]

A convention not discussed by Tribble but probably pertinent is the one governing asides and soliloquies. Humphrey Gyde showed that it is possible to construct a single simple rule-based convention for both forms of address, only the former of which (asides) were so named in early modern theater scripts.[28] Gyde argued that the implied injunction to a character who steps onto the Renaissance stage is 'tell us how you feel' and in responding to this call – especially in soliloquy where deception is not possible – dramatic personality is created. Gyde produced an elegantly simple model of the audibility of asides and soliloquies that he called 'represented awareness', and I have found no violations of it in the extant early modern drama. Gyde claimed that a character speaking an aside or soliloquy can, as it were, deafen those characters whom she knows are on the stage at the time, invoking the convention of deafening by taking a step toward one of the edges of the stage ('aside') to confide in the audience. However, a character cannot deafen those of whose presence she is unaware, and hence the fear of being crept upon shown by soliloquizers and their explicit silencing of themselves lest the speech that was privileged address to the audience be overheard in the world of the play:

[RICHARD GLOUCESTER]
Enter George Duke of Clarence, guarded, and Sir Robert Brackenbury
Dive, thoughts, down to my soul: here Clarence comes

(*Richard 3* 1.1.41)

[HAMLET]
... Soft you, now,
The fair Ophelia!

(*Hamlet* 3.1.90–91)

[BANQUO]
Thou hast it now: King, Cawdor, Glamis, all ...
... But hush, no more.
Sennet sounded. Enter Macbeth as King, Lady Macbeth as Queen, Lennox, Ross, lords, and attendants

(*Macbeth* 3.1.1–10)

In response to these cues, and others that answer the 'tell us how you feel' injunction, the audience or reader infers a relatively unified and stable (albeit changeable) dramatic self, and as Bruce R. Smith showed – and Gyde independently confirmed – this process is most obviously logocentric (in Jacques Derrida's sense) in the soliloquies, and more acoustically social in the ensemble scenes.[29] Smith, however, was concerned with the disorder – what he called 'green' sound – that encroaches upon the edges of logocentric order, and did not address the more philosophically fundamental problem of what we are doing when we infer a character from its utterances.

Even the most postmodern of us does not feel entitled to treat what a character says as genuinely self-contradictory; we habitually infer coherence of mind and seek alternative explanations for contradiction. Take the simple problem of how many men Miranda thinks she has encountered before meeting Ferdinand in *The Tempest*. In an aside Miranda implicitly counts Caliban among humankind: 'This [Ferdinand] | Is the third man that e'er I saw' (1.2.447–448). Obviously, Prospero and Caliban were the first and second. However, speaking to Ferdinand she excludes Caliban from the count: 'nor have I seen | More that I may call men than you, good friend, | And my dear father' (3.1.50–52). Rather than treat this as a fundamental rupture in the coherence or consistency of the dramatic world, we quite naturally treat such discrepancies as examples of a singular, unified character being in possession of imperfect knowledge of herself, of her selfhood perhaps changing over time, and, whether consciously or not, of a self shaping its account of reality to suit the hearer. Thus, we might say that in her own mind – and presumably her near-rape is significant in this regard – Miranda considers Caliban a potential, albeit revolting, sexual partner, but she would not want Ferdinand to think so.

We treat ourselves as likewise partially self-divided when we say 'I forced myself to do it', or 'I did not mean it'. Such moments of self-division are among the markers of realistic dramatic character, and this fact itself corroborates the claim that consciousness – in dramatic characters as much as in real people – is an overarching phenomenological illusion of mental continuity that sutures the differences in our behaviors across time and space. Thus, although I do not feel that I am wholly the same person I was twenty years ago, I can hardly claim to be someone else entirely; at the very least the law may hold me to account for what I did back then. And yet there is a distinct fear of such continuities in much modern criticism and theorizing about Shakespeare. Foucauldians such as Margreta de Grazia would have us believe that before the Enlightenment people were simply more comfortable than we are with the discontinuous, the contradictory, and hence that (to take her example) when in Shakespeare the same letter is read aloud twice to produce seemingly different words, a problem emerges for us (with our unity-loving post-Enlightenment minds) that simply did not exist for them.[30] As I argue elsewhere, de Grazia misread the historical evidence and an enabling fiction of characterological and textual unity was as important for them as for us, and this should inform our textual theory.[31] At the very least, both the singularity of the performance script (normally only one text was licensed by the Master of the Revels) and, press correction notwithstanding, the singularity of the printed text, belie the textual multiplicities favored by postmodernism.

In our current dramatic practices, we grant coherent singularity to dramatic characters within a single text or performance, but withhold it when there exist multiple, seemingly distinct, early printings. We no longer feel entitled to consider the Hamlet found in Q1 to be the same as the Hamlet found in Q2 or in F, and hence the new Arden 3 edition of the play contains edited texts of all three versions.[32] And yet we treat the Hamlet at the end of the play (in each version) as effectively the same as the Hamlet at the beginning, despite him having quite a different outlook and, as Roger Lewis pointed out, having answered his own question 'To be or not to be' (3.1.58) with a definitive and conclusive 'Let be'.[33] This might well be considered something of a contradiction, and perhaps we should draw an analogy with our habitual refusal to construct a different person to account for how the character has altered over the course of the play (and we do not, as Dennett showed, do that with real people either) in order to say that we will not construct a different play each time we come across a variant version. The difficulty lies in overdetermination. There are two ways that a singular text may become a pair of variant texts, by revision or by corruption, and although either is sufficient to explain variation both may be operating as the cause of difference between two versions of a play. The problem was neatly stated by John Jowett, who observed that we used to think that Q1 and Folio *King Lear* were imperfect witnesses to a singular antecedent authorial version, and now we are in danger of deluding ourselves that they are perfect witnesses to two equally viable authorial versions, but in fact the truth lies between these positions: authorial revision and corruption separate these printings.[34]

Assuming that corruption will always be present in the printing of early modern plays, how high should we set the threshold for variation before we say that revision must also be brought in as an explanation? In a preface to Blackmore's book on memes, Dawkins gave a cogent reason for distinguishing between variations that matter and those that do not, using Blackmore's distinction of reproductions that copy-the-product and those (much more important, and mostly human) imitations that copy-the-instructions. Imagine a Japanese master carpenter teaching an English apprentice:

> The apprentice would not copy obvious mistakes. If the master hit his thumb with a hammer, the apprentice would correctly guess, even without understanding the Japanese expletive '** **** **!', that

he meant to hit the nail. He would not make a Lamarckian copy of the precise details of every hammer blow, but copy instead the inferred Weismannian instruction: drive the nail in with as many blows of your hammer as it takes your arm to achieve the same idealized end result as the master achieved with his – a nail head flush with the wood.[35]

This is pure Platonism, even down to the use of an analogy from woodwork, allied with a view of genetics that rightly dispenses with Jean-Baptiste de Lamarck's notion of inheritance – that each generation's particularities (the ironsmith's large biceps, the bicycle-courier's powerful calves) are passed on to descendants – in favor of August Weismann's assertion of the continuity of the germ line: your genes are not, in fact, altered by your behavior. Just as the apprentice copies not the master's actual practice but what is inferred to be the ideal behind it, so in genetic reproduction the 'recipe' for an organism, not a particular reading of that 'recipe' (the phenotype or bodily particularities), is what is passed on.

A version of this distinction was clearly in Philip Sidney's mind when he responded to Plato's famous attack on visual and poetic art, which attack claimed that because any real-world object, say a bed, is only an imitation of a perfect Idea or Form of 'bed-ness', a painting of, or a poem about, a bed necessarily is only an imitation of an imitation.[36] According to Sidney, '… the skill of ech Artificer standeth in that *Idea*, or fore conceit of the worke, and not in the worke itself'. Through imitating the idea, not the actual behavior, the prentice may surpass the master and the poet can make 'things either better than nature bringeth foorth, or quite a new, formes such as neuer were in nature'.[37] Using Blackmore's example, we may notice that a person's recipe for a new soup could be disseminated by repeated imitation of the soup by those who tasted it (Lamarckian, copy-the-product dissemination), in which case slight alterations might accumulate quite quickly as each chef chooses to add more salt.[38] But if the recipe circled the world as a text on the internet, such local variations would not accumulate because the text (like the germ line, DNA) is not altered by the variations (thus Weismannian, copy-the-instructions dissemination). The analogy works well for Shakespeare: most of the early textual reproduction was monogenetic (Q1 was copy for Q2, which was copy for Q3 and so on) and, now that the memorial reconstruction theory is largely discredited (in which theory actors created a complete script by recalling their individual lines), we can say that there was relatively little copy-the-product dissemination.

If we think the Dawkins/Blackmore/Dennett line on Platonism and the dissemination of cultural knowledge is reasonable in its distinction between variations that matter and those that do not, and if we think that human character is an example of where we should permit considerable latitude (such as variations of behavior and beliefs) before we conclude that we are dealing with more than one person, there seems little reason to accept the current textual theorizing that finds in each textual variant (apart from those that are egregious errors) the branching off of a new version of the play. Recast in terms of human labor, playwrighting tends toward the singular not the plural and to accord a distinct line to each textualization that happens to survive is as mistaken as asserting that there is no singularity called Coca-Cola because (contrary to the corporation's official line) we know that its sweetness is varied when it is sold in different markets across the world. It is true that we do not possess a manuscript recipe for Shakespeare's *Hamlet*, only three copies of it that differ markedly. But we are entitled to treat these as three approximations of one thing, the Platonic Ideal of *Hamlet* as it existed (in material form, as configurations of neurons) in the mind of Shakespeare. If, over time, Shakespeare changed his mind about *Hamlet*, it is still conceptually *Hamlet* even if a text closely representing the conceptual state at time T_1 is quite different from a text closely representing the conceptual state at time T_2.

It is not that there was actually a pure and unembodied form of the play in the mind of Shakespeare and that all textualizations are fallings-off from this perfected state. Ideas are not quite like that. According to Dennett, even in our minds experiences and intentions exist in multiple and inchoate forms, never coming together at one place (the now-dismissed Cartesian Theater):

> We don't directly experience what happens on our retinas, in our ears, on the surface of our skin. What we actually experience is a product of many processes of interpretation – editorial processes, in effect. They take in relatively raw and one-sided representations, and yield collated, revised, enhanced representations, and they take place in the streams of activity occurring in various parts of the brain.[39]

The written state of the plays existing in multiple textualizations has an analogy in the neurological state inside our heads, the 'multiple-drafts' model of consciousness. The feeling that the experiences and intentions adhere together, are coherent and persistent over time, comes from the combination of memes that we call consciousness: the simplified version of oneself that the various modules of the brain collectively generate for the purpose of managing the whole. (A useful analogy would be the 'My Computer' icon that the Microsoft Windows operating system presents to its user: like the Cartesian homuncule this exists within the hardware that it purports to represent, and it hides from the user the messy detail of the modules that make the real system.) The editorial work performed by the mind to generate and perpetuate the useful fiction of me (to me) or you (to you) is precisely like the editorial work formerly undertaken by editors to generate and perpetuate the useful fiction of a singular, coherent Shakespeare play for his readership and theater practitioners. As that singularity and coherence breaks down in modern editorial practice, the plays become at worst impossible to speak of at all, or at best they become identified with, and constrained by, their extant early textualizations. This is not how Shakespeare would have thought of them, nor how the early actors would have thought of them, because it is not really how the human mind works. Conflating multiple drafts to create usable fictions of self, of textuality, of existence, are what our minds have evolved to do.

Notes

1. Terry Eagleton, *Ideology: An Introduction* (London: Verso, 1991), 33–159.
2. Slavoj Žižek, *The Sublime Object of Ideology* (London: Verso, 1989).
3. Clifford Geertz, 'The Impact of the Concept of Culture on the Concept of Man', in John R. Platt, ed., *New Views of the Nature of Man* (Chicago: University of Chicago Press, 1965), 93–118 (107).
4. Ibid., 112.
5. Ibid., 111.
6. Stephen Greenblatt, *Renaissance Self-fashioning: From More to Shakespeare* (Chicago: University of Chicago Press, 1980), 3.
7. Bertrand Russell, *History of Western Philosophy and Its Connections with Political and Social Circumstances from the Earliest Times to the Present Day* (London: George Allen & Unwin, 1946), 146.
8. René Descartes, *Meditationes de Prima Philosophia* (Paris: Michael Soly, 1641).
9. Gilbert Ryle, *The Concept of Mind* (London: Hutchinson's University Library, 1949).
10. Daniel C. Dennett, *The Intentional Stance* (Cambridge, MA: MIT Press, 1987).
11. Fritz Heider and Mary-Ann Simmel, 'An Experimental Study of Apparent Behaviour', *American Journal of Psychology* 57 (1944), 243–249.
12. William Shakespeare, *Venus and Adonis: A Masque for Puppets* (work-in-progress), directed by Gregory Doran at the Royal Shakespeare Theatre, Stratford-upon-Avon on 10 October 2004 as part of the New Work Festival, 29 September–17 October.

13. Daniel C. Dennett, *Consciousness Explained*, illustrated by Paul Weiner (London: Penguin, 1993).

14. Richard Dawkins, *The Selfish Gene* (Oxford: Oxford University Press, 1976), 189–201.

15. Susan Blackmore, *The Meme Machine* (Oxford: Oxford University Press, 1999), 67–81.

16. Geertz, 'The Impact of the Concept of Culture', 110–111.

17. Vittorio Gallese, Luciano Fadiga, Leonardo Fogassi, and Giacomo Rizzolatti, 'Action Recognition in the Premotor Cortex', *Brain* 119 (1996), 593–609.

18. Michael A. Arbib, 'From Monkey-like Action Recognition to Human Language: An Evolutionary Framework for Neurolinguistics', *Behavioral and Brain Sciences* 28 (2005), 105–167.

19. Claude Lévi-Strauss, *The Raw and the Cooked*, trans. John and Doreen Weightman, in volume 1 of *Mythologiques: Introduction to the Science of Mythology* (London: Cape, 1970), 20.

20. F. Elizabeth Hart, 'The Renaissance Theory of "Things" and "Words", or What Iago Knows That Othello Doesn't', paper for the seminar 'Theorizing the Mind in English Renaissance Literature' at the National Convention of the Modern Language Association of America held in Washington, DC, 27–30 December 2005.

21. David Premack and Guy Woodruff, 'Does the Chimpanzee Have a "Theory of Mind"?' *Behavioral and Brain Sciences* 4 (1978), 515–526.

22. Simon Baron-Cohen, Alan M. Leslie, and Uta Frith, 'Does the Autistic Child Have a "Theory of Mind"?' *Cognition* 21 (1985), 37–46.

23. Unless otherwise stated all quotations of plays are from William Shakespeare, *The Complete Works*, ed. Stanley Wells, Gary Taylor, John Jowett, and William Montgomery, 1989. Electronic edition prepared by William Montgomery and Lou Burnard (Oxford: Oxford University Press).

24. Tiffany Stern, *Rehearsal from Shakespeare to Sheridan* (Oxford: Clarendon Press, 2000).

25. Evelyn Tribble, 'Distributing Cognition in the Globe', *Shakespeare Quarterly* 56 (2005), 135–55 (140).

26. Ibid., 150.

27. Ibid., 143; Andrew Gurr and Mariko Ichikawa, *Staging in Shakespeare's Theatres*, Oxford Shakespeare Topics (Oxford: Oxford University Press, 2000).

28. Robert Humphrey Gyde, 'Speaking Apart: The Formation and Exploration of Character Through the Aside and Soliloquy in Elizabethan and Jacobean Drama', unpublished PhD thesis, Stanford University, California, 1990.

29. Bruce R. Smith, 'Hearing Green: Logomarginality in Hamlet', 1.1–2, *Early Modern Literary Studies* 7:1 (2001), n.p., online at http://purl.oclc.org/emls/07-1/logomarg/intro.htm, accessed 9 August 2001.

30. Margreta de Grazia, *Shakespeare Verbatim: The Reproduction of Authenticity and the 1790 Apparatus* (Oxford: Clarendon Press, 1991), 222–226.

31. Gabriel Egan, 'Foucault's Epistemic Shift and Verbatim Repetition in Shakespeare', in Richard Meek, Jane Rickard, and Richard Wilson, eds, *Shakespeare's Book* (Manchester: Manchester University Press, 2008), 123–139.

32. William Shakespeare, *Hamlet*, ed. Ann Thompson and Neil Taylor, The Arden Shakespeare (London: Thomson Learning, 2006); *Hamlet: The Texts of*

1603 and 1623, ed. Ann Thompson and Neil Taylor, The Arden Shakespeare (London: Thomson Learning, 2006).

33. Roger Lewis, 'The Hortatory Hamlet', *Papers of the Bibliographical Society of America* 72 (1978), 59–60; William Shakespeare, *[Hamlet] The Tragicall Historie of Hamlet, Prince of Denmarke. Newly Imprinted and Enlarged*, STC 22276 (Q2a) (London: [J. Roberts] for N. L[ing], 1604), N3v.

34. John Jowett, *Shakespeare and Text*, Oxford Shakespeare Topics (Oxford: Oxford University Press, 2007), 3.

35. Blackmore, *The Meme Machine*, 12.

36. Plato, *The Republic*, trans. and ed. Francis Macdonald Cornford (Oxford: Clarendon Press, 1941), 10: 595a–608b, pp. 314–32.

37. Philip Sidney Sidney, *The Defence of Poesie*, STC 22535 (London: [Thomas Creede] for William Ponsonby, 1595), B4v–Clr.

38. Blackmore, *The Meme Machine*, 60–61.

39. Dennett, *Consciousness Explained*, 112.

13
Theater and the Scriptural Economy in *Doctor Faustus*

Ian Munro

> 'O, it strikes, it strikes! Now, body, turn to air'
> (*Doctor Faustus*, A-text, 5.2.116)[1]

'One bare hour' (5.2.66)

At the end of Christopher Marlowe's *Doctor Faustus* an hour passes in less than sixty lines. Immediately after the scholars who wish to pray for Faustus are dismissed, 'The clock strikes eleven' (5.2.64, s.d.); Faustus announces, 'Ah Faustus, / Now hast thou but one bare hour to live, / And then thou must be damned perpetually', and begins his impassioned plea to stop time and what is coming for him:

> Stand still, you ever-moving spheres of heaven,
> That time may cease and midnight never come!
> Fair Nature's eye, rise, rise again, and make
> Perpetual day; or let this hour be but
> A year, a month, a week, a natural day,
> That Faustus may repent, and save his soul.
> *O lente lente currite noctis equi!*
> The stars move still; time runs; the clock will strike;
> The devil will come, and Faustus must be damned.
>
> (65–76)

Thirty-one lines after the first bell, 'The watch strikes' (95, s.d.), and Faustus declares, 'Ah, half the hour is past! / 'Twill all be past anon' (96–97). The second half hour is even shorter: in only twenty lines 'The clock striketh twelve' (115, s.d.), there is 'thunder and lightning'

296

(117, s.d.), and the devils enter and remove Faustus, who is still scream-
ing 'let me breathe awhile' (121).

How might the accelerating fate of Faustus be useful for talking about
a 'return to theory' for early modern theatrical criticism? This essay uses
the play, and especially its ending, as a location for examining certain
ideas about theory, text, and embodiment. *Doctor Faustus* can be understood as a kind of theoretical *exemplum* for relations between script and
body, I would argue, with influences that extend far beyond the direct
study of the play. What follows is less a reading of the play *per se* than
a kind of case study built around its complex modeling of theatrical
embodiment. I am particularly interested in one conceptual negotiation
of *Doctor Faustus*, albeit a perhaps unconscious one: the chapter called
'The Scriptural Economy' in Michel de Certeau's *The Practice of Everyday
Life*. This analysis of how bodies are written in Western culture, and
written upon, proceeds via a kind of historical allegory that connects
strongly to the specifics of *Doctor Faustus* – so much so that the play
seems an unacknowledged guide for the discussion. By turning back
to this seminal theoretical text, I am in part suggesting that a 'return
to theory' involves a re-reading of our own theoretical past, as part of
an exploration of unremarked conceptual paths. In this case, what has
remained unremarked is the extent to which de Certeau's modeling
of embodied orality and literacy depends upon the apparatus of theatrical representation. Reading *Doctor Faustus* through 'The Scriptural
Economy', and vice versa, offers an opportunity to examine the ontological and epistemological conditions of early modern theatricality.

In order to facilitate this examination, I first want to use the play to
survey, in brief terms, some dominant threads in the critical discourse
surrounding theatricality in the past generation. I begin with two fundamentally opposed ways of thinking about theater's relation to theory,
whose conjunction offers a miniature illustration of how theory became
something that would need to *return* to criticism. The first is a centrally
important book in performance studies: Stanton Garner's 1994 *Bodied
Spaces*, which attempted 'to return performance theory to the body and
its perceptual worlds', particularly by challenging 'the apparent ease with
which contemporary theory has dispatched the phenomenal (or lived)
body in favor of the representational (or signifying) body'.[2] Garner frames
his project in terms that have become a prominent dynamic in performance studies, and which seem especially suggestive of *Doctor Faustus*:

> To the extent that theory involves an unquestioned flight from
> the corporeal subject and the materiality of its phenomenal

fields, the phenomenology of a realm as penetrated with actuality as the theater will necessarily and strategically invoke *resistance*, in the technical sense of opposing or retarding motion. Against the transparency of theoretical disembodiment, it points to that which challenges such erasure.[3]

In this reading, the phenomenological becomes that which fights against a totalizing power that promises, too soon, to theorize away the body, to reduce it to transparent semiotic meaning. This desire for resistance, to stop the abstract consumption of the body, is exactly what we see Faustus attempting in the final speech of the play; even his call for his body to 'turn to air' (5.2.116) is less about disembodiment than a desire to perpetuate embodiment by making it immaterial. Faustus resists motion, resists time, resists being taken away. 'Not yet' would seem to be the central dynamic of the scene. In the context of Garner's rhetoric, we might thus imagine the final scene of *Doctor Faustus* as staging a confrontation between phenomenological and semiotic approaches to the theater. On one level, the final scene is all about the lived body of Faustus, the fear he is experiencing, the panic at the passage of time. On another level, the scene is all about the significance, in quite stark terms, of Faustus himself: in the A-text, immediately after his departure the Chorus returns to declare, 'Faustus is gone. Regard his hellish fall' (Epilogue, 4) and propounds the concluding moral. For the audience, empathy with Faustus is produced through experiential association and intimacy; judgment of Faustus is produced through imagining his body as part of a signifying system.[4]

This opposition between the phenomenal body and the signifying body is especially notable in the A-text because of its intense focus on Faustus in the final scene: after the departure of the scholars we don't *see* anything except his body. In the B-text the final speech is prefaced by the reappearance of the Good and Bad Angels, who show Faustus the descent and ascent of God's throne and reveal the mouth of Hell, a spectacle that perhaps remains visible throughout his final speech. Also, at the close of the play the scholars return and observe the scattered limbs and entrails that are the grisly remainder of Faustus. None of this is present in the A-text: despite the lavish theatrical pageants found earlier in the play, at the end every opportunity for visual spectacle is avoided. In fact, we might imagine that the terse final stage directions of the A-text – 'Enter devils' (120, s.d.); 'Exeunt with him' (123, s.d.) – should be understood as pointing to a grudging necessity: that somehow the physical body of Faustus must be taken from us, dragged down through a trapdoor in the bare, daylit space of the early modern stage.[5] Because of the physical apparatus of the theater, a simple disappearance of the body is not possible.

The significance of Faustus' body is scriptural, in both literal terms – the logic and the language of biblical scripture pervade the speech – and figurative terms. In Marjorie Garber's seminal 1984 analysis of the play, Faustus is described as 'a figure, a representation, a terminable fiction, a dramatic character who perversely tries to turn blood into ink as an act of willful self-inscription – the antithesis of a creating Word who self-lessly transforms wine into blood as an act of grace'.[6] Characteristically, Garber employs a light hand with theoretical language, but her reading of 'the trope of writing and unwriting' (301) is clearly deconstructive, presenting the play as a Derridean meditation on 'the riddle of the creating Word' (318) and the paradox of incarnation: 'The verb after which [Faustus] strives, that which he wishes to *do*, is God's verb, the verb intrinsic and implicit in Jehovah's name, the verb "to be", "to pre-exist", which has neither beginning nor end' (317). In his impossible act of self-inscription, Faustus competes with 'two other authors, or authorities – the author of the play, and the Author of the Universe, who is also the Author of the Scriptures' (308), who have been inscribing him all along. And as Faustus' body is rendered (or revealed) as only a text, so *Doctor Faustus* as a performed play becomes expressly textual, operating under the signature of 'Marlowe's written supplement to his play, *terminat Author opus*' (318). The value of theatrical performance, in Garber's analysis, is that it highlights the semiological conundrum: the theater offers 'a special opportunity for articulating – or disarticulating – semiotic problems of presence and absence' (315). The physical incarnation of *Doctor Faustus* on the stage only helps to demonstrate its fundamental disembodiment.

Garber's reading is thus exemplary of the kind of 'theoretical disembodiment' that Garner would oppose with a return to the embodied space of the theater, and the tension between these two methodologies offers one perspective on what has happened to 'theory' over the past generation in early modern scholarship. Garner's Faustian resistance to the dominating apparatus of theory is an interesting rhetorical maneuver in part because it inverts the terms under which poststructuralist analysis has typically presented itself. This methodological inversion can be neatly expressed through de Certeau's opposition between 'strategic' and 'tactical' social practices, introduced in *The Practice of Everyday Life*.[7] To rehearse briefly this well-known pairing, strategies are dominant practices that operate by establishing a bounded space as a proper location for their actions; tactics, by contrast, must operate within this strategic space and have at their root a kind of fugitive mobility that allows them to destabilize the assumptive logic of strategies. Thus, while Derrida's deconstruction of Husserlian phenomenology (and, more broadly, the metaphysics of presence) positioned itself as tactical, showing the occluded impossibilities grounding a widely accepted theoretical model, for Garner poststructuralism has become strategic, a totalizing discourse whose own blind spots must be explored. At the same time, there is something implicitly strategic in Garner's focus on the theater as 'a realm … penetrated with actuality'. Garner specifically cites Derrida's work on Husserl, but somewhere in the background is surely Derrida's two essays on Antonin Artaud's Theater of Cruelty, which argue (put crudely) that Artaud is in search of an impossibility: a theater of pure presence, beyond representation and thus beyond textuality.[8] The resistance that Garner supports is therefore also legible as the resistance of the material theater to Derrida's conceptual theater. In effect, Garner's rhetoric claims the embodied space of the stage as a demarcated territory from which theory's encroachments might be beaten back.

'We'll canvass every quiddity thereof' (1.2.166)

Theory is thus associated with an 'unquestioned flight' from theatricality, which is associated with materiality, corporeality, and 'actuality', the practical and real matters of the stage. In this respect the methodological turn signaled by Garner's book dovetails with developments in early modern theater scholarship over the past two decades – although the resistance to semiotic transparency that Garner valorizes has tended to get lost along the way.[9] Two years after Garner published *Bodied Spaces*, Kier Elam surveyed the ongoing 'corporeal turn' in Shakespearean criticism and observed that despite claims to the contrary the critical discourse of the early modern body was continuing the semiotic project under different terms: 'the early modern body turns out to be more bookish than corporeal, its readability guaranteed by the fact that it is already constituted by the play of discourses and intertexts'.[10] Elam concluded that it was necessary to turn again to the material early modern stage and the material early modern body: 'in order to be both fully historicized and fully materialized', a post-semiotics of Shakespearean drama 'can only set out from the one historical and material body we have, the actor's'.[11] Work in this area, which might be loosely grouped under Bruce Smith's term 'historical phenomenology', has proliferated in recent years – perhaps most prominently in Robert Weimann's exploration, across a series of books, of issues of authority and representation in the early modern theater.[12] Although its methodological framing tends to be more implicit than explicit, this range of work could be described as proceeding under the influence of Pierre Bourdieu, in that it understands the theater as a place of embodied social practices and the body (especially the actor's body) as a site of durable social dispositions that can be inventoried through their interactions with the socio-cultural field in which they find expression.[13] In this way, the space of theatrical performance, as distinct from the dramatic representation being performed, becomes representational in another register; as Henry Turner declares, 'performance itself functioned as a practical model of knowledge'.[14] Correspondingly, the idea of the script as master-text (as in Garber's reading) has increasingly been undermined, as scholars have explored how early modern actors used scripts as a platform for improvisation, or how the scripts themselves are patchwork cultural artifacts, loci of cultural translation and negotiation.[15]

This is a critical movement that has produced extraordinary rewards, giving us a much stronger sense of the complexities of early modern performance than was previously available. Furthermore, attention to the sociology of the early modern stage forms part of a larger conceptual shift from drama as literature to drama as theater – a redressing of the relationship of 'author's pen' and 'actor's voice', in Robert Weimann's resonant pairing – that is (in my opinion) wholly salutary. At the same time, it is perhaps appropriate to think about tactical responses to the strategic reach of historical phenomenology. What often eludes such methodologies, in my opinion, is the inherent strangeness of theatrical experience – a strangeness that generally comes through most clearly in moments when our expectations are shifted, as in the final soliloquy of *Doctor Faustus*. The idea of theatrical experience that generally emerges from this discourse is profoundly lucid: the stage is a space in which we can explore not only the fictions presented but the legible means whereby that presentation occurs. The limits of our understanding of the space of the stage are principally historical and archival: the evidence surviving from the period and our ability to extrapolate from it.

In talking about the 'strangeness' of theater, I am thinking specifically of what Garber calls (in a later book that clearly builds on her Marlowe essay) 'uncanny causality', our sense of an unknowable agency driving the dramatic action.[16] For Garber, again, this causality operates under the sign of literature and is general to language. But uncanniness is an experience proper and specific to theater: it is inseparable from what we might label the theatricality of theater. As Anthony Kubiak has noted:

> The closer one comes to the seemingly real in theater, the more one is reminded that what one sees is not what one is seeing. I am, in other words, never fooled that what I am looking at is 'real', or that it mimics the Real, but I *am* aware that what I am looking at is really happening ... and that despite its realness ... what is 'really' happening is never the substance of what I am 'really' seeing.[17]

The theater of early modern England shows a special interest in such uncanny reality effects, particularly in its fascination with dramatic scriptedness; it is commonplace in early modern drama for characters to register the limitations of the representational mode in which they find themselves, either through direct metatheatrical allusion or more indirect means. In the midst of 'The Mousetrap', to take one famous example, Hamlet remarks to Ophelia, 'I could interpret between you and your love if I could see the puppets dallying.'[18] For Hamlet to see the puppets (meaning, I assume, Ophelia and himself) would be to establish a position of privileged knowledge outside of himself, from which accurate interpretation could take place. But who, or what, is the puppeteer? Hamlet's theoretical leap stands at some distance from Garber's focus on script as only a form of *writing*; the authorial figure Hamlet imagines is not the author but the operator of the theatrical machinery. In a broad sense, the strangeness of the theater is thus intimately bound up with its material manifestation.

The close of *Doctor Faustus* provides an excellent example of this material uncanniness through its manipulation of time. It has long been commonplace to observe that *Doctor Faustus* is a play *about* time, on several levels.[19] Less attention has been paid, however, to the play's complex temporal phenomenology, especially in its final minutes – or perhaps its final hour. The trick with the clock that begins this essay involves a dislocation of mind and environment of a sort we are all familiar with: what we think of as our internal clock, in a revealing expression, is shown to be out of sync with the passing of time.[20] It's later than we think. The striking of 'the watch' surprises us, and in that surprise creates an identification between us and the space of performance. Paradoxically, we are both dislocated and drawn in by it. Time is the inevitable motor of theater, as both a form of narrative and as a performing art, but our typical experience of theatrical time is so strongly governed by convention that we remain largely unaware of it: twenty-four years of Faustus' life pass by in about two hours, yet a series of familiar theatrical devices – clearing the stage, interposing the comic plot, regular choric emplacements – means that this compression raises no alarms. The final scene of the play is phenomenologically different, because there are no breaks: we enter into another time that moves more quickly than our own.

Time is the buckle of this scene, in both meanings of the word: it breaks the environmental relation between stage and audience and it binds that relation. To take a modern analogy, consider a hackneyed cinematic device: a clock with rapidly spinning hands, illustrating the passage of an extended period of time. Even if we accept the convention we are brought face to face with the artifice involved, jolted out of the normal representational space of the movie. And considered outside of its convention, a spinning clock is a somewhat disturbing idea, in that it breaks the illusion that a clock *reflects* something (the passage of time) rather than *produces* something (the motion of the hands). Rather than being something that merely passes, time seems to be something that is propelled: what is powering those hands?[21] In the theater, as in a movie, the answer is somehow both prosaic and uncanny. 'To be that scared of a bell!', Jean exclaims at the end of August Strindberg's *Miss Julie*, in a scene that self-consciously recalls the conclusion of *Doctor Faustus*.[22] Jean explains his terror: 'it isn't only the bell – there is something behind it – a hand that makes it move – and something else that makes the hand move' (36). The representational hand that Jean directly fears is that of the unseen Count, the returning authority figure who will call both him and Julie to judgment. But as in Hamlet's imagined puppet theater, the material hand belongs to someone behind the scenes, a 'stagehand' who pulls the cord that rings the physical bell. And as in Kubiak's observations on the 'reality' of theater, both the representational hand and the material hand are related to, but distinct from, the 'something else' that Jean fears, the unnamable motive power of the play.

Theatricality is thus something more, or something other, than representation. In a brilliant analysis of Plato's Cave as a foundational theorization of theatrical experience, Samuel Weber focuses his attention not on the fettered spectators, forced to watch the play of shadows that they take for reality, but on the other inhabitants of the cave, those who carry the objects that cast the shadows:

What is their ontological or, for that matter, political status? How do they relate to that spellbound, enthralled audience of spectator-prisoners? How do they relate to the organization and significance of the 'spectacle' itself? Plato does not respond to these questions, though his own scenario stages, and thus implicitly raises, them. The question of theater and theatricality thereby remains unaddressed by the ontological condemnation Plato reserves for emphatically mimetic practices. But that condemnation sets the scene, as it were, for all successive attempts to determine the precise place – ontologically, epistemologically, ethically, politically – of theater and its 'special' effects, including spectators and actors, stages and their 'props', lighting, sounds, and perhaps *effectiveness* in general.[23]

In order to arrive at his condemnation of representation, Plato must elide the machinery of the scene he stages. But the machinery of the theater is never invisible: as Weber concludes, 'theater marks the spot where the spot reveals itself to be an ineradicable macula, a stigma or stain that cannot be cleansed or otherwise rendered transparent, diaphanous'.[24] While Garner describes the opaque *body* as that which exceeds and resists representation, for Weber *theater* is opaque, exactly because its material structures exist in a complex and nebulous relation to its project of representation. And while these structures – 'spectators and actors, stages and their 'props', lighting, sounds', and so on – have often been the focus of the discourses of historical phenomenology, the usual result has been to incorporate theatricality into the paradigm of representation, in effect making it into another species of cultural performance.

'The machine is the *primum mobile*, the solitary god from which all the action proceeds', writes de Certeau in *The Practice of Everyday Life*: 'It not only divides spectators and beings, but also connects them; it is a mobile symbol between them' (113). The context of this statement is railway travel, a social practice which de Certeau describes in terms remarkably similar to the representational experience of Plato's Cave: modern travelers are kept immobile in a train carriage as an untouchable landscape flows in front of them, 'a module of imprisonment that makes possible the production of an order, a closed and autonomous insularity – that is what can traverse space and make itself independent of local roots' (111). It is the *sound* of train travel, tellingly, that communicates the panoptic experience most effectively:

These sounds also indicate … the Principle responsible for all the action taken away from both travelers and nature: the machine. As invisible as all theatrical machinery, the locomotive organizes from afar all the echoes of its work ... There is also an accidental element in it. Jolts, brakings, surprises arise from this motor of the system. This residue of events depend on an invisible and single actor, recognizable only by the regularity of the rumbling or by the sudden miracles that disturb the order. (113)

De Certeau's description of the locomotive as the orchestrator of the spectacle of travel connects strongly with Weber's vision of theatricality as a medium, something 'relational and situational, depending decisively on alien or extraneous instances that, in the case of theater, are generally identified with the spectators or audience' (43). It also connects strongly to the striking clock which concludes *Doctor Faustus*, one of the 'sudden miracles that disturb the order' and reinforce the power of the order by our acquiescence to their evanescent presence. It is appropriate, too, that time is at issue: like the accelerating clock, the jolts and brakings of the train signal increases or reductions in speed, reminding us that something is driving our sensory experience. As throughout *The Practice of Everyday Life*, de Certeau links this spatial practice to textuality and storytelling, but the textual model is supported by a largely unacknowledged theatricality, elided – 'as invisible as all theatrical machinery' – even as it utilized. The fantasy of theater as a world-making device, traversing space, presenting any imaginable scene, depends upon a machinery that we can only fitfully notice, in the same way that the 'residue of events' noticed by the traveler can only gesture toward the 'solitary god from which all action proceeds'.

One can certainly see a clear evolutionary line from *The Practice of Everyday Life*, dominantly concerned with how tactical social practices resist incorporation into the strategies of established powers, to Garner's vision of the body as an irreducible location of asignifying power. At the same time, the paradoxes that underpin the operation of the scriptural economy have the potential to lead in different directions. By 'the scriptural', de Certeau means something that stands at some distance from either God's 'creating Word' or poststructuralist *différance*. For de Certeau, writing is always material, a social practice involving physical tools; the scriptural can never be disembodied because it involves an interface between idea and flesh.[25] At the same time, it begins with the abstract territory of the blank page, a place 'where the ambiguities of the world have been exorcized', which 'delimits a place of production for the subject' (134), and which can be used to capture and dominate that which has been deemed its exterior: 'the scriptural enterprise transforms or retains within itself what it receives from its outside and creates internally the instruments for an appropriation of the external space' (135). The scriptural stands as the book's central instance of a strategic practice, the example to which all of de Certeau's other discussions eventually turn. Yet as in the example of railway travel, the orchestrating power of the scriptural economy is subtended by an occluded theatricality. In fact, I would argue, de Certeau's examination of the scriptural economy's relation to the writing of the law addresses the same problematic exteriorities of theatrical representation noted by Weber.

'Ecce signum!' (3.2.3)

In the context of the larger project of *The Practice of Everyday Life*, 'The Scriptural Economy' is positioned as a first step toward an investigation of orality. 'I am trying to hear these fragile ways in which the body makes itself heard in the language', de Certeau writes, but 'these voices can no longer be heard except within the interior of the scriptural systems where they recur' (131); as a result, he proposes to explore the 'historical implantation' and 'rules and instruments' of the scriptural system in order to 'begin to locate the points at which voices slip into the great book of our law' (132). De Certeau is specifically interested in a new kind of writing that he argues instantiates itself in the early modern period and quickly spreads. Catalyzed by the loss of 'Holy Scripture' as the 'Spoken Word' of God, 'another writing is imposed little by little in scientific, erudite or political forms: it is no longer something that speaks, but something that is made' (136–137). This shift from hearing the word to making words characterizes the basic problematic of modern subjectivity:

> It is because he loses his position that the individual comes into being as a *subject*. The place a cosmological language formerly assigned to him and which was understood as a 'vocation' and a placement in the order of the world, becomes a 'nothing', a sort of void, which drives the subject to make himself the master of a space and to set himself up as a producer of writing. (138)

The blank page is thus invested with the promise of both a remarkable new power and a recuperation of what has been lost: 'the disappearance of the First Speaker' (138) creates a pressing need to write ourselves and to write our world, to make a world of writing, to make writing the world. Or, as that prescient scriptural worker Doctor Faustus puts it: 'O, what a world of profit and delight, / Of power, of honour, of omnipotence, / Is promised to the studious artisan! / All things that move between the quiet poles / Shall be at my command' (1.1.55–59).

The thematic relevance of *Doctor Faustus* to the story de Certeau lays out is inescapable. At the start of the play, we join the scriptural subject already in the midst of his expansive appropriations: the space that Faustus has been driven to make himself master of is physically represented by his study, and the play begins with an inventory of the territories that study (and the study) has conquered, albeit colored by a desperate longing for new worlds to dominate. Writing is fundamentally productive in the opening scene of the play: the question Faustus asks of all his books, including the Bible, is what they are *for*, what they *do*. And throughout the play, scholarship – a dominant engine of the scriptural economy, then as now – is described as having acquired strange new powers to transform people and places, to contain 'all nature's treasury' (1.1.77), to resolve all ambiguities. As Stephen Greenblatt has observed, 'Marlowe writes in the period in which European man embarked on his extraordinary career of consumption, his eager pursuit of knowledge, with one intellectual model after another seized, squeezed dry, and discarded, and his frenzied exhaustion of the world's resources.'[26] In de Certeau's account, this endless succession of intellectual models is facilitated by the new *mythos* of scriptural production: 'one can read above the portals of modernity such inscriptions as "Here, to work is to write", or "Here only what is written is understood"' (134). Although *Doctor Faustus* makes an orthodox distinction between the legitimate practice of learning and damnable magic, from the perspective of the practitioners the first leads directly to the next: 'He that is grounded in astrology, / Enriched with tongues, well seen in minerals, / Hath all the principles magic doth require' (1.1.140–142).

The parallels between the rise of the scriptural economy and the fortunes of Faustus compound in the context of the next part of de Certeau's essay, which connects the new scriptural practice to the 'virtually immemorial effort to place the (social and/or individual) body under the law of writing' (139).[27] 'There is no law that is not inscribed on bodies', de Certeau declares:

> It engraves itself on parchments made from the skin of its subjects. It articulates them in a juridical corpus. It makes a book out of them. These writings carry out two complementary operations: through them, living beings are 'packed into a text' (in the sense that products are canned or packed), transformed into signifiers of rules (a sort

of 'intertextuation') and, on the other hand, the reason or *Logos* of a society 'becomes flesh' (an incarnation). (139–140)

This process, he goes on to explain, requires a machine of some sort, an 'apparatus ... that can mediate the relation between' the law and the body. This apparatus includes the specific tools of the law – ropes, knives, handcuffs, cells, billyclubs, and so on – but also, and more significantly, 'the panoply of orthopedic instruments and means of treatment', physical and ideological, with which our bodies are disciplined (142). Surveying the discourse of the body from the Renaissance anatomy theater to modern surgery, de Certeau observes that transforming 'the opaque carnal reality' of flesh into the readable fiction of a unified body requires an ever-proliferating apparatus of articulation: 'Between the tool and the flesh, there is thus an interaction that shows itself on the one hand by a change in the fiction (a correction of knowledge) and, on the other by the cry, which shrieks an inarticulable pain and constitutes the unthought part of bodily difference' (145). Tellingly, de Certeau titles these operations 'the machinery of representation', in that they have as their aim 'making the body tell its code', the 'immense task of "machining" bodies to make them spell out an order' (147–148). And, finally, it is the ability to use bodies to incarnate this fictional order that gives credibility to the entire enterprise: 'From initiation ceremonies to tortures, every social orthodoxy makes use of instruments to give itself the form of a story and to produce the credibility attached to a discourse articulated by bodies' (149). If the scriptural economy once presented its myth as that of Robinson Crusoe, master of the blank page of his island, 'in which he can produce what he wants' (136), now the guiding myth is Franz Kafka's penal colony, which houses a writing machine that inscribes the words of the law on the body of the offender.

In de Certeau's account, the movement from Defoe's island to Kafka's is a historical progression, grounded in massive political and economic alterations in Western society over several centuries. But this simple causality is troubled by the sense that the myth of Crusoe persists today (albeit with less assurance) and was always already subtended by the myth of the penal colony. To move from the first to the second – to appreciate one's emplacement within the scriptural apparatus – involves an act of recognition, or perhaps disillusionment: to look behind the curtain, to observe the apparatus, to see the puppets dallying. In this regard, it is unsurprising that *Doctor Faustus* – a play that stands (with the rest of Marlowe's corpus) as a kind of degree zero for modern drama, the definitive break with the sacral theater of medieval drama – demonstrates the entire range of de Certeau's progression, articulating both the new practice of writing that grants Faustus his extraordinary power and the inscription of Faustus' body with the machinery of representation. If within the scope of the play Faustus' fortune is presented as a rebuke to 'forward wits', as the Chorus calls those who 'practice more than heavenly power permits' (Epilogue 6–7), when placed alongside de Certeau's account the semiotic inscription of Faustus' body is something exactly catalyzed by the practice of writing the self, the incarnation of a writing that exceeds the category of either God or the author.

There is a theatrical implication to all of this as well, of course, and not only in that the movement from *Mankind* to Marlowe coincides so neatly with de Certeau's imagined origin point for the scriptural economy. The disciplining of the body under the law described so vividly by de Certeau connects not only to the character of Faustus, forced to become a scriptural sign, but also to the disciplines of the theater, where actors' bodies are made the canvas for the law that we call the script, are used to incarnate the articulate fiction that we call a 'character', are made to speak with words that are not their own – an extraordinarily laborious practice, one requiring years of training, months of dedicated practice, and a complete surrender to the imperatives of an alien power. In a less thorough sense, the theater audience is also disciplined by the machinery of representation, both in our acquiesence to the (increasingly strict) decorums of theatrical attendance and more basically in our submission to the conventions through which the imaginary dramatic world is presented. Beyond these analogies, the theater is itself a fundamental aspect of (or institution within) the scriptural economy, especially in its articulation of the fictions of the human body. The 'blank page' of the empty stage is the *sine qua non* of the strategic incorporation of the world, the location which assimilates human experience and repackages it as product, eventually producing (through a movement into new media of dissemination) what Raymond Williams bleakly termed a 'dramatized society', in which representations of reality take primacy over reality itself.[28]

'Read, read the Scriptures. That is blasphemy' (1.1.75)

The theater that emerges from these conceptualizations may seem caught in its own trap – unable to be anything *but* a trap, a totalizing strategic practice. As Derrida comments, in the context of Artaud, 'The classical Western stage defines a theater of the organ ... a theater of deviation from the groundwork of a preestablished text, a table written by a God-Author who is the sold wielder of the primal word.'[29] One tactical response, then, would be to attend to what de Certeau terms 'the unthought part of bodily difference', demonstrated by 'the cry, which shrieks with inarticulable pain'. In the larger context of *The Practice of Everyday Life*, embodied orality is indeed that which resists the empire of signification, what de Certeau will refer to in a subsequent chapter as 'white pebbles dropped through the forest of signs' (163); it has no *place* in the scriptural economy and yet persists within it, troubling its domination.[30] De Certeau's rhetorical framing of 'the cry' owes a great deal to Artaud – specifically to his ecstatic, excruciating call in *To Have Done with the Judgment of God* to halt the proliferation of representation via a 'body without organs', disarticulate and antiscriptural.[31] In Jerzy Grotowski's Artaudian production of *Doctor Faustus* in 1963, the spectacle of Faustus' body provides not an illustration of the judgment of God but a fundamental resistance to it: 'He is in a rapture, his body is shaken by spasms. The ecstatic failure of his voice becomes at the moment of his Passion a series of inarticulate cries – the piercing, pitiable shrieks of an animal caught in a trap.'[32] Reconfigured as the Passion, the moment of the Law's inscription on the body of Faustus becomes the moment at which he escapes it. Variations on this theme have also been the dominant strand in performance studies for as long as performance studies has existed as a discipline, through a persistent valorizing of performance over theater, body over text, improvisation over recitation, repertoire over archive, and so on.[33] And yet there is a kind of repetition compulsion involved, whereby the extra-textuality of the body must be continually reaffirmed, and the influence of the script continually eluded. In part, this is because this mode of performance studies is founded on a contradiction, whereby writing about the body perpetually reinscribes it within the scriptural economy, perpetually makes it transparent, a representation – perpetually makes it tell its code.[34] In conclusion, I want to suggest a somewhat different methodological tactic, one that seems to me to have specific relevance to both *Doctor Faustus* and to the project of understanding early modern theatricality, by returning to that *other* opacity, the 'ineradicable macula' of theatricality proposed by Weber.

In de Certeau's account, the apotheosis of the scriptural economy in Kafka's penal colony is also an omen of a further historical development: 'Today ... the scriptural system moves forward on its own; it is becoming self-moving and technocratic; it transforms the subjects that controlled it into operators of the writing machine that orders and uses them' (136). De Certeau calls this a 'celibate machine', after the model of Michel Carrouges' *Les Machines célibataires*, which observed profound resemblances between Kafka's writing machine and Marcel Duchamp's *Grand Verre: La mariée mise à nu par ses célibataires, même.*[35] Connecting Duchamp and Kafka to a host of other autonomous machines in literature, art, and theater, Carrouges imagined the 'celibate machine' as an important modern mythic structure, one in which humanity is inescapably embedded within a totalizing (or perhaps totalitarian) machine.[36] But for de Certeau, the paradoxical effect of the advent of the celibate machine is that the myth that originally sustained the scriptural economy is evacuated: 'Stripped naked (*mise à nu*) by a mechanically organized deterioration, the bride (*la mariée*) is never married to a reality of meaning' (151). Although it begins in service to the law, facilitating its bodily incarnation, the celibate machine destroys the fiction of the law by dismantling the idea of representation as a mediation between *Logos* and flesh. As de Certeau says in conclusion to the chapter:

> It is through this stripping naked of the modern myth of writing that the celibate machine becomes, in a derisive mode, blasphemy. It takes away the *appearance* of being (i.e., of content, of meaning) that was the sacred secret of the Bible, transformed by four centuries of bourgeois writing into the power of the letter and the numeral. Perhaps this anti-myth is still ahead of our history ... Or perhaps it has simply been placed 'alongside' a galloping technocratization, like a suggestive para-dox, a little white pebble. (153)

The conceptual paradox that this conclusion presents is that the place outside the law ('para-dox') which elsewhere in de Certeau's writing is associated with the asignifying body, 'white pebbles dropped through the forest of signs' (163), is here occupied by the very apparatus of signification. Rather than seeking a point of escape or resistance *exterior* to the machinery of the scriptural economy, the potential failure of that economy is catalyzed by its overwhelming *success*, in an action that shows the distance between the fiction and the machine that incarnates it. De Certeau's discussion of *les machines célibataires* in 'The Scriptural Economy' derives from an earlier essay, 'The Arts of Dying: Celibatory Machines', and in a number of respects the implications of the 'celibate machine' can only be imperfectly adapted to the conceptual project of *The Practice of Everyday Life.*[37] In the context of ways of thinking about the relation of theater to the law, it points to another path, another line of flight, than the reiteration of the opacity of the body.

For de Certeau the celibate machine is a modern invention, and yet, as with the historical dimensions of the scriptural economy as a whole, its temporal moment is hard to locate: at what point did the fiery-footed steeds of 'technocratization' begin to gallop apace? Reading modernist art (Kafka, Duchamp, Alfred Jarry) as an avatar of the postmodern condition, de Certeau claims 'a poetics, once again, has preceded theory', and yet cannot quite commit to its presence in the present: the anti-myth is either 'still ahead', or perhaps already 'alongside'.[38] We might imagine that the idea of the celibate machine is itself atemporal, is always already alongside the moment that it would characterize – even in the early modern period. Amid the proliferation of Renaissance automata, we might especially place Kafka's disintegrating writing machine against the Brazen Head of Friar Bacon, powered by demonic forces, which notes the passing of time and which fails to encircle England with brass, and thus create a strategic space from which England may defend itself.[39] In a less overtly machinic fashion, the last hour of Doctor Faustus also presents a strategic breakdown. Richard Proudfoot has recently commented on his initial exposure to Marlowe, through reading 'the final speech of Faustus' in a poetry anthology: 'Its impact was immediate ... I longed (for many years) to see the play, imagining the power that speech must have in context. Some seven theatrical experiences later I am still longing, still imagining.'[40] What Proudfoot positions as a theatrical problem of the play – 'in the case of *Doctor Faustus*, the inevitable gap between text and performance yawns into a gulf'[41] – might also be seen as the action of the celibate machine, which fails through its prolific effectiveness, which can only succeed through failure.

The principal 'failure' of the final scene, I want to suggest, concerns to its relationship to the supernatural framework that ostensibly gives it meaning. This framework receives its most desperate articulation in the lines leading up to the first striking of the clock:

> See, see where Christ's blood streams in the firmament!
> One drop would save my soul, half a drop. Ah, my Christ!
> Ah, rend not my heart for naming of my Christ!
> Yet will I call on him. O, spare me, Lucifer!
> Where is it now? 'Tis gone, and see where God
> Stretcheth out his arm and bends his ireful brows!
> Mountains and hills, come, come, and fall on me
> And hide me from the heavy wrath of God!
> No, no!

Then will I headlong run into the earth.
Earth, gape! O, no, it will not harbour me.
You stars that reigned at my nativity,
Whose influence hath allotted death and hell,
Now draw up Faustus like a foggy mist
Into the entrails of yon labouring cloud,
That when you vomit forth into the air,
My limbs may issue from your smoky mouths,
So that my soul may but ascend to heaven.

(5.2.78–95)

Twice we are told to *see*. 'See, see, where Christ's blood streams in the firmament!'; and then, in response, 'See where God / Stretcheth out his arm and bends his ireful brows'. These two visual moments enact the *agon* of the play, the dialectic between salvation and damnation. But crucially, we can't actually *see* either of them. The most eerie moments of the play are these twin commands: what are we supposed to be looking at? On one level, we are meant to look at Faustus looking. The scene presents us with a reference to a larger realm of significance that we cannot perceive directly, only through Faustus' body. But because of that mediation, we also cannot commit to them fully – at least in the A-text. Without a gaudy vision of heaven and hell to guide us, the meaning of Faustus' removal remains indeterminate. The only thing we see is the theater, less a metaphor for a God-Author than the thing itself. Faustus' appeal to 'You stars that reigned at my nativity / Whose influence hath allotted death and hell' surely involves looking up at the playhouse roof, the overarching canopy that he cannot move out from under. After all, it is this theatrical ceiling that reigned at Faustus' nativity. What we see at the end of the play is a body caught within the machinery of the theater, unable to escape, unable to find a point of reference that exceeds it. But in that theatrical domination lies the specific blasphemy of the piece: that this is *only* theater, that there is no law beyond the script.

   The divine significance of the scene is deliberately presented through the vehicle of Faustus' body. This body is not merely a sign, pointing to a divine signified, but a medium – and by medium I mean not a surface inscribed with meaning, but the intermediary apparatus that facilitates meaning. With his frenzied command that we see the hand of an angry god inscribing its judgment, Faustus' body becomes part of the autonomous theatrical machine: the opacity of the body, its inability to communicate this meaning to us unequivocally, is not Garner's embodied resistance to signification but the macula of Weber's theatricality. And thus at exactly the moment when the play would have us be transported, captivated by our close encounter with divinity, the real artificiality of the material theater is brought before us, taking away 'the appearance of being'. In theatrical performance, *Doctor Faustus* offers us a demystified mystery play, in which the controlling context of the judgment of God is removed, replaced by only a practice. On another level, it demystifies the scriptural economy by failing at representation, incarnating its *Logos* in a manner that strips bare the fiction of incarnation. Excepting Faustus' final vision, the single direct presence of the divine in the A-text is the biblical inscription that appears on Faustus' arm as he prepares to sign away his soul:

> 'Homo, fuge!' Whither should I fly?
> If unto God, he'll throw thee down to hell –
> My senses are deceived; here's nothing writ. –
> It see it plain. Here in this place is writ
> 'Homo, fuge!' Yet shall not Faustus fly.

> (2.1.78–81)[42]

This is also the most direct example of the action of the scriptural economy, understood in its most explicit form: the theater as a writing machine, the body of the actor as the surface on which the law is inscribed. No other moment shows the play reaching outside of the machinery of representation. And yet 'here's nothing writ' is the literal truth in the context of the early modern theater: with its limited spectacular technologies, the early modern theater would have communicated the appearance and disappearance of the scriptural admonition only through words. 'Homo fuge' is as invisible to us, as immaterial, as the blood of Christ streaming in the firmament.

In this regard, the inscription scene stands in sharp contrast to the final striking of the watch: a real device, requiring an actual hand. It is the actuality of the clock's bell that places it 'alongside' *Doctor Faustus*, a point of suture between the space of representation and the space of performance. Time is the voice of the scriptural economy of *Doctor Faustus*; more than anything else, especially in the A-text, it is the striking of the clock that marks the inexorability of the law. Faustus says, 'it strikes, it strikes! Now body turn to air' (5.2.116), as if the clock is striking *him*, physically, disciplining his body with its measured strokes. If we witness the theatrical disciplining of Faustus' body, our bodies are disciplined, too: we recognize and accept the slippage of time, we recognize and accept the imposition of new theatrical rules, and we recognize and accept the approaching conclusion of the play. Our experience of the striking clock does not involve a simple opposition, whereby we can easily oppose spaces of representation and performance, or imagine the body as 'actual' and the signification as not, as if our spectating bodies occupied a privileged position of unfettered agency. Instead, we are reminded, viscerally, of the subjection and disciplining of our own bodies, existing as well within scriptural economies in which signification and 'actuality' are an impossible opposition. The striking clock is a celibate machine, a paradoxical engine subjecting us to time while alerting us to its fictiveness. 'Where is it now? 'Tis gone.' The space is empty, the body is down, and time is up.

Notes

1. *Doctor Faustus: A- and B- texts (1604, 1616)*, ed. David Bevington and Eric Rasmussen, The Revels Plays (Manchester: Manchester University Press, 1993). All subsequent citations are to this edition's A-text of the play, except as noted.
2. Stanton B. Garner, Jr., *Bodied Spaces: Phenomenology and Performance in Contemporary Drama* (Ithaca: Cornell University Press, 1994), 17, 13.
3. Ibid., 14.
4. See especially Johannes H. Birringer, 'Between Body and Language: "Writing" the Damnation of Faustus', *Theatre Journal* 36:3 (October, 1984), 335–355.
5. In quoting these stage directions, I have ignored Bevington and Rasmussen's editorial expansions.
6. Marjorie Garber, '"Here's Nothing Writ": Scribe, Script, and Circumscription in Marlowe's Plays', *Theatre Journal* 36:3 (October, 1984), 301–320 (316). This remains one of the finest readings of the play I have encountered. Subsequent references are cited parenthetically in the text.
7. Michel de Certeau, *The Practice of Everyday Life*, trans. Steven Rendall (Berkeley: University of California Press, 1984). Subsequent references are cited parenthetically in the text. For an excellent discussion of the complex interactions between tactical and strategic methodologies in de Certeau's own work, see chapter 3 of Bryan Reynolds' *Transversal Subjects: From Montaigne to Deleuze after Derrida* (Basingstoke: Palgrave Macmillan, 2009).
8. See 'La parole soufflée' and 'The Theater of Cruelty and the Closure of Representation', both in *Writing and Difference*, trans. Alan Bass (Chicago: University of Chicago Press, 1978).
9. I use the word 'dovetails' because it is generally difficult to assess the *direct* influence of performance studies as a discipline on early modern criticism, although a number of early modern scholars (such as Anthony Dawson, Paul Yachnin, Kier Elam, Barbara Hodgson, Bruce Smith, and others) make their interdisciplinary connections clear. Critical debates around the relations between performance and text, and performativity and textuality, cannot

be adequately captured in my brief summary. W. B. Worthen's extensive work on this topic is invaluable; see especially *Shakespeare and the Authority of Performance* (Cambridge: Cambridge University Press, 1997); 'Drama, Performativity, and Performance', *PMLA* 113:5 (October 1998), 1093–1107; and 'Antigone's Bones', *TDR: The Drama Review* 52:3 (Fall 2008), 10–33.

10. Keir Elam, '"In what chapter of his bosom?" Reading Shakespeare's Bodies', in Terence Hawkes, ed., *Alternative Shakespeares, Volume 2* (New York: Routledge, 1996), 152–153.

11. Ibid., 163.

12. For Weimann, see especially *Actor's Pen and Author's Voice: Playing and Writing in Shakespeare's Theatre* (Cambridge: Cambridge University Press, 2000) and *Shakespeare and the Power of Performance: Stage and Page in the Elizabethan Theatre*, co-authored with Douglas Bruster (Cambridge: Cambridge University Press, 2008). Bruce Smith introduced the term 'historical phenomenology' in his introduction to a roundtable discussion of 'the early modern body' in *Shakespeare Studies* 29 (2001), principally in reference to his earlier book, *The Acoustic World of Early Modern England: Attending to the O-Factor* (Chicago: University of Chicago Press, 1999). See also Mary Floyd-Wilson and Garrett Sullivan's introduction to *Renaissance Drama* 35 (2006), a special issue devoted to 'Embodiment and Environment in Early Modern Drama and Performance' (vii), and the essays that follow. I lack the space even to list an adequate selection of important criticism in this general area. Despite methodological contiguities, this criticism has many mansions; see Anthony Dawson and Paul Yachnin, *The Culture of Playgoing in Shakespeare's England: A Collaborative Debate* (Cambridge: Cambridge University Press, 2001), for two distinct approaches.

13. For Bourdieu, see especially *The Logic of Practice*, trans. Richard Nice (Stanford: Stanford University Press, 1990). On theatrical performance as cultural performance, see *Shakespeare and the Cultures of Performance*, ed. Paul Yachnin and Patricia Badir (Aldershot: Ashgate, 2008), especially in the introduction.

14. Henry Turner, *The English Renaissance Stage: Geometry, Poetics, and the Practical Spatial Arts, 1580–1630* (Oxford: Oxford University Press, 2006). Turner's theoretically sophisticated analysis of how new spatial arts and developing technologies shaped the early modern theater is always attentive to the conflicts and tensions that arise in theatrical representation and the complexities of signification that result; nevertheless, I think it would be fair to characterize the book as a semiotic project.

15. See David Mann, *The Elizabethan Player: Contemporary Stage Representation* (New York: Routledge, 1991), and Tiffany Stern, 'Re-patching the Play', in Peter Holland and Stephen Orgel, eds, *From Script to Stage in Early Modern England* (Basingstoke: Palgrave Macmillan, 2004).

16. Marjorie Garber, *Shakespeare's Ghost Writers: Literature as Uncanny Causality* (New York: Methuen, 1987).

17. Anthony Kubiak, *Agitated States: Performance in the American Theater of Cruelty* (Ann Arbor: University of Michigan Press, 2002), 22–23.

18. *Hamlet*, 3.2.225–26, in *The Norton Shakespeare*, ed. Stephen Greenblatt et al. (New York: W. W. Norton & Company, 2008). The meaning of Hamlet's offer is, of course, subject to much debate.

19. See especially Stephen Greenblatt's foundational analysis of Marlowe's plays in *Renaissance Self-Fashioning: From More to Shakespeare* (Chicago: University of Chicago Press, 1980), and Angus Fletcher's *Time, Space, and Motion in the Age of Shakespeare* (Cambridge, MA: Harvard University Press, 2007), both of which position Marlowe's representations of time as a key factor in the 'modernity' of his plays.

20. I acknowledge that the 'we' I employ to discuss this phenomenological moment is vulnerable to various criticisms, not least of universalizing a theatrical experience that is in many ways historically inflected and bounded; see, for example, Kristen Poole, 'The Devil's in the Archive: *Doctor Faustus* and Ovidian Physics', *Renaissance Drama* 35 (2006), 191–219. My intention is not to ventriloquize the thought processes of early modern theatergoers, but to provide a more expansive understanding of what theatrical experience might inherently involve.

21. It is unsurprising that the uncanniness of this moment has produced a common comic subtheme, where the camera pulls back to show a character manipulating the clock, returning control to the space of representation.

22. August Strindberg, *Miss Julie* (New York: Dover Publications, 1992), 36.

23. Samuel Weber, *Theatricality as Medium* (New York: Fordham University Press, 2004), 6.

24. Ibid., 7.

25. This is not to say that poststructuralism has been uninterested in the implements (and implementation) of the scriptural; see, for example, Jonathan Goldberg's *Writing Matter: From the Hands of the English Renaissance* (Stanford: Stanford University Press, 1990). Nevertheless, I think it is clear that de Certeau intends this distinction.

26. Greenblatt, *Renaissance Self-Fashioning*, 199. Greenblatt's reading of *Doctor Faustus* is remarkably congruent with de Certeau's description of the masterful phase of the scriptural economy.

27. Although he does not directly acknowledge it, de Certeau's discussion of the relation between law and the body is clearly indebted to Gilles Deleuze and Félix Guattari's *Anti-Oedipus: Capitalism and Schizophrenia* (Minneapolis: University of Minnesota Press, 1992), especially the chapter 'The Barbarian Despotic Machine', which (like de Certeau's text) takes Franz Kafka's 'In the Penal Colony' as a guiding narrative.

28. Raymond Williams, 'Drama in a Dramatized Society', in *Writing in Society* (London: Verso, 1991).

29. Jacques Derrida, 'La parole soufflée', *Writing and Difference*, 185.

30. On this point, see Donovan Sherman, 'The Absent Elegy: Performing Trauma in *The Winter's Tale*', *Shakespeare Bulletin* 27:2 (2009), 197–221 (205–206).

31. Antonin Artaud, *To Have Done with the Judgment of God*, trans. Clayton Esbleman and Norman Glass (Los Angeles: Black Sparrow Press, 1975).

32. Jerzy Grotowski, *Towards a Poor Theatre*, ed. Eugenio Barba (New York: Routledge, 2002), 86. The quotations are from Barba's production notes.

33. See, for example, Diana Taylor, *The Archive and the Repertoire: Performing Cultural Memory in the Americas* (Durham: Duke University Press, 2003).

34. My sense of my own painful complicity in this system, as I strive to complete this scriptural product (under pressing deadline, naturally), has been profound.

35. Note the complexities of 'célibataire', which could be translated as both 'celibate' and 'bachelor' and connotes both sterility and autonomy.

36. The impact of Carrouges' book on French theory has been diverse, and de Certeau's celibate machine should probably be separated from the nihilistic celibate machine of Jean Baudrillard and the psychosexual celibate desiring machine of Deleuze and Guattari.

37. Published in English in the essay collection *Heterologies: Discourse on the Other* (Minneapolis: University of Minnesota Press, 1986).

38. Kafka's machine, of course, is not what it once was; we encounter it in the fitful throes of its lingering, ironized failure.

39. Robert Greene, *Friar Bacon and Friar Bungay*, ed. Daniel Seltzer (Lincoln, NE: University of Nebraska Press, 1963). On early modern machines, see Jonathan Sawday's fascinating *Engines of the Imagination: Renaissance Culture and the Rise of the Machine* (New York: Routledge, 2007).

40. Richard Proudfoot, 'Marlowe and the Editors', in J. A. Downie and J. T. Parnell, eds, *Constructing Christopher Marlowe* (Cambridge: Cambridge University Press, 2000), 46.

41. Ibid.

42. I would argue that the Good Angel operates in a different register of meaning than the divine.

Index